Advances in Delays and Dynamics

Volume 11

Delay systems are largely encountered in modeling propagation and transportation phenomena, population dynamics and representing interactions between interconnected dynamics through material, energy and communication flows. Thought as an open library on delays and dynamics, this series is devoted to publish basic and advanced textbooks, explorative research monographs as well as proceedings volumes focusing on delays from modeling to analysis, optimization, control with a particular emphasis on applications spanning biology, ecology, economy and engineering. Topics covering interactions between delays and modeling (from engineering to biology and economic sciences), control strategies (including also control structure and robustness issues), optimization and computation (including also numerical approaches and related algorithms) by creating links and bridges between fields and areas in a delay setting are particularly encouraged.

More information about this series at http://www.springer.com/series/11914

Kun Liu · Emilia Fridman · Yuanqing Xia

Networked Control Under Communication Constraints

A Time-Delay Approach

Springer

Kun Liu
Beijing Institute of Technology
Beijing, China

Yuanqing Xia
School of Automation
Beijing Institute of Technology
Beijing, China

Emilia Fridman
School of Electrical Engineering
Tel Aviv University
Tel Aviv, Israel

ISSN 2197-117X ISSN 2197-1161 (electronic)
Advances in Delays and Dynamics
ISBN 978-981-15-4229-9 ISBN 978-981-15-4230-5 (eBook)
https://doi.org/10.1007/978-981-15-4230-5

This Springer imprint is published by the registered company Springer Nature Singapore Pte Ltd.
The registered company address is: 152 Beach Road, #21-01/04 Gateway East, Singapore 189721, Singapore

This work is dedicated to

My wife, Yuxuan, and
My parents, Deyou Liu and Chunying Yin, and
My parents-in-law, Yi Sun and Yaru Zhang

—Kun Liu

Boris, Eugenii and Leonid

—Emilia Fridman

My family with tolerance, patience, encouragement, support and wonderful frame of mind

—Yuanqing Xia

Preface

With the development of communication techniques, network topologies and control methods, networked control systems (NCSs) have received increasing attention in the past decades due to its widespread applications. The rapidly developing wireless communication technology enables NCSs with increased flexibility, ease of installation and reduced costs. A drawback of networking the control system is, however, that it leads to communication delays, packet dropouts, quantization errors in the signals transmitted over the network and communication constraints (i.e., it is no longer possible to transmit all sensor and actuator signals at each transmission instant). Three main approaches have been applied to the sampled-data control and later to NCSs. The first one is based on the exact discretization of the continuous-time linear plant over a sampling interval. The second one is based on the representation of the sampled-data system in the form of impulsive/hybrid model. The third approach is the time-delay approach, where the system is modeled as a continuous-time system with the delayed control input. The time-delay approach via time-independent Lyapunov–Krasovskii functionals or Lyapunov–Razumikhin functions leads to linear matrix inequalities in the analysis and the design of linear (uncertain) NCSs.

Most of the existing methods in the framework of time-delay approach are based on some Lyapunov-based analysis of systems with uncertain and bounded fast-varying delays. Therefore, the stability cannot be guaranteed by these methods for delays with upper bound that is greater than the analytical upper bound on the constant delay that preserves the stability. However, it is well known (see examples in [149] and discussions on quenching in [170]) that in many particular systems, the upper bound on the sampling interval that preserves the stability may be larger than the corresponding bound for the constant delay. In [46], piecewise-continuous (in time) Lyapunov functionals have been proposed for the analysis of sampled-data systems in the framework of the time-delay approach. The proposed time-dependent terms of the Lyapunov functionals lead to essentially new results, allowing a performance under the sampling that is superior to the one under the constant delay. Therefore, it is of both theoretical significance and practical importance to develop new Lyapunov-based methods for networked control under communication constraints.

This research monograph is summarizing our work on networked control by virtue of the discontinuous Lyapunov functional approach. Chapter 1 introduces the fundamental issues in NCSs and the three main approaches adopted for the analysis of this type of control systems. Detailed literature review is addressed in this chapter. In particular, the time-delay approach is emphasized in Chap. 1. In Chaps. 2–8, we study networked control for continuous-time plants. Chapters 2 and 3 develop the discontinuous Lyapunov functional approach, which is based on the upper bound of the network-induced delays and the vector extension of Wirtinger's inequality, respectively. A time-delay approach is developed for the modeling of NCSs under scheduling protocols in Chaps. 4–6, where Round-Robin, try-once-discard (TOD)/Round-Robin and stochastic protocols are discussed, respectively. Note that Chap. 5 presents a unified hybrid system model with time-varying delays in the continuous dynamics and in the reset equations for the closed-loop system under both TOD and Round-Robin protocols. Chapter 7 considers the decentralized exponential stabilization of larger-scale NCSs with local networks under TOD/Round-Robin protocol. Chapter 8 develops a time-delay approach to uncertain linear NCSs with dynamic quantization. Chapters 9–12 present our analysis on networked control for discrete-time plants. In Chap. 9, we focus on the stability analysis of discrete-time NCSs in the presence of actuator constraints under Round-Robin or under TOD protocol. A Lyapunov-based method has been proposed for finding the domain of attraction under both scheduling protocols. Chapter 10 presents the discrete-time counterpart of the continuous-time results of Chap. 8 with inclusion of Round-Robin protocol. The discrete-time counterpart of the continuous-time results of Chap. 5 for TOD protocol and Chap. 6 for stochastic protocol is demonstrated in Chap. 11 with an improvement achieved by a newly constructed augmented Lyapunov functional and the discrete-time counterpart of the second-order Bessel–Legendre integral inequality. In Chap. 12, we develop a time-delay approach to the decentralized exponential stabilization and l_2 gain analysis of large-scale NCSs with local networks in the presence of TOD/Round-Robin protocol.

This book is intended to scientists and engineers with interest in sampled-data and networked control, time-delay systems, and to graduate students in automatic control and systems theory. The book will lead the readers to recent developments on Lyapunov-based analysis of networked control under communication constraints.

The results in this book would not have been possible without the efforts and support of our colleagues and students. In particular, it is a great pleasure to acknowledge Professor Karl Henrik Johansson, who supervised the first author when he did his postdoctoral research at KTH Royal Institute of Technology. The first author is grateful for the guidance, scientific insights, great ideas, warm encouragement and all the support that Professor Johansson has always provided. We are also indebted to Professor Laurentiu Hetel at CNRS, Centrale Lille, and Professor Alexandre Seuret at LAAS-CNRS, for their many fruitful discussions on research ideas. The first author would like to use this opportunity to thank Professor

Liying Zhao at University of Science and Technology Beijing for her ongoing support and encouragement. The second author is grateful to her former students Dror Freirich and Anton Selivanov whose results are partially included in the book. Our special thanks to current and former students Aoyun Ma, Shuhe Ma, Dongyu Han, Xia Pan, Qirui Zhang and Yeming Lin for their great help. In addition, we acknowledge Prof. Silviu-Iulian Niculescu from CNRS, Gif-sur-Yvette, for handling and supporting this book. We also acknowledge Wiley, Elsevier, SIAM and IEEE for granting us the permission to reuse materials from our publications copyrighted by these publishers in this book. Finally, we gratefully acknowledge the support of the National Natural Science Foundation of China under Grant 61873034 and Grant 61503026, the Beijing Natural Science Foundation under Grant 4182057, the Major International (Regional) Joint Research Project of the National Natural Science Foundation of China under Grant 61720106010, the Open Subject of Beijing Intelligent Logistics System Collaborative Innovation Center under Grant BILSCIC-2019KF-13, the Israel Science Foundation under Grant 754/10, Grant 1128/14 and Grant 673/19, Chana and Heinrich Manderman chair at Tel Aviv University, the Knut and Alice Wallenberg Foundation under Grant Dnr KAW 2009.088 and the Swedish Research Council under Grant VR 621-2014-6282.

Beijing, China — Kun Liu
Tel Aviv, Israel — Emilia Fridman
Beijing, China — Yuanqing Xia
February 2020

Contents

Symbols and Acronyms

$\mathbb{R}$	Set of real numbers
$\mathbb{R}^n$	Set of n-dimensional real vectors
$\mathbb{R}^{n\times m}$	Set of $n \times m$ real matrices
$\mathbb{R}^+$	Set of non-negative real numbers
$\mathbb{Z}$	Set of integers
$\mathbb{Z}^+$	Set of non-negative integers
$\mathbb{N}$	Set of positive integers
$\mathbb{S}^n_+$	Set of symmetric positive definite matrices
$\mathbb{Z}[a, b]$	All integers in the interval $[a, b]$ with a, b integers and $b > a$
I, I_n	Identity matrix
$0, 0_{m\times n}$	Zero matrix
$P > 0$	The symmetric matrix P is positive definite
$P \geq 0$	The symmetric matrix P is positive semi-definite
$P < 0$	The symmetric matrix P is negative definite
$P \leq 0$	The symmetric matrix P is negative semi-definite
P^T	Transpose of matrix P
P^{-1}	Inverse of matrix P
$\lambda_{\max}(P)$	Maximum eigenvalue of matrix P
$\lambda_{\min}(P)$	Minimum eigenvalue of matrix P
$\mathrm{He}(A)$	$A + A^T$ for any square matrix A
$\mathrm{diag}\{\cdots\}$	Block-diagonal matrix
$\lvert\cdot\rvert$	Euclidean norm of a vector
$\lVert\cdot\rVert_\infty$	Essential supremum of a vector and the induced infinity norm of a matrix
sup	Supremum
$L_p(a, b)$, $p \in \mathbb{N}$	Space of functions $\phi : [a, b] \to \mathbb{R}^n$ with the norm $\lVert\phi\rVert_{L_p} = \left[\int_a^b \lvert\phi(s)\rvert^p ds\right]^{\frac{1}{p}}$
$L_2[0, \infty)$	Space of functions $\phi : \mathbb{R}^+ \to \mathbb{R}^n$ with the norm $\lVert\phi\rVert_{L_2} = \left[\int_0^\infty \lvert\phi(s)\rvert^2 ds\right]^{\frac{1}{2}}$

$l_2[0,\infty)$	Space of functions $\phi : \mathbb{Z}^+ \to \mathbb{R}^n$ with the norm $\Vert\phi\Vert_{l_2} = \left[\sum_{k=0}^{\infty} \vert\varphi(k)\vert^2\right]^{\frac{1}{2}}$
$W[a,b]$	Space of absolutely continuous functions $\phi : [a,b] \to \mathbb{R}^n$ with $\dot{\phi} \in L_2(a,b)$ and with the norm $\Vert\phi\Vert_W = \max_{\theta \in [a,b]} \vert\phi(\theta)\vert + \Vert\dot{\phi}\Vert_{L_2}$
$\mathrm{Prob}\{x\}$	Probability of x
$\mathbb{E}\{x\}$	Expectation of x
$\mathbb{E}\{x \vert y\}$	Expectation of x conditional on y
$x_t : [-h, 0] \to \mathbb{R}^n$	$x_t(\theta) = x(t+\theta)$, $\theta \in [-h, 0]$
$\lfloor x \rfloor$	Largest integer k such that $k < x$, i.e., $\lfloor x \rfloor = \max\{k \in \mathbb{Z} : k < x\}$
$\lceil x \rceil$	Smallest integer k such that $k \geq x$, i.e., $\lceil x \rceil = \min\{k \in \mathbb{Z} : k \geq x\}$
$\mathrm{col}\{\mathrm{x}, \mathrm{y}\}$	Vector $[x^T \; y^T]^T$
$\begin{bmatrix} X & Y \\ * & Z \end{bmatrix}, \begin{bmatrix} X & * \\ Y^T & Z \end{bmatrix}$	Symmetric matrix $\begin{bmatrix} X & Y \\ Y^T & Z \end{bmatrix}$
Iid	Independent and identically distributed
ISS	Input-to-state stability
LKF	Lyapunov–Krasovskii functional
LMI	Linear matrix inequality
LTI	Linear time-invariant
MAD	Maximum allowable delay
MATI	Maximum allowable transmission interval
NCS	Networked control system
TOD	Try-once-discard
ZOH	Zero-order-hold

Chapter 1
Introduction

1.1 Networked Control Systems

The point-to-point architecture is the traditional communication architecture for control systems, that is, sensors and/or actuators are connected to controllers via wires. Due to the expansion of physical setups and functionality, a traditional point-to-point architecture is no longer able to meet new requirements, such as modularity, integrated diagnostics, quick and easy maintenance, and low cost. Such requirements are particularly demanding in the control of complex control systems [18, 33] and remote control systems [8, 113, 116, 179].

To satisfy these new requirements, the common-bus network architectures have been introduced. The common-bus network architectures can improve the efficiency, flexibility and reliability of integrated applications, and reduce installation, reconfiguration and maintenance time and costs [169]. It gives rise to the so-called networked control systems (NCSs) [5, 88, 94, 229, 259].

In general, NCSs are a type of distributed control systems where sensors, actuators and controllers are interconnected through a communication network as shown in Fig. 1.1. The sensors measure the states of the plant and transmit these states over the communication network to the controllers. The controllers receive these states, calculate appropriate control actions and send them to the actuators over the communication network. The actuators receive control actions and control the plant appropriately. Due to its low cost, flexibility and less wiring, NCSs are rapidly increasing in industrial applications, including telecommunications, remote process control, altitude control of airplanes and so on [8, 33, 113, 116, 179].

It can be seen from the block diagram in Fig. 1.1 that in NCSs, the closed loops are closed via communication networks. The insertion of the communication network in the feedback control loop makes the analysis and design of systems more complex than the traditional point-to-point architecture. The network can introduce unreliable and time-dependent levels of service in terms of, for example, delays, jitter or losses. In general, the network-induced imperfections can jeopardize the stability, safety and performance of the units in a physical environment [103].

K. Liu et al., *Networked Control Under Communication Constraints*, Advances in Delays and Dynamics 11, https://doi.org/10.1007/978-981-15-4230-5_1

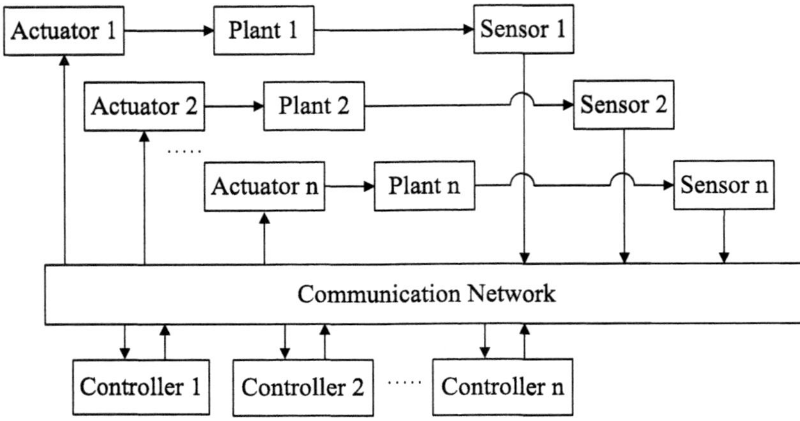

Fig. 1.1 Typical setup of NCSs

1.2 Fundamental Issues in NCSs

The main network-induced imperfections and constraints can be categorized in the following five types [84, 144, 264]:

(i) Variable Sampling/Transmission Intervals

The conventional computer-controlled systems theories assume equal-distance sampling of the plant outputs, which means the samples are taken periodically at the time instants kh, $k \in \mathbb{Z}^+$, where $h > 0$ is the constant sampling period. This assumption leads to linear time-invariant (LTI) sampled-data systems and greatly simplifies the stability and performance analysis [258].

However, the assumption of equal-distance sampling should not be imposed on the NCSs analysis. To transmit a continuous-time signal over a network, the signal must be sampled, encoded in a digital format, transmitted over the network, and finally the data must be decoded at the receiver side. This process is significantly different from the usual periodic sampling in digital control. A significant number of results have attempted to characterize maximum allowable transmission interval (MATI) for which stability can be guaranteed [84, 92, 163].

Moreover, in contrast to periodic sampling control, the event-based control aims at minimizing the bandwidth utilization while still guaranteeing the desired level of control performance. There are two main triggering strategies, one is event-triggering strategy (see, e.g., [13, 153, 221]) and the other is self-triggering strategy (see, e.g., [82, 155, 232]). The difference between event-triggering strategy and self-triggering strategy is that the former is reactive, while the latter is proactive. In event-triggering strategy, a triggering condition based on current measurements is monitored and when violated, an event is triggered. In self-triggering strategy, the next updating time is pre-computed at a control updating time based on predictions using previously received data and knowledge on the plant dynamics [82]. Thus, growing attention is paid to event-based control for NCSs, e.g., see [175–177, 188, 233, 248] for event-triggered control over networks, and see [2, 4, 168, 212, 255] for self-triggered control over networks.

(ii) Communication Delay/Network-Induced Delay

The network-induced delay, including sensor-to-controller delay and controller-to-actuator delay, happens when the data exchange among devices connected by the communication network, which will deteriorate the system performance as well as the stability. This delay, depending on the network characteristics such as network load, topologies, routing schemes, can be constant, time-varying or even random. In the literature, two ways of modeling network-induced uncertainties can be distinguished.

- The first approach bounds the network-induced delay and considers maximum allowable delay (MAD). The NCSs are modeled as discrete-time (uncertain) systems [22, 259], time-delay systems [63, 65, 128, 129, 214, 246], hybrid systems [83, 84] or switched systems [214, 261].
- The second one is the stochastic modeling approach. By assuming that the network-induced delay has a known probability distribution function [99, 139, 147, 167, 180, 211], or incorporating the network-induced delays as Markov process [102, 207, 208, 256], the resulting closed loop is modeled as a stochastic system.

For some systems, the presence of communication delay may have the positive effect on system performance [54, 55, 190], e.g., see [129, 191] for sampled-data stabilization, and see [25, 245] for consensus of multi-agent systems.

(iii) Packet Dropouts Caused by the Unreliability of the Network

Another significant difference between NCSs and the standard digital control is the possibility that some packets not only suffer communication delays but, even worse, may be lost while in transit through the network. Typically, the packet dropouts result from the transmission errors in physical network links (which is far more common in wireless than in wired networks) or from the buffer overflows due to congestion. Thus, how much packet dropouts affect the stability and the performance of NCSs is an issue that must be considered [96, 242, 243].

In general, in most of the literature two different strategies are considered for dealing with packet dropouts. The first one is zero-input, i.e., the actuator input to the plant is set to zero when the control packet from the controller to the actuator is lost [104]. The second one is hold-input, i.e., the latest control input stored in the actuator buffer is used when a packet is lost [259]. By studying the linear quadratic performance of NCSs where the control packets are subject to loss, it was shown in [185] that none of these two control schemes can be claimed to be superior to the other.

Different control techniques have been proposed for the modeling of NCSs with data packet dropouts. They can be roughly categorized into the following types based on the resulting closed-loop systems:

- Switching systems [105, 260], asynchronous dynamical systems [259] and jump linear systems with Markov chains [185, 186]. Note that the packet dropouts defined in the aforementioned references have two cases, dropped or sent successfully, which are modeled as a Bernoulli or a two-state Markov chain process based on zero-input or hold-input.

- Another type of the resulting closed-loop system is in terms of time-delay systems [65, 244]. By modeling the dropouts as prolongations of the variable sampling intervals or the communication delays, the NCSs with data packet dropouts are modeled as linear systems with time-varying input delays based on hold-input strategy. Then, the delay-dependent approach can be applied to the resulting time-delay systems, and the maximum allowable value of the successive packet dropouts can be determined by solving a set of linear matrix inequalities (LMIs).

(iv) Quantization Errors in the Signals Transmitted over the Network Due to the Finite Word Length of the Packets

Due to the limited transmission capacity of the network, the data transmitted in practical NCSs should be quantized before they are sent to the next network node [50, 163, 240]. A quantizer is a function that maps a real-valued function into a piecewise constant function taking values on a finite set. At present, there exist two kinds of quantizers, which are uniform quantizers [16] and logarithmic ones [32].

- The uniform quantizer maps the real-valued function to a finite number of quantization regions with rectilinear shape [16] or arbitrary shape [123, 125, 137, 163]. The study of system affected by the uniform quantizer is usually based on "zoom" strategy, which is composed of two stages, i.e., "zooming-out" and "zooming-in." In the first stage, the range of quantizer is increased to guarantee the states of system can be adequately measured. In the second stage, the quantization error is decreased to drive the states to the origin.
- When the system is affected by the logarithmic quantizer, in which the quantization levels are linear in logarithmic scale, a simple classical approach to the analysis and mitigation of the quantization effects is to treat the quantization error as uncertainty or nonlinearity and bound it using a sector bound [60, 65, 247].

(v) Communication Scheduling

In NCSs, the performance of control loops not only depends on the design of the control algorithms, but also relies on the scheduling of the shared network resource. The communication constraints impose that, per transmission, only one node can access the network and send its information. Hence, many existing works are focused on that how often a plant should schedule to transmit the data and with what priority the packet should be sent out. There are three main classes of network protocols, namely:

- The class of static protocols, of which Round-Robin protocol is a special case [29, 84, 133, 140, 163, 164]. In the Round-Robin protocol, the node j is transmitted periodically with period l, where l is the total number of nodes. The transmission order is decided in advance, and this order is repeated indefinitely.
- The class of dynamic protocols, which includes the well known try-once-discard (TOD) protocol [29, 84, 134, 163, 164]. In the TOD protocol, the node that has

the largest network-induced error, i.e., the largest difference between the latest transmitted values and the current values of the signals corresponding to the node, is granted access to the network.
It was observed in some examples that the TOD protocol stabilized the system for larger MATI than the Round-Robin protocol whenever $l > 1$ [164, 230]. However, in some examples opposite conclusions were made [39, 134].

- The stochastic protocol, which was introduced in [28, 220]. The stochastic protocol determines the transmitted node through a Bernoulli or a two-state Markov chain process [138, 265]. The quadratic and stochastic protocols belong to dynamic protocols.

1.3 Three Main Approaches to NCSs

Consider a generic schematic diagram of NCSs as shown in Fig. 1.2. The LTI continuous-time plant is given by

$$\dot{x}(t) = Ax(t) + Bu(t), \tag{1.1}$$

where $x(t) \in \mathbb{R}^n$ is the state vector, $u(t) \in \mathbb{R}^{n_u}$ is the control input, $A \in \mathbb{R}^{n\times n}$ and $B \in \mathbb{R}^{n\times n_u}$ are system matrices with appropriate dimensions. Consider, for simplicity, the state-feedback case.

Let s_k denote the unbounded monotonously increasing sequence of sampling instants, i.e.,

$$0 = s_0 < s_1 < \cdots < s_k < \cdots, \quad k \in \mathbb{Z}^+, \ \lim_{k\to\infty} s_k = \infty \tag{1.2}$$

with the time-varying sampling intervals $h_k = s_{k+1} - s_k > 0$. There are two sources of delays from the network: sensor-to controller η_k^{sc} and controller-to-actuator η_k^{ca}. It is assumed that the sensor acts in a time-driven fashion (i.e., sampling occurs at

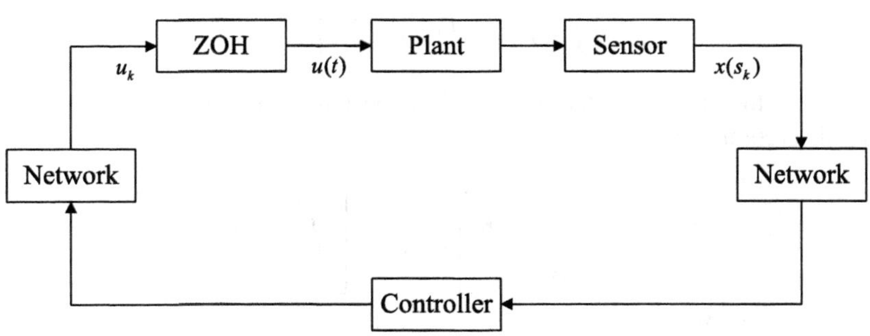

Fig. 1.2 Schematic diagram of a NCS

the times s_k, $k \in \mathbb{Z}^+$) and that both the controller and the actuator act in an event-driven fashion (i.e., they respond instantaneously to newly arrived data). Under these assumptions, these two delays can be captured by a single delay $\eta_k = \eta_k^{sc} + \eta_k^{ca}$. Assume that $\eta_k \in [\eta_m, \text{ MAD}]$, where η_m and MAD denote the lower and upper delay bounds on the network-induced delays η_k, respectively. Denote by $t_k = s_k + \eta_k$ the updating instant time of the zero-order-hold (ZOH), finally the ZOH function transforms the discrete-time control input u_k to a continuous-time control input

$$u(t) = u_k = Kx(s_k),\ t \in [t_k, t_{k+1}),\ k \in \mathbb{Z}^+. \tag{1.3}$$

The resulting closed-loop system is (1.1), (1.3):

$$\dot{x}(t) = Ax(t) + BKx(s_k),\ t \in [t_k, t_{k+1}),\ k \in \mathbb{Z}^+. \tag{1.4}$$

In the literature, three main approaches have been used to the sampled-data control (see, e.g., [6, 9, 19, 31, 47, 91, 156, 209]) and later to the NCSs.

1.3.1 Discrete-Time Modeling Approach

For the sake of clarity, we focus here on the case of small network-induced delays, where these delays are smaller than the sampling intervals, i.e., $\eta_k = t_k - s_k < h_k$, $k \in \mathbb{Z}^+$. The most common discrete-time NCS model is explained in, e.g., [166, 259]. The discrete-time uncertain system can be obtained by describing the evolution of the states between s_k and $s_{k+1} = s_k + h_k$. Discretizing the linear plant (1.1) at the sampling time s_k, $k \in \mathbb{Z}^+$, we obtain (see, e.g., [22, 29])

$$x(s_{k+1}) = e^{Ah_k}x(s_k) + \int_0^{h_k-\eta_k} e^{As}ds\,Bu_k + \int_{h_k-\eta_k}^{h_k} e^{As}ds\,Bu_{k-1}. \tag{1.5}$$

Using now the state vector

$$\xi(s_k) = \left[x^T(s_k)\ \ u_{k-1}^T\right]^T \tag{1.6}$$

that includes the current system state and the past system input, we obtain the following discrete model

$$\begin{aligned}\xi(s_{k+1}) &= \begin{bmatrix} e^{Ah_k} & \int_{h_k-\eta_k}^{h_k} e^{As}ds\,B \\ 0 & 0 \end{bmatrix} \xi(s_k) + \begin{bmatrix} \int_0^{h_k-\eta_k} e^{As}ds\,B \\ I \end{bmatrix} u_k \\ &= \tilde{A}_{h_k,\eta_k}\xi(s_k),\end{aligned} \tag{1.7}$$

where

$$\tilde{A}_{h_k,\eta_k} = \begin{bmatrix} e^{Ah_k} + \int_0^{h_k-\eta_k} e^{As} ds BK & \int_{h_k-\eta_k}^{h_k} e^{As} ds B \\ K & 0 \end{bmatrix}.$$

Hence, the stability analysis for the uncertain system (1.7) with the uncertainty $\eta_k \in [\eta_m, \text{ MAD}]$, $h_k \in (0, \text{ MATI}]$ is essentially a robust stability analysis problem. Then, the obstruction to apply existing robust stability analysis techniques directly is that the uncertainty appears in an exponential fashion in $\tilde{A}_{h_k,\eta_k}$ of (1.7). To make the formulation (1.7) suitable for robust stability analysis, the over-approximation techniques are employed in the literature to embed the original model (as tight as possible) in a larger model that has nice structural properties suitable for the application of robust stability methods. The adopted over-approximation techniques are based on the real Jordan form [22–24, 227], the Taylor series [89], gridding and norm-bounding [61, 210, 213] and the Cayley–Hamilton theorem [69]. The over-approximation techniques typically result in discrete-time polytopic models [12, 85, 115] with (or without) additive norm-bounded uncertainties. These models are amendable for robust stability assessment using LMIs. In [85], a comparison was presented between the different over-approximation methods and the subsequent LMI-based stability analysis.

The most general and complete modeling based on discrete-time approach was provided in [22] (see also [23, 24]) that includes imperfection types (i), (ii), (iii) of network-induced uncertainties, i.e., time-varying sampling intervals, time-varying communication delays (both smaller and larger than the sampling interval) and explicit modeling of the dropouts.

In [30], the discrete-time modeling approach was further applied to the network-based stabilization including imperfection types (i), (iii), (v). In [29], the small communication delays were taken into account. The work of [29] was extended in [228] by including quantization, where the quantization-induced disturbances were incorporated as one of the stability and performance limiting factors. A state-dependent sampling approach was introduced in [37] for the stabilization of sampled-data systems to reduce the number of computations.

For LTI systems, as shown in, e.g., [29], the discrete-time approach can lead to less conservative results in terms of the so-called MATI and MAD. However, the discrete-time methods become complicated for systems with uncertain coefficients. Moreover, it is tedious to include large delays (that are larger than the sampling intervals) in such models and the stability analysis methods may fail when the interval between two transmissions takes small values.

1.3.2 Impulsive System Approach

The second approach is based on the representation of the system in the form of hybrid/impulsive system [209]. The impulsive dynamical systems exhibit continuous evolutions described by ordinary differential equations and instantaneous state jumps or impulses (see, e.g., [70]).

The idea of the impulsive system approach is to rewrite its closed-loop system (1.4) as the following delay impulsive system

$$\begin{aligned} \dot{\xi}(t) &= \begin{bmatrix} A & BK \\ 0 & 0 \end{bmatrix} \xi(t), \quad t \in [t_k, t_{k+1}), \\ \xi(t_k) &= \begin{bmatrix} x(t_k^-) \\ x(s_k) \end{bmatrix}, \quad k \in \mathbb{Z}^+, \end{aligned} \tag{1.8}$$

where $\xi(t) = [x^T(t) \;\; z^T(t)]^T$, $z(t) = x(s_k)$, $t \in [t_k, t_{k+1})$.

In [162], the impulsive system approach was proposed to the analysis and control of sampled-data systems ($\eta_k \equiv 0$, i.e., $t_k = s_k$, $k \in \mathbb{Z}^+$) with variable sampling. A discontinuous Lyapunov functional, which is discontinuous at input update instants and is decreasing between discontinuities, was introduced. Later on in [161], the impulsive system approach was further extended to the modeling and analysis of NCSs with variable sampling and communication delays. This discontinuous Lyapunov function method improved the existing Lyapunov-based results in, e.g., [43, 63, 65, 246]. The main advantage of this modeling approach is the possibility to incorporate time-delays larger than the sampling interval without increasing model complexity, as is the case in the discrete-time modeling approach [20, 24].

Based on the impulsive system approach, the input–output stability properties of nonlinear NCSs were studied in [164] for NCSs with imperfection types (i), (iii), (v). In [163], the stabilization of nonlinear NCSs with dynamic quantization was studied. However, the delays are not included in the analysis. In [84], the imperfection types (i), (ii), (iii), (v) were considered and the methods for computing MATI and MAD were provided, for which the stability of a nonlinear system is ensured. A unifying modeling framework was provided in [83] to incorporate all the five types of networked-induced effects. Note that some of the mentioned results that study varying transmission intervals and/or varying communication delays can be extended to include type (iii), i.e., the data packet dropouts phenomena as well by modeling the dropouts as prolongations of the MATI [83]. In the above works, only small communication delays were considered.

1.3.3 Time-Delay Approach

The modeling of continuous-time systems with digital control in the form of continuous-time systems with delayed control input was introduced by [40, 156] (see also [150]). The digital control law for sampled-data systems ($\eta_k \equiv 0$, i.e., $t_k = s_k$, $k \in \mathbb{Z}^+$) may be represented as delayed control as follows:

$$\begin{aligned} u(t) &= Kx(t_k) = Kx(t - (t - t_k)) = Kx(t - \tau(t)), \\ \tau(t) &= t - t_k, \quad t_k \le t < t_{k+1}, \end{aligned} \tag{1.9}$$

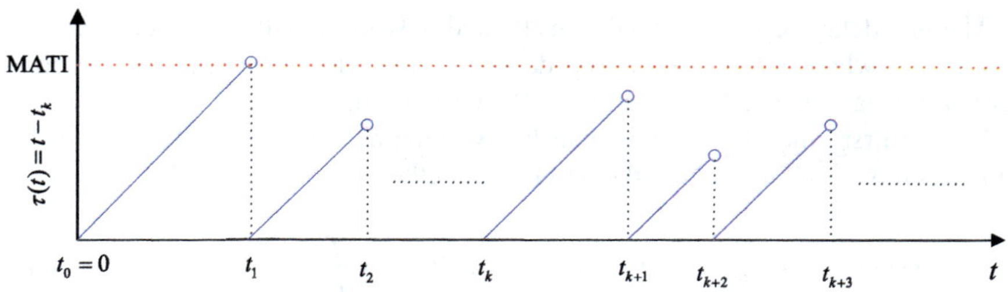

Fig. 1.3 Sampled-data systems: piecewise-continuous time-delay

In this case, the closed-loop system becomes an infinite-dimensional delay differential equation

$$\dot{x}(t) = Ax(t) + A_1 x(t - \tau(t)),\ t_k \le t < t_{k+1},\ k \in \mathbb{Z}^+, \tag{1.10}$$

where $A_1 = BK$, the time-varying delay $\tau(t) = t - t_k$ is piecewise linear with derivative $\dot{\tau}(t) = 1$ for $t \neq t_k$ (see Fig. 1.3). Moreover, $\tau(t) \le t_{k+1} - t_k = s_{k+1} - s_k = \text{MATI}$ for $t_k \le t < t_{k+1}$. The stability of (1.10) can be established using Lyapunov–Razumikhin or Lyapunov–Krasovskii theorems. The time-delay approach was applied to robust sampled-data stabilization via Lyapunov–Krasovskii technique in [53].

1.3.3.1 Delay-Dependent Analysis

The choice of Lyapunov–Krasovskii functional (LKF) (that we will call also Lyapunov functional) is crucial for deriving stability criteria [48]. The first delay-dependent (both Krasovskii- and Razumikhin-based) conditions were derived by using the relation

$$x(t - \tau(t)) = x(t) - \int_{t-\tau(t)}^{t} \dot{x}(s)ds \tag{1.11}$$

via different model transformations and by bounding the cross-terms [112, 120, 171]. The widely used first model transformation, where (1.11) is substituted into (1.10) with $\dot{x}(s)$ substituted by the right-hand side of (1.10), has the form

$$\dot{x}(t) = (A + A_1)x(t) - A_1 \int_{t-\tau(t)}^{t} [Ax(s) + A_1 x(s - \tau(s))]ds. \tag{1.12}$$

Note that this transformation is valid for $t - \tau(t) \ge t_0$. The latter system is not equivalent to the original one possessing some additional dynamics [74, 109]. The stability of the transformed system (1.12) guarantees the stability of the original one, but not vice versa.

The first delay-dependent conditions treated only the slowly varying delays with $\dot{\tau} \le d < 1$, whereas the fast-varying delay (without any constraints on the delay derivative) was analyzed via Lyapunov–Razumikhin functions.

For the first time, the systems with fast-varying delays were analyzed by using Krasovskii method in [56], via the descriptor method introduced in [41]:

$$\dot{x}(t) = y(t),\ 0 = -y(t) + (A + A_1)x(t) - A_1 \int_{t-\tau(t)}^{t} y(s)ds. \tag{1.13}$$

The descriptor system (1.13) is equivalent to (1.10) in the sense of stability. In the descriptor approach, $\dot{x}(t)$ is not substituted by the right-hand side of the differential equation. Instead, it is considered as an additional state variable of the resulting descriptor system (1.13). Therefore, the novelty of the descriptor approach is not in $V = x^T(t)Px(t) + \cdots\ (P > 0)$, but in $\dot{V}$, where $\frac{d}{dt}\left[x^T(t)Px(t)\right]$ is found as

$$\begin{aligned}\frac{d}{dt}\left[x^T(t)Px(t)\right] &= 2x^T(t)P\dot{x}(t) + 2[x^T(t)P_2^T + \dot{x}^T(t)P_3^T] \\ &\quad \times[-\dot{x}(t) + (A + A_1)x(t) - A_1 \int_{t-\tau(t)}^{t} \dot{x}(s)ds],\end{aligned} \tag{1.14}$$

and where $P_2 \in \mathbb{R}^{n\times n}$ and $P_3 \in \mathbb{R}^{n\times n}$ are "slack variables." This leads to $\dot{V} \le -\gamma(|x(t)|^2 + |\dot{x}(t)|^2),\ \gamma > 0$.

The advantages of the descriptor method are:

- Less conservative conditions (even without delay) for uncertain systems,
- "Unifying" LMIs for the discrete-time and for the continuous-time systems, having almost the same form and the same advantages [58],
- Simple conditions for neutral type systems can be derived (where the stability of the difference operator follows from LMIs) [42],
- Efficient design is obtained for systems with state, input and output delays by choosing $P_3 = \varepsilon P_2$ with a tuning scalar parameter ε [217],
- Simple delay-dependent conditions can be derived for diffusion partial differential equations [51].

Most of the recent Krasovskii-based results do not use model transformations and cross-terms bounding. They are based on the application of Jensen's inequality (see, e.g., [73]).

1.3.3.2 Simple Delay-Dependent Conditions

The first Krasovskii-based LMI conditions for systems with fast-varying delays were derived in [56] via the descriptor method. We differentiate $x^T(t)Px(t)$ as in (1.14) along system (1.10) with MATI $\overset{\Delta}{=} h$. To "compensate" $\int_{t-\tau(t)}^{t} \dot{x}(s)ds$, consider the double integral term [56]:

$$V_R(\dot{x}_t) = \int_{-h}^{0}\int_{t+\theta}^{t} \dot{x}^T(s)R\dot{x}(s)dsd\theta,\ R > 0.$$

The term V_R can be rewritten equivalently as

$$V_R(\dot{x}_t) = \int_{t-h}^{t} (h + s - t)\dot{x}^T(s)R\dot{x}(s)ds.$$

Differentiating $V_R(\dot{x}_t)$, we obtain

$$\begin{aligned}\frac{d}{dt}V_R(\dot{x}_t) &= -\int_{t-h}^{t} \dot{x}^T(s)R\dot{x}(s)ds + h\dot{x}^T(t)R\dot{x}(t)\\ &= -\int_{t-\tau(t)}^{t} \dot{x}^T(s)R\dot{x}(s)ds + h\dot{x}^T(t)R\dot{x}(t) - \underbrace{\int_{t-h}^{t-\tau(t)} \dot{x}^T(s)R\dot{x}(s)ds}_{\text{will be ignored}}.\end{aligned} \tag{1.15}$$

We apply further Jensen's inequality

$$-\int_{t-\tau(t)}^{t} \dot{x}^T(s)R\dot{x}(s)ds \le -\frac{1}{h}\int_{t-\tau(t)}^{t} \dot{x}^T(s)dsR\int_{t-\tau(t)}^{t} \dot{x}(s)ds.$$

Then, for Lyapunov functional

$$V(x(t), \dot{x}_t) = x^T(t)Px(t) + V_R(\dot{x}_t),$$

we find

$$\begin{aligned}\frac{d}{dt}V(x(t), \dot{x}_t) &\le 2x^T(t)P\dot{x}(t) + h\dot{x}^T(t)R\dot{x}(t) - \frac{1}{h}\int_{t-\tau(t)}^{t} \dot{x}^T(s)dsR\int_{t-\tau(t)}^{t} \dot{x}(s)ds\\ &\quad +2[x^T(t)P_2^T + \dot{x}^T(t)P_3^T][(A + A_1)x(t) - A_1\int_{t-\tau}^{t} \dot{x}(s)ds - \dot{x}(t)]\\ &\le \eta^T(t)\Psi\eta(t) \le -\varepsilon(|x(t)|^2 + |\dot{x}(t)|^2),\ \varepsilon > 0,\end{aligned}$$

where $\eta(t) = \text{col}\{x(t), \dot{x}(t), \frac{1}{h}\int_{t-\tau}^{t} \dot{x}(s)ds\}$, if

$$\begin{aligned}\Psi &= \begin{bmatrix} \Phi & P - P_2^T + (A + A_1)^T P_3 & -hP_2^T A_1 \\ * & -P_3 - P_3^T + hR & -hP_3^T A_1 \\ * & * & -hR \end{bmatrix} < 0,\\ \Phi &= P_2^T(A + A_1) + (A + A_1)^T P_2.\end{aligned} \tag{1.16}$$

As it was understood later [51, 216], the equivalent delay-dependent conditions can be derived without the descriptor method, where $\dot{x}$ is substituted by the right-hand side of (1.10) and the Schur complements is applied further.

Note that $\Psi < 0$ yields that the eigenvalues of hA_1 are inside of the unit circle. In the example $\dot{x}(t) = -x(t - \tau(t))$ with $A_1 = -1$, the simple delay-dependent conditions cannot guarantee the stability for $h \geq 1$, which is far from the analytical bound 1.5. This illustrates the conservatism of the simple conditions.

1.3.3.3 Improved Delay-Dependent Conditions

The relation between $x(t-\tau(t))$ and $x(t-h)$ (and not only between $x(t-\tau(t))$ and $x(t)$) has been taken into account in [80]. The widely used by now LKF for delay-dependent stability is state-derivative dependent one of the form

$$\begin{aligned} V(t, x_t, \dot{x}_t) &= x^T(t)Px(t) + \int_{t-h}^{t} x^T(s)Sx(s)ds \\ &+h \int_{-h}^{0}\int_{t+\theta}^{t} \dot{x}^T(s)R\dot{x}(s)dsd\theta + \int_{t-\tau(t)}^{t} x^T(s)Qx(s)ds, \end{aligned} \tag{1.17}$$

where $P > 0,\ R \geq 0,\ S \geq 0,\ Q \geq 0$. This functional with $Q = 0$ leads to delay-dependent conditions for systems with fast-varying delays, whereas for $R = S = 0$ it leads to delay-independent conditions (for systems with slowly varying delays). The above V with $S = 0$ was introduced in [56], whereas the S-dependent term was added in [80].

Differentiating V given by (1.17), we find

$$\begin{aligned} \frac{d}{dt}V &\leq 2x^T(t)P\dot{x}(t) + h^2\dot{x}^T(t)R\dot{x}(t) \\ &-h\int_{t-h}^{t} \dot{x}^T(s)R\dot{x}(s)ds + x^T(t)(S+Q)x(t) \\ &-x^T(t-h)Sx(t-h) - (1-d)x^T(t-\tau(t))Qx(t-\tau(t)) \end{aligned} \tag{1.18}$$

and employ the representation

$$-h\int_{t-h}^{t} \dot{x}^T(s)R\dot{x}(s)ds = -h\int_{t-h}^{t-\tau(t)} \dot{x}^T(s)R\dot{x}(s)ds - h\int_{t-\tau(t)}^{t} \dot{x}^T(s)R\dot{x}(s)ds. \tag{1.19}$$

Applying Jensen's inequality to both terms on the right-hand side of (1.19), we arrive at

$$-h\int_{t-h}^{t} \dot{x}^T(s)R\dot{x}(s)ds \leq -\frac{h}{\tau(t)}e_1^T Re_1 - \frac{h}{h-\tau(t)}e_2^T Re_2, \tag{1.20}$$

where

$$e_1 = x(t) - x(t-\tau(t)),\ e_2 = x(t-\tau(t)) - x(t-h).$$

Here for $\tau = 0$ and $\tau = h$, we mean the following limits:

$$\lim_{\tau(t)\to 0} \frac{h}{\tau(t)}e_1^T Re_1 = h \lim_{\tau(t)\to 0} \tau(t)\dot{x}^T(t)R\dot{x}(t) = 0$$

and

$$\lim_{\tau(t)\to h} \frac{h}{h-\tau(t)} e_2^T R e_2 = 0.$$

In [80], the right-hand side of (1.20) was upper-bounded by $-e_1^T Re_1 - e_2^T Re_2$ that was conservative. The convex analysis of [173] allowed to avoid the latter restrictive bounding. We reformulate the reciprocally convex combination lemma proposed in [173] in a more convenient for the Lyapunov-based analysis form:

Lemma 1.1 *Let $R_1 \in \mathbb{R}^{n_1\times n_1}, \ldots, R_N \in \mathbb{R}^{n_N\times n_N}$ be positive definite matrices. For $e_1 \in \mathbb{R}^{n_1}, \ldots, e_N \in \mathbb{R}^{n_N}$, and for all $\alpha_i > 0$ with $\sum_{i=1}^N \alpha_i = 1$, define a reciprocally convex combination as a function of the form $\sum_{i=1}^N \frac{1}{\alpha_i} e_i^T R_i e_i$. Then for all $S_{ij} \in \mathbb{R}^{n_i\times n_j}$, $j = 2, \ldots, N$, $i = 1, \ldots, j-1$, such that*

$$\begin{bmatrix} R_i & S_{ij} \\ * & R_j \end{bmatrix} \geq 0,$$

the following inequality holds:

$$\sum_{i=1}^{N} \frac{1}{\alpha_i} e_i^T R_i e_i \geq \begin{bmatrix} e_1 \\ e_2 \\ \vdots \\ e_N \end{bmatrix}^T \begin{bmatrix} R_1 & S_{12} & \cdots & S_{1N} \\ * & R_2 & \cdots & S_{2N} \\ * & * & \ddots & \vdots \\ * & * & \cdots & R_N \end{bmatrix} \begin{bmatrix} e_1 \\ e_2 \\ \vdots \\ e_N \end{bmatrix}.$$

The novelty of this method consists in merging the non-convex terms into a single expression to derive an accurate convex inequality. It was notably shown in [145] that the reciprocally convex combination lemma leads to the same conservatism as the Moon et al.'s inequality [160] when considering Jensen-based stability criteria, but with a lower computational burden. A relaxed reciprocally convex combination lemma was proposed in [253] without requiring any extra decision variable. More insights on the relationship between some existing matrix inequalities were provided in [203] that revealed strong links between the existing inequalities of [160, 173, 206, 251, 253].

1.3.3.4 Stability Analysis of Systems with Interval or Non-small Delay

The time-delay approach became popular in NCSs, being applied to uncertain systems under uncertain sampling with the known upper bound on the sampling intervals [63, 65, 246]. The time-delay approach was further extended to event-triggered networked control [188, 248], distributed networked control of partial differential equations [7, 49, 107, 187], etc.

By defining

$$\tau(t) = t - t_k + \eta_k, \ t_k \le t < t_{k+1}, \ k \in \mathbb{Z}^+, \tag{1.21}$$

the digital control law has the following form:

$$u(t) = Kx(t - \tau(t)), \ t_k \le t < t_{k+1}. \tag{1.22}$$

Hence, the LTI system (1.1) under (1.22) is then modeled as a time-delay system with time-varying delay

$$\dot{x}(t) = Ax(t) + BKx(t - \tau(t)), \ t_k \le t < t_{k+1}, \ k \in \mathbb{Z}^+, \tag{1.23}$$

where the time-varying delay $\tau(t)$ is piecewise linear with derivative $\dot{\tau}(t) = 1$ for $t \neq t_k$. Moreover, we have

$$\eta_m \le \eta_k \le \tau(t) < t_{k+1} - t_k + \eta_k \le \tau_M,$$

where η_m is a lower bound on the network-induced delay, τ_M denotes the maximum time span between the time $s_k = t_k - \eta_k$ at which the state is sampled and the time t_{k+1} at which the next update arrives at the ZOH and $\tau_M = \text{MATI} + \text{MAD}$. See Fig. 1.4 for an example of $\tau(t)$.

This is one of the applications that motivate the stability analysis of systems with interval (or non-small) delay $\tau(t) \in [h_0, h_1]$ with $h_0 > 0$ (see, e.g., [45, 80, 110]). Keeping in mind that (1.23) can be represented as

$$\dot{x}(t) = Ax(t) + A_1 x(t - h_0) - A_1 \int_{t-\tau(t)}^{t-h_0} \dot{x}(s)ds, \ h_0 \triangleq \eta_m,$$

the stability of (1.23) can be analyzed via Lyapunov functionals of the form [45]:

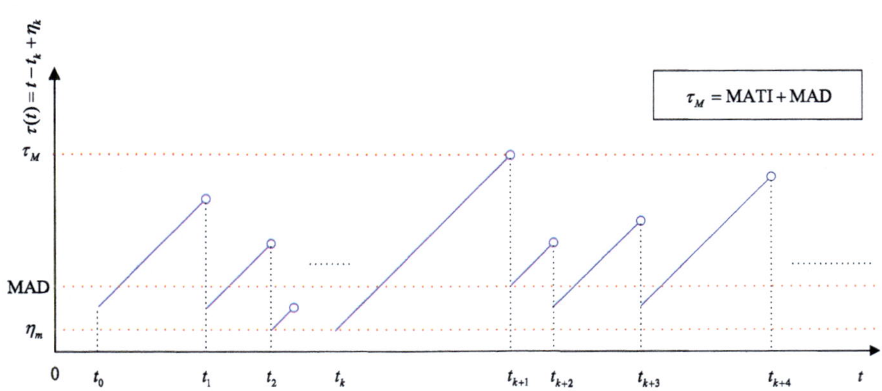

Fig. 1.4 NCSs: piecewise-continuous time-delay

$$V(t, x_t, \dot{x}_t) = V_n(x_t, \dot{x}_t) + V_1(t, x_t, \dot{x}_t),$$

where V_n is a "nominal" functional for the "nominal" system with constant delay

$$\dot{x}(t) = Ax(t) + A_1x(t - h_0)$$

and where

$$\begin{aligned} V_1 = &\int_{t-h_1}^{t-h_0} x^T(s)S_1x(s)ds + \int_{t-\tau(t)}^{t-h_0} x^T(s)Q_1x(s)ds \\ &+(h_1 - h_0)\int_{-h_1}^{-h_0}\int_{t+\theta}^{t} \dot{x}^T(s)R_1\dot{x}(s)dsd\theta, \ h_1 \triangleq \tau_M \end{aligned}$$

with $S_1 > 0,\ Q_1 > 0,\ R_1 > 0$.

In the case where the nominal system is stable for all constant delays from $[0, h_0]$, V_n can be chosen in the form of (1.17), where $h = h_0$ and $Q = 0$. Then, the stability conditions in terms of LMIs can be derived by using the standard arguments for delay-dependent analysis, e.g., in [47, 173].

1.3.3.5 Time-Dependent Lyapunov Functionals for Sampled-Data Systems

Note that the existing methods in the framework of time-delay approach are based on some Lyapunov-based analysis of systems with uncertain and bounded fast-varying delays. Therefore, these methods cannot guarantee the stability if the delay is not smaller than the analytical upper bound on the constant delay that preserves the stability. However, it is well known that in many systems the upper bound on the sampling that preserves the stability may be higher than the one for the constant delay, see examples in [149], as well as the following example.

Example 1.1 Consider the following simple and much-studied problem (see, e.g., [75, 170] and the references therein):

$$\dot{x}(t) = -x(s_k),\ t_k \le t < t_{k+1},\ k \in \mathbb{Z}^+. \tag{1.24}$$

It is well known that the equation

$$\dot{x}(t) = -x(t - \tau) \tag{1.25}$$

with constant delay τ is asymptotically stable for $\tau \le \pi/2$ and unstable for $\tau > \pi/2$, whereas for the fast-varying delay, it is stable for $\tau < 1.5$ and there exists a destabilizing delay with an upper bound greater than 1.5. This means that all the existing methods via time-independent Lyapunov functionals cannot guarantee the stability of system (1.25) for the sampling intervals that may be greater than $\pi/2$. It is easy to check that in the case of pure (uniform) sampling, the system remains stable for all constant samplings less than 2 and becomes unstable for samplings greater than 2.

Therefore, it is necessary to develop new Lyapunov functional-based techniques for sampled-data control to improve the results. Inspired by the construction of discontinuous Lyapunov functions in [162] for impulsive systems, the time-dependent Lyapunov functionals were proposed in [46] for the analysis of sampled-data systems in the framework of time-delay approach. The main idea is that for (1.10), the standard time-independent term

$$\int_{-h}^{0}\int_{t+\theta}^{t} \dot{x}^T(s)R\dot{x}(s)dsd\theta, \ R > 0 \tag{1.26}$$

with $h \triangleq$ MATI can be advantageously replaced by the term

$$(h - t + t_k)\int_{t_k}^{t} \dot{x}^T(s)R\dot{x}(s)ds, \ t_k \le t < t_{k+1}, \tag{1.27}$$

which provides time-dependent LKFs $\bar{V}(t) = V(t, x_t, \dot{x}_t)$. The function $\bar{V}(t)$ may be discontinuous in time, but it is not allowed to grow in the jumps as shown in Fig. 1.5. The introduced time-dependent Lyapunov functionals lead to qualitatively new results for time-delay systems, allowing a superior performance under the sampling, than the one under the constant delay. The stability of system (1.10) is based on the following lemma

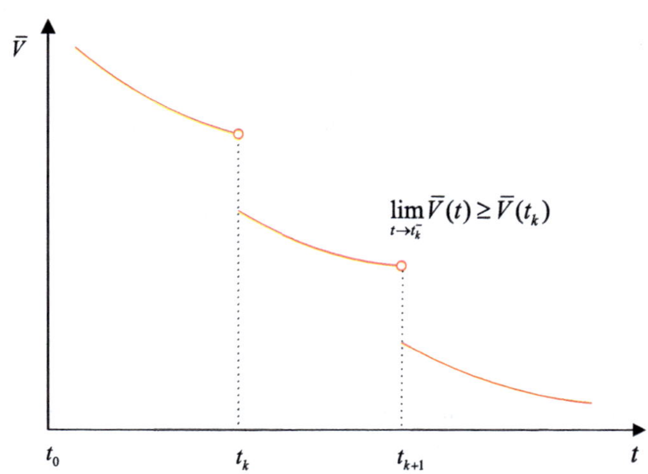

Fig. 1.5 Discontinuous in time Lyapunov functional

Lemma 1.2 *Let there exist positive numbers* α, β, δ *and a functional* $V : \mathbb{R}^+ \times W[-h, 0] \times L_2[-h, 0] \to \mathbb{R}^+$ *such that*

$$\beta|\phi(0)|^2 \leq V(t, \phi, \dot{\phi}) \leq \delta\|\phi\|_W^2. \tag{1.28}$$

Let the function $\bar{V}(t) = V(t, x_t, \dot{x}_t)$ *be continuous from the right for* $x(t)$ *satisfying (1.10), absolutely continuous for* $t \neq t_k$ *and satisfy*

$$\lim_{t\to t_k^-} \bar{V}(t) \geq \bar{V}(t_k). \tag{1.29}$$

(i) If along (1.10)

$$\dot{\bar{V}}(t) \leq -\tilde{\beta}|x(t)|^2 \ for\ t \neq t_k\ and\ for\ some\ scalar\ \tilde{\beta} > 0,$$

then (1.10) is asymptotically stable.

(ii) If along (1.10)

$$\dot{\bar{V}}(t) + 2\alpha\bar{V}(t) \leq 0, \quad for\ t \neq t_k,$$

then $\bar{V}(t) \leq e^{-2\alpha t}\bar{V}(0)$, *which implies that*

$$|x(t)|^2 \leq e^{-2\alpha t}\frac{\delta}{\beta}\|x_0\|_W^2, \tag{1.30}$$

and thus, system (1.10) is exponentially stable with the decay rate α.

Remark 1.1 The above discontinuous Lyapunov constructions and their extensions [46, 192] give efficient tool for different control problems; see, e.g., [128] for stabilization of NCSs with large network-induced delays, [38, 39, 133, 134, 137] for stabilization of NCSs under scheduling protocols, [263] for H_∞ filter of sampled-data systems, [101, 118, 151, 238] for synchronization of complex systems.

1.3.3.6 General and Augmented Lyapunov Functional Method

A necessary condition for the application of the simple LKFs considered in the previous sections is the asymptotic stability of (1.10) or (1.23) with $\tau(t) = 0$. Consider e.g., the following system with a constant delay

$$\dot{x}(t) = \begin{bmatrix} 0 & 1 \\ -2 & 0.1 \end{bmatrix} x(t) + \begin{bmatrix} 0 & 0 \\ 1 & 0 \end{bmatrix} x(t-h), \ x(t) \in \mathbb{R}^2.$$

This system is unstable for $h = 0$ and is asymptotically stable for the constant delay $h \in [0.10017, 1.7178]$ [73]. For analysis of such systems (particularly, for using delay for stabilization), the simple Lyapunov functionals considered in the previous sections are not suitable. For stability conditions of system (1.10) or (1.23) with constant $\tau(t) \equiv h$ and with $A + A_1$ not necessary to be Hurwitz, one can use a general quadratic Lyapunov functional:

$$
\begin{aligned}
V(x_t) = {} & x^T(t)Px(t) + 2x^T(t)\int_{-h}^{0} Q(\xi)x(t+\xi)d\xi \\
& + \int_{-h}^{0}\int_{-h}^{0} x^T(t+s)R(s,\xi)dsx(t+\xi)d\xi \\
& + \int_{-h}^{0} x^T(t+\xi)S(\xi)x(t+\xi)d\xi,
\end{aligned}
\tag{1.31}
$$

where $0 < P \in \mathbb{R}^n$ and where $n \times n$ matrix functions $Q(\xi)$, $R(\xi, \eta) = R^T(\eta, \xi)$ and $S(\xi) = S^T(\xi)$ are absolutely continuous. For the sufficiency of (1.31), one has to formulate conditions for $V \geq \alpha_0|x(t)|^2$, $\alpha_0 > 0$ and $\dot{V} \leq -\alpha|x(t)|^2$, $\alpha > 0$.

The LMI sufficient conditions via general Lyapunov functional of (1.31) and discretization were found in [71], where $Q(\xi)$, $R(\xi, \eta) = R^T(\eta, \xi)$ and $S(\xi) = S^T(\xi) \in \mathbb{R}^{n\times n}$ are continuous and piecewise-linear matrix functions. The resulting LMI stability conditions appeared to be very efficient, leading in some examples to results close to analytical ones. For the discretized Lyapunov functional method, see Sect. 5.7 of [73].

Till [44], no design problems were solved by this method due to bilinear terms in the resulting matrix inequalities. The latter terms arise from the substitution of $\dot{x}(t)$ by the right-hand side of the differential equation in $\dot{V}$. The descriptor discretized method suggested in [44] avoids this substitution. The descriptor discretized method was applied to state-feedback design of H_∞ controllers for neutral-type systems with discrete and distributed delays [59] and to dynamic output-feedback H_∞ controller of retarded systems with state, input and output delays [218]. For differential-algebraic systems with delay, the corresponding general LKFs were studied in [72].

A more general quadratic Lyapunov functional has the form

$$
\begin{aligned}
V(x_t) = {} & x^T(t)Px(t) + 2x^T(t)\int_{-h}^{0} Q(\xi)x(t+\xi)d\xi \\
& + \int_{-h}^{0}\int_{-h}^{0} x^T(t+s)R(s,\xi)ds\, x(t+\xi)d\xi + \int_{t-h}^{t} x^T(\xi)Sx(\xi)d\xi \\
& + \int_{-h}^{0}\int_{t+\theta}^{t} \dot{x}^T(s)R_0\dot{x}(s)dsd\theta,
\end{aligned}
\tag{1.32}
$$

where $P > 0$, $S > 0$, $R_0 > 0$. The matrix functions $Q(\xi) \in \mathbb{R}^{n\times n}$ and $R(\xi, \eta) = R^T(\eta, \xi) \in \mathbb{R}^{n\times n}$ are absolutely continuous. For the sufficiency of (1.32), one has to formulate conditions for $V \geq \beta|x(t)|^2$, $\beta > 0$ and $\dot{V} \leq -\alpha|x(t)|^2$, $\alpha > 0$.

Choosing in (1.32) $R = Q = 0$ and replacing R_0 by hR, we arrive at the simple Lyapunov functional (1.17), where $Q = 0$. Consider now (1.32) with constant $R \equiv Z$ and Q, and replace R_0 by hR. Then, we arrive at the augmented Lyapunov functional of the form

$$V(x_t, \dot{x}_t) = \begin{bmatrix} x(t) \\ \int_{t-h}^{t} x(s)ds \end{bmatrix}^T \begin{bmatrix} P & Q \\ * & Z \end{bmatrix} \begin{bmatrix} x(t) \\ \int_{t-h}^{t} x(s)ds \end{bmatrix} + \int_{t-h}^{t} x^T(s)Sx(s)ds$$
$$+h \int_{-h}^{0} \int_{t+\theta}^{t} \dot{x}^T(s)R\dot{x}(s)dsd\theta, \tag{1.33}$$

where

$$\begin{bmatrix} P & Q \\ * & Z \end{bmatrix} > 0, \ S > 0, \ R > 0.$$

Note that the term $Q \neq 0$ in (1.33) allows to derive non-convex in h conditions that do not imply the stability of the original system with $h = 0$. A remarkable result was obtained in [197] for systems with constant discrete and distributed delays and in [201] for systems with fast-varying discrete delays: LMI conditions that may guarantee the stability of systems which are unstable with the zero delay (i.e., in the case of "stabilizing delay") were derived by the Wirtinger-based integral inequality, which includes Jensen's inequality as a particular case, and by the augmented Lyapunov functional (1.33).

In recent years, several attempts have been done concerning the extension of Jensen's inequality such as auxiliary-based [174], Bessel–Legendre inequality [198–200] or polynomials-based inequality [117]. By construction of more general augmented LKFs and the application of these developed integral inequalities, a series of less conservative stability conditions was achieved [146, 198–200, 251, 252].

Remark 1.2 In the literature, there are many other methods that have been proposed to stability analysis and/or control synthesis of NCSs. Networked predictive method is effective for NCSs with communication delays and packet dropouts [127, 231, 240, 241]. A looped-functional approach is for robust analysis of sampled-data system either considering directly the sampled-data system formulation [192] or the impulsive system formulation [15, 193]. Model predictive control plays an important role in dealing with constraints, such as actuator or physical limitations [78, 79, 142, 154]. Sliding model control, as an effective robust control strategy, has also been applied to NCSs [35, 76, 205]. With the development of cloud computing technologies, the cloud-based networked control has exhibited potential advantages [1, 239]. Furthermore, due to the openness of the network, NCSs are more vulnerable to malicious threats such as data tampering, eavesdropping and interception. Therefore, the issues of network security have attracted ever-increasing attention in recent years; see, e.g., [3, 27, 86, 98, 141, 158, 184, 257] and the references therein.

The objective of this research monograph is summarizing our work on linear NCSs under communication constraints by virtue of the time-delay approach. Note that various types of NCSs and control problems have been considered in existing monographs. To mention a few, the stabilization and optimization of stochastic NCSs under information constraints have been addressed in [250], where the issue of quantization is specifically dealt with and the existence of optimal quantizers and control policies is established. The synchronization control for large-scale network systems with non-identical nodes has been studied in [237].

1.4 Monograph Overview

Besides the Introduction in this chapter, this monograph is organized as follows:

Chapter 2 extends the discontinuous Lyapunov functional approach to NCSs, where variable network-induced delays, variable sampling intervals and data packet dropouts are taken into account. The LKFs in this chapter depend on the time and on the upper bound of the network-induced delays. The new analysis is applied to the state-feedback and to a novel network-based static output-feedback H_∞ control problems.

Chapter 3 provides new discontinuous Lyapunov functionals for sampled-data control in the presence of a constant input delay. The construction of these functionals is based on the vector extension of Wirtinger's inequality. These functionals lead to simplified and efficient stability conditions in terms of LMIs. The new stability analysis is applied to sampled-data state-feedback stabilization and to a novel sampled-data static output-feedback problem, where the delayed measurements are used for stabilization.

Chapter 4 develops a time-delay approach to the exponential stability and the induced L_2-gain of NCSs that are subject to time-varying transmission intervals, time-varying transmission delays and communication constraints. The system sensor nodes are supposed to be distributed over a network. The scheduling of sensor information toward the controller is ruled by the classical Round-Robin protocol. The closed-loop system is presented as a switched system with multiple and ordered time-varying delays. The efficiency of the method is illustrated on the batch reactor and on the cart–pendulum benchmark problems. The results essentially improve the hybrid system-based existing ones and allow treating the case of non-small network-induced delay.

Chapter 5 investigates the stability of NCSs with distributed sensor nodes under TOD or under Round-Robin protocols. A unified hybrid system model under both protocols for the closed-loop system is presented; it contains time-varying delays in the continuous dynamics and in the reset conditions. A new Lyapunov–Krasovskii method, which is based on discontinuous in time Lyapunov functional, is introduced for the stability analysis of the delayed hybrid systems. The efficiency of the time-delay approach is illustrated on the examples of uncertain cart–pendulum and of batch reactor.

Chapter 6 is devoted to the modeling and analysis of NCSs under stochastic protocols. Two classes of protocols are considered. The first one is defined by an independent and identically distributed (iid) stochastic process. The activation probability of each sensor node for this protocol is a given constant, whereas it is assumed that collisions occur with a certain probability. The resulting closed-loop system is a stochastic impulsive system with delays both in the continuous dynamics and in the reset equations, where the system matrices have stochastic parameters with Bernoulli distributions. The second scheduling protocol is defined by a discrete-time Markov chain with a known transition probability matrix taking into account collisions. The resulting closed-loop system is a Markovian jump impulsive system with delays both in the continuous dynamics and in the reset equations. Sufficient conditions for the exponential mean-square stability of the resulting closed-loop system are derived via a Lyapunov–Krasovskii-based method. The efficiency of the method is illustrated on an example of a batch reactor. It is demonstrated how the time-delay approach allows treating network-induced delays larger than the sampling intervals in the presence of collisions.

Chapter 7 develops the time-delay approach to large-scale NCSs with multiple local communication networks connecting sensors, controllers and actuators. The local networks operate asynchronously and independently of each other in the presence of variable sampling intervals, transmission delays and scheduling protocols. The communication delays are allowed to be greater than the sampling intervals. A novel Lyapunov–Krasovskii method is proposed for the exponential stability analysis of the closed-loop large-scale system. In the case of networked control of a single plant, the presented results lead to simplified conditions in terms of reduced-order LMIs comparatively to the results in Chaps. 4 and 5. Numerical examples from the literature illustrate the efficiency of the results.

Chapter 8 is concerned with the stability analysis of NCSs with dynamic quantization, variable sampling intervals and communication delays. A time-triggered zooming algorithm for the dynamic quantization at the sensor side is proposed leading to an exponentially stable closed-loop system. The algorithm includes proper initialization of the zoom parameter. More precisely, given a bound on the state initial conditions and the values of the quantizer range and error, we derive conditions for finding the initial value of the zoom parameter, starting from which the exponential stability is guaranteed by using "zooming-in" only. Polytopic-type uncertainties in the system model can be easily included in the analysis. The efficiency of the method is illustrated on an example of an uncertain cart–pendulum system.

Chapter 9 focuses on the solution bounds for discrete-time NCSs via delay-dependent Lyapunov–Krasovskii methods. The solution bounds are widely used for systems with input saturation caused by actuator saturation or by the quantizers with saturation. In the presence of actuators saturation, the time-delay approach is extended to the stability analysis of NCSs with discrete-time plant under Round-Robin or under TOD protocols. The communication delays are allowed to be larger than the sampling intervals. A novel Lyapunov-based method is proposed for finding the domain of attraction.

Chapter 10 analyzes the exponential stability of a discrete-time linear plant in feedback control over a communication network with distributed sensor nodes, dynamic quantization, large communication delays, variable sampling intervals and Round-Robin protocol. The closed-loop system is modeled as a switched system with multiple ordered time-varying delays and bounded disturbances. We propose a time-triggered zooming algorithm implemented at the sensors that preserves the exponential stability of the closed-loop system. A direct Lyapunov approach is proposed for the initialization of the zoom variable. The effectiveness of the method is illustrated on cart–pendulum and quadruple-tank processes.

Chapter 11 deals with the stability of discrete-time networked systems with multiple sensor nodes under dynamic scheduling protocols. The access to the communication medium is orchestrated by TOD or by iid stochastic protocol that determines which sensor node can access the network at each sampling instant and transmit its corresponding data. Through a time-delay approach, a unified discrete-time hybrid system with time-varying delays in the dynamics and in the reset conditions is formulated under both scheduling protocols. A new stability criterion for discrete-time systems with time-varying delays is proposed by the discrete-time counterpart of the second-order Bessel–Legendre integral inequality. The proposed approach is applied to guarantee the stability of the resulting discrete-time hybrid system model with respect to the full state under TOD or iid scheduling protocol.

Chapter 12 considers discrete-time large-scale networked control systems with multiple local communication networks connecting sensors, controllers and actuators. The local networks operate asynchronously and independently of each other in the presence of variable sampling intervals, transmission delays and scheduling protocols. The time-delay approach is extended to H_∞ decentralized control in the discrete-time. An appropriate Lyapunov–Krasovskii method is proposed leading to efficient LMI conditions for the exponential stability and l_2-gain analysis of the closed-loop large-scale system. The differences from the continuous-time results are discussed. A numerical example of decentralized control of two coupled cart–pendulum systems illustrates the efficiency of the results.

1.5 Notes

The results in this monograph are based mainly on the following journal publications. Acknowledgment is given to

Chapter 2
Stabilization of NCSs via Discontinuous Lyapunov Functionals

More general constructions of the time-dependent Lyapunov functionals addressed in Chap. 1 can be found in [46] for the analysis of sampled-data systems in the framework of time-delay approach. The introduced time-dependent terms of Lyapunov functionals lead to qualitatively new results, allowing a superior performance under the sawtooth delay, than the one under the constant delay. In this chapter, we extend the discontinuous Lyapunov functional method (in the framework of time-delay approach) for sampled-data systems to network-based H_∞ control, where data packet dropouts and variable network-induced delays are taken into account. The proposed Lyapunov functional depends on the time and on the upper bound of the network-induced delay, and it does not grow along the input update times. We apply our analysis results to the state-feedback and to a novel static output-feedback H_∞ control. It is noted that the observer-based control via network is usually encountered with some waiting strategy and buffers [204]. The implementation of the network-based static output-feedback controller is simple, provided that the system is stabilizable by such a controller. The sufficient conditions for the stabilization via the continuous static output feedback can be found in the survey [219]. Following the sampled-data H_∞ control [217], we consider an H_∞ performance index that takes into account the updating rates of the measurement. This index is related to the energy of the measurement noise. Numerical examples show that the novel discontinuous terms in the Lyapunov functional essentially reduce the conservatism.

2.1 Problem Formulation

Consider the system

$$\begin{aligned} \dot{x}(t) &= Ax(t) + B_2u(t) + B_1w(t), \\ z(t) &= C_1x(t) + D_{12}u(t), \end{aligned} \tag{2.1}$$

K. Liu et al., *Networked Control Under Communication Constraints*, Advances in Delays and Dynamics 11, https://doi.org/10.1007/978-981-15-4230-5_2

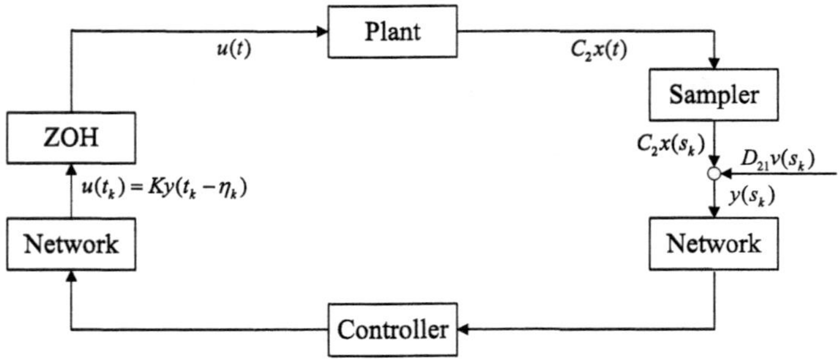

Fig. 2.1 Networked static output-feedback control system

where $x(t) \in \mathbb{R}^n$ is the state vector, $w(t) \in \mathbb{R}^{n_w}$ is the disturbance, $u(t) \in \mathbb{R}^{n_u}$ is the control input and $z(t) \in \mathbb{R}^{n_z}$ is the signal to be controlled or estimated, A, B_1, B_2, C_1 and D_{12} are system matrices with appropriate dimensions. We will consider the exponential stabilization (for $w = 0$) and H_∞ control of (2.1) via state feedback or static output feedback.

2.1.1 *Static Output-Feedback Control*

Consider the static output-feedback controller of NCSs shown in Fig. 2.1. The sampler is time-driven, whereas the controller and the ZOH are event-driven (in the sense that the controller and the ZOH update their outputs as soon as they receive a new sample). We assume that the measurement output $y(s_k) \in \mathbb{R}^{n_y}$ is available at discrete sampling instants

$$0 = s_0 < s_1 < \cdots < s_k < \cdots, \quad k \in \mathbb{Z}^+, \quad \lim_{k\to\infty} s_k = \infty \tag{2.2}$$

and it may be corrupted by a measurement noise signal $v(s_k)$ (see Fig. 2.1):

$$y(s_k) = C_2x(s_k) + D_{21}v(s_k). \tag{2.3}$$

We take into account the data packet dropouts by allowing the sampling to be nonuniform. Thus, in our formulation $y(s_k)$, $k \in \mathbb{Z}^+$, correspond to the measurements that are not lost.

Denote by t_k the updating instant time of the ZOH, and suppose that the updating signal at the instant t_k has experienced a signal transmission delay η_k. The timing diagram of the considered NCSs with both delay and packet dropout is shown in Fig. 2.2, where $s_k = t_k - \eta_k$ denotes the sampling time of the data that has not been lost. Following [161], we allow the delays η_k to grow larger than $s_{k+1} - s_k$, provided that the sequence of input update times t_k remains strictly increasing. This means that if an old sample gets to the destination after the most recent one, it should be dropped.

The static output-feedback controller has the form $u(t_k) = Ky(t_k - \eta_k)$, where K is the controller gain. Thus, considering the behavior of the ZOH, we have

$$u(t) = Ky(t_k - \eta_k), \quad t_k \le t < t_{k+1}, \quad k \in \mathbb{Z}^+ \tag{2.4}$$

with t_{k+1} being the next updating instant time of the ZOH after t_k.

As in [63, 65, 161, 246], we assume that

$$t_{k+1} - t_k + \eta_k \le \tau_M, \quad 0 \le \eta_k \le \eta_M, \quad k \in \mathbb{Z}^+, \tag{2.5}$$

where η_M, i.e., MAD is a known upper bound on the network-induced delays η_k. As a Corollary from the main result, we will formulate the sufficient conditions for the stabilization of NCSs with constant delay $\eta_k \equiv \eta_M$.

Remark 2.1 The assumption (2.5) is equivalent to

$$s_{k+1} - s_k + \eta_{k+1} \le \tau_M, \quad 0 \le \eta_{k+1} \le \eta_M, \quad k \in \mathbb{Z}^+. \tag{2.6}$$

The latter implies that $s_{k+1} - s_k \le \tau_M$, i.e., the sampling intervals and the numbers of successive packet dropouts are uniformly bounded.

Remark 2.2 Consider now a more general situation, where the older sample can get to the destination later than the most recent one and where the older data packet is not discarded. In this more general case, our results will remain true provided that τ_M satisfying (2.5) can be found and that $\lim_{k\to\infty} t_k = \infty$.

Defining

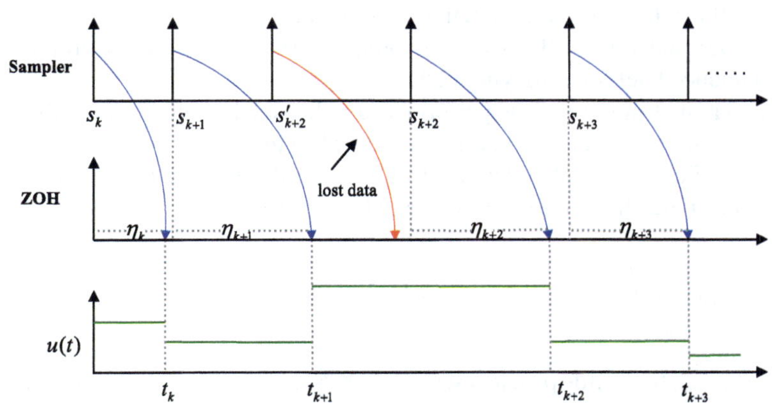

Fig. 2.2 Timing diagram of the NCSs (s'_k denotes the sampling instant, where the measurement is lost)

$$\tau(t) = t - t_k + \eta_k, \; t_k \le t < t_{k+1}, \tag{2.7}$$

we obtain the following closed-loop system (2.1), (2.4):

$$\begin{aligned}\dot{x}(t) &= Ax(t) + A_1 x(t-\tau(t)) + A_2 v(t-\tau(t)) + B_1 w(t),\\ z(t) &= C_1 x(t) + D_1 x(t-\tau(t)) + D_2 v(t-\tau(t)),\end{aligned} \tag{2.8}$$

where

$$\begin{aligned}A_1 &= B_2 K C_2, \quad A_2 = B_2 K D_{21},\\ D_1 &= D_{12} K C_2, \quad D_2 = D_{12} K D_{21}.\end{aligned} \tag{2.9}$$

Under (2.5) and (2.7), we have $0 \le \tau(t) < t_{k+1} - t_k + \eta_k \le \tau_M$ and $\dot{\tau}(t) = 1$ for $t \ne t_k$.

Denote $\bar{v}(t) = v(t-\tau(t)), \; t \ge t_0$. Then, (2.8) has two disturbances $\bar{v} \in L_2[t_0, \infty)$ and $w \in L_2[t_0, \infty)$, where

$$\begin{aligned}\|\bar{v}\|^2_{L_2} &= \int_{t_0}^{\infty} v^T(t-\tau(t))v(t-\tau(t))dt\\ &= \sum_{k=0}^{\infty}(t_{k+1} - t_k)v^T(t_k - \eta_k)v(t_k - \eta_k).\end{aligned} \tag{2.10}$$

For a prescribed scalar $\gamma > 0$, we thus define the following performance index [217]:

$$\begin{aligned}J &= \|z\|^2_{L_2} - \gamma^2(\|\bar{v}\|^2_{L_2} + \|w\|^2_{L_2})\\ &= \int_{t_0}^{\infty}[z^T(s)z(s) - \gamma^2 w^T(s)w(s)]ds - \gamma^2\sum_{k=0}^{\infty}(t_{k+1} - t_k)v^T(t_k - \eta_k)v(t_k - \eta_k).\end{aligned} \tag{2.11}$$

The objective is to find a controller of (2.4) that internally exponentially stabilizes the system and that leads to L_2-gain of (2.8) less than γ. The latter means that along (2.8) $J < 0$ for the zero initial function $x_{t_0} \equiv 0$ and for all nonzero $w \in L_2[t_0, \infty)$, $v \in l_2[t_0, \infty)$ and for all allowable sampling intervals, data packet dropouts and network-induced delays, satisfying (2.5).

Note that the last term of the performance index J takes into account the updating rates of the measurement and is thus related to the energy of the measurement noise [217]. For the sampled-data control under uniform sampling, a conventional performance index has the form (see, e.g., [183, 209])

$$J_{samp} = \int_{t_0}^{\infty}[z^T(s)z(s) - \gamma^2 w^T(s)w(s)]ds - \gamma^2\sum_{k=0}^{\infty}v^T(t_k)v(t_k).$$

The index J_{samp} has a little physical sense for NCSs since it does not take the updating rates into account.

2.1.2 State-Feedback Control

For the state-feedback case, we consider the static output-feedback formulation, where $C_2 = I$, $D_{21} = 0$ and $v(t_k - \eta_k) \equiv 0$. Thus, the resulting state-feedback controller has the form

$$u(t) = Kx(t_k - \eta_k), \; t_k \leq t < t_{k+1} \tag{2.12}$$

and leads to the following closed-loop system (2.1), (2.12):

$$\begin{aligned} \dot{x}(t) &= Ax(t) + A_1 x(t - \tau(t)) + B_1 w(t), \\ z(t) &= C_1 x(t) + D_1 x(t - \tau(t)), \end{aligned} \tag{2.13}$$

where $\tau(t)$ is defined by (2.7) and where

$$A_1 = B_2 K, \; D_1 = D_{12} K. \tag{2.14}$$

The corresponding performance index has the form

$$J_1 = J_{|v(t_k - \eta_k)=0} = \int_{t_0}^{\infty} [z^T(s)z(s) - \gamma^2 w^T(s)w(s)]ds. \tag{2.15}$$

The objective is to find a state feedback (2.12) that internally exponentially stabilizes the system and that leads to $J_1 < 0$ for the zero initial function and for all nonzero $w \in L_2[t_0, \infty)$ and for all allowable sampling intervals, data packet dropouts and network-induced delays, satisfying (2.5). For the sake of brevity, in the sequel, throughout the chapter, the notation τ stands for the time-varying delay $\tau(t)$.

2.2 Exponential Stability and L_2-Gain Analysis

In this section, we analyze the closed-loop systems (2.8) (and its particular case (2.13), where $C_2 = I$ and $v(t_k - \eta_k) \equiv 0$). The exponential stability of (2.8) with $w = v = 0$, i.e., of

$$\dot{x}(t) = Ax(t) + A_1 x(t - \tau), \;\; \tau = t - t_k + \eta_k, \;\; t_k \leq t < t_{k+1}, \tag{2.16}$$

as well as the L_2-gain analysis of (2.8) will be based on the following

Lemma 2.1 *Assume that there exist positive numbers α, β, δ and a functional $V : \mathbb{R}^+ \times W[t_0 - \tau_M, t_0] \times L_2[t_0 - \tau_M, t_0] \to \mathbb{R}^+$ such that*

$$\beta|\phi(t_0)|^2 \leq V(t, \phi, \dot{\phi}) \leq \delta\|\phi\|_W^2. \tag{2.17}$$

Let the function $\bar{V}(t) = V(t, x_t, \dot{x}_t)$ be continuous from the right for $x(t)$ satisfying (2.8), absolutely continuous for $t \neq t_k$ and satisfy

$$\lim_{t \to t_k^-} \bar{V}(t) \geq \bar{V}(t_k). \tag{2.18}$$

(i) If along (2.16)

$$\dot{\bar{V}}(t) \leq -\tilde{\beta}|x(t)|^2 \ for\ t \neq t_k\ and\ for\ some\ scalar\ \tilde{\beta} > 0, \tag{2.19}$$

then (2.16) is asymptotically stable.

(ii) If along (2.16)

$$\dot{\bar{V}}(t) + 2\alpha \bar{V}(t) \leq 0, \quad for\ t \neq t_k, \tag{2.20}$$

then $\bar{V}(t) \leq e^{-2\alpha t}\bar{V}(t_0)$, i.e., $|x(t)|^2 \leq e^{-2\alpha t}\frac{\delta}{\beta}\|x_{t_0}\|_W^2$, and thus, system (2.16) is exponentially stable with the decay rate α.

(iii) For a prescribed $\gamma > 0$, if along (2.8)

$$\begin{aligned} &\dot{\bar{V}}(t) + z^T(t)z(t) - \gamma^2 w^T(t)w(t) - \gamma^2 v^T(t_k - \eta_k)v(t_k - \eta_k) < 0, \\ &\forall w \neq 0,\ v \neq 0,\ t_k \leq t < t_{k+1}, \end{aligned} \tag{2.21}$$

then the performance index (2.11) achieves $J < 0$ for all nonzero $w \in L_2$, $v \in l_2$ and for the zero initial function.

Proof For the proof of (ii), see [46]. We now give the proof of (iii). Given $N >> 1$, we integrate the first inequality (2.21) from t_0 till t_N. We have

$$\begin{aligned} &\bar{V}(t_N) - \bar{V}(t_{N-1}) + \bar{V}(t_{N-1}^-) - \bar{V}(t_{N-2}) \cdots + \bar{V}(t_1^-) - \bar{V}(t_0) \\ &+ \int_{t_0}^{t_N} [z^T(t)z(t) - \gamma^2 w^T(t)w(t)]dt - \gamma^2 \sum_{k=0}^{N-1} (t_{k+1} - t_k)v^T(t_k - \eta_k)v(t_k - \eta_k) < 0. \end{aligned}$$

Since $\bar{V}(t_N) \geq 0$, $\bar{V}(t_{k-1}^-) - \bar{V}(t_{k-1}) \geq 0$ for $k = 2, \ldots, N$ and $\bar{V}(t_0) = 0$, we find

$$\int_{t_0}^{t_N} [z^T(t)z(t) - \gamma^2 w^T(t)w(t)]dt - \gamma^2 \sum_{k=0}^{N-1} (t_{k+1} - t_k)v^T(t_k - \eta_k)v(t_k - \eta_k) < 0.$$

Thus, for $N \to \infty$ we arrive at $J < 0$.

A standard time-independent functional for delay-dependent stability of (2.16) with fast-varying delay $\tau \in [0, \tau_M]$ has the form (see, e.g., [36, 56, 80, 172])

$$V_0(x_t, \dot{x}_t) = x^T(t)Px(t) + \int_{t-\tau_M}^{t} e^{2\alpha(s-t)}x^T(s)Sx(s)ds + \frac{1}{\tau_M}\int_{-\tau_M}^{0}\int_{t+\theta}^{t} e^{2\alpha(s-t)}\dot{x}^T(s)R\dot{x}(s)dsd\theta, \ P > 0, \ S > 0, \ R > 0, \tag{2.22}$$

where $\alpha > 0$ corresponds to exponential stability with the decay rate α. In the existing papers [43, 53, 63, 65, 246] in the framework of time-delay approach, the time-independent Lyapunov functionals are usually involved.

For the case of pure sampling with $\eta_k \equiv 0$ and $\tau = t - t_k$, the following time-dependent functional has been introduced in [46]

$$V_s(t, x_t, \dot{x}_t) = \bar{V}_s(t) = x^T(t)Px(t) + \sum_{i=1}^{2} V_{is}(t, x_t, \dot{x}_t),$$

where the discontinuous terms V_{1s} and V_{2s} have the form

$$\begin{aligned} V_{1s}(t, x_t, \dot{x}_t) &= \frac{\tau_M - \tau}{\tau_M}[x(t) - x(t-\tau)]^T X[x(t) - x(t-\tau)], \ X > 0, \\ V_{2s}(t, x_t, \dot{x}_t) &= \frac{\tau_M - \tau}{\tau_M}\int_{t-\tau}^{t} e^{2\alpha(s-t)}\dot{x}^T(s)U\dot{x}(s)ds, \ U > 0, \ \alpha > 0. \end{aligned} \tag{2.23}$$

For $\eta_k \equiv 0$, V_{1s} and V_{2s} do not increase along the jumps, since these terms are non-negative before the jumps at t_k and become zero just after the jumps (because $t_{|t=t_k} = (t - \tau)_{|t=t_k}$). Thus, $\bar{V}_s(t)$ does not increase along the jumps and the condition $\lim_{t \to t_k^-} \bar{V}_s(t) \geq \bar{V}_s(t_k)$ holds.

In the case of $\eta_k \neq 0$ and $\tau = t - t_k + \eta_k$, the discontinuous terms (2.23) cannot be used, because $t_{|t=t_k} \neq (t - \tau)_{|t=t_k} = t_k - \eta_k$. One can modify V_{1s} and V_{2s} as follows

$$\begin{aligned} \tilde{V}_{1s}(t, x_t, \dot{x}_t) &= (t_{k+1} - t)[x(t) - x(t_k)]^T X[x(t) - x(t_k)], \ X > 0, \\ \tilde{V}_{2s}(t, x_t, \dot{x}_t) &= (t_{k+1} - t)\int_{t_k}^{t} e^{2\alpha(s-t)}\dot{x}^T(s)U\dot{x}(s)ds, \ U > 0, \ \alpha > 0 \end{aligned} \tag{2.24}$$

and use the bounds $0 \leq \eta_k \leq \eta_M$ and $t_{k+1} - t_k = t_{k+1} - s_k - \eta_k \leq \tau_M$. However, this leads to the overall bound $\tau_M + \eta_M$ on the delay τ, which is greater than τ_M, because

$$\tau = t - t_k + \eta_k < t_{k+1} - t_k + \eta_k \leq \tau_M + \eta_M. \tag{2.25}$$

Such an extension can be efficient only for $\eta_M \to 0$, whereas for bigger η_M it can lead to more conservative results than the standard results for the uncertain time-varying delays $\tau \in [0, \tau_M]$.

In the case of $\eta_k \neq 0$, the upper bound τ_M, that preserves the stability, is between the corresponding upper bounds on the (arbitrary) fast-varying delay and on the pure sampling (with $\eta_k \equiv 0$). Since the biggest bound is the one for the pure sampling, we construct the discontinuous terms of Lyapunov functional that correspond to the "worst case," where we have the maximum network-induced delay η_M. Defining

$$\tau_1 = \max\{0,\ \tau - \eta_M\} = \max\{0,\ t - t_k - \eta_M + \eta_k\},\quad t_k \le t < t_{k+1}, \tag{2.26}$$

we note that $\tau_{1|t=t_k} = 0$ and that $\tau_1 \le \tau_M - \eta_M$. Consider the functional of the form

$$V(t, x_t, \dot{x}_t) = \bar{V}(t) = V_0(x_t, \dot{x}_t) + \sum_{i=1}^{2} V_i(t, x_t, \dot{x}_t), \tag{2.27}$$

where V_0 is defined by (2.22) and

$$\begin{aligned} V_1(t, x_t, \dot{x}_t) &= \frac{\tau_M - \tau}{\tau_M - \eta_M}[x(t) - x(t-\tau_1)]^T X [x(t) - x(t-\tau_1)],\ X > 0,\\ V_2(t, x_t, \dot{x}_t) &= \frac{\tau_M - \tau}{\tau_M - \eta_M}\int_{t-\tau_1}^{t} e^{2\alpha(s-t)}\dot{x}^T(s)U\dot{x}(s)ds,\ U > 0,\ \alpha > 0. \end{aligned} \tag{2.28}$$

Along the input update times $t = t_k$, V_1 and V_2 do not increase since these terms are non-negative before t_k and become zero just after t_k (because $t_{|t=t_k} = (t - \tau_1)_{|t=t_k}$). Thus, $\bar{V}$ does not increase along the input update times and the condition $\lim_{t\to t_k^-} \bar{V}(t) \ge \bar{V}(t_k)$ holds.

For the simplicity of presentation, we will use the notations:

$$\begin{aligned} &F_{10} = [I\ \ 0_{n\times 5n}],\ \ F_{11} = [-I\ \ 0\ \ \tau I\ \ 0_{n\times 2n}\ \ I],\ \ F_{12} = [0_{n\times 3n}\ (\tau_M - \tau)I\ \ I\ \ -I],\\ &F_{13} = [A\ \ -I\ \ 0_{n\times 3n}\ A_1],\ \ F_{14} = [0\ \ I\ \ 0_{n\times 4n}],\\ &F_{20} = [I\ \ 0_{n\times 7n}],\ \ F_{21} = [-I\ \ 0_{n\times 2n}\ (\tau - \eta_M)I\ \ 0_{n\times 2n}\ I\ 0],\\ &F_{22} = [0_{n\times 2n}\ \ I\ \ 0\ (\tau_M - \tau)I\ \ 0_{n\times 2n}\ \ -I],\\ &F_{23} = [0_{n\times 5n}\ \eta_M I\ \ -I\ \ I],\ \ F_{24} = [A\ \ -I\ \ 0_{n\times 5n}\ A_1],\\ &F_{25} = [0\ \ I\ \ 0_{n\times 6n}],\ \ F_{26} = [I\ \ 0_{n\times 5n}\ \ -I\ \ 0]. \end{aligned} \tag{2.29}$$

By the discontinuous Lyapunov functional (2.27), we obtain the following sufficient conditions.

Theorem 2.1 *(i) Given $\alpha > 0$, assume that there exist $n \times n$ matrices $P > 0$, $R > 0$, $U > 0$, $X > 0$, $S > 0$, T_{11}, P_{2i}, P_{3i}, T_{2i}, M_{2i}, Y_{ij} and Z_{ij}, $i, j = 1, 2$, such that the following four LMIs*

$$\begin{aligned} \Psi_{11} &= \Psi|_{\tau<\eta_M}(0) < 0, & \Psi_{12} &= \Psi|_{\tau<\eta_M}(\eta_M) < 0,\\ \Psi_{21} &= \Psi|_{\tau\ge\eta_M}(\eta_M) < 0, & \Psi_{22} &= \Psi|_{\tau\ge\eta_M}(\tau_M) < 0 \end{aligned} \tag{2.30}$$

hold with the zero row and the zero column deleted, where

$$
\begin{aligned}
\Psi|_{\tau<\eta_M}(\tau) &= \Upsilon_1 + \mathrm{He}(F_{10}^T P F_{14} + \sum_{j=1}^{3} \Sigma_{1j}^T F_{1j}) < 0, \\
\Psi|_{\tau\geq\eta_M}(\tau) &= \Upsilon_2 + \mathrm{He}(F_{20}^T P F_{25} + \sum_{j=1}^{4} \Sigma_{2j}^T F_{2j} + \frac{\tau_M-\tau}{\tau_M-\eta_M} F_{25}^T X F_{26}) \\
&\quad - \frac{1-2\alpha(\tau_M-\tau)}{\tau_M-\eta_M} F_{26}^T X F_{26} < 0
\end{aligned} \tag{2.31}
$$

with

$$
\begin{aligned}
\Upsilon_1 &= \mathrm{diag}\{2\alpha P + S,\ R,\ -e^{-2\alpha\tau_M}\frac{\tau}{\tau_M}R,\ -e^{-2\alpha\tau_M}\frac{\tau_M-\tau}{\tau_M}R,\ -e^{-2\alpha\tau_M}S\}, \\
\Upsilon_2 &= \mathrm{diag}\{2\alpha P + S,\ R + \frac{\tau_M-\tau}{\tau_M-\eta_M}U,\ -e^{-2\alpha\tau_M}S,\ \Omega_{44},\ \Omega_{55},\ \Omega_{66},\}, \\
\Omega_{44} &= -(\tau-\eta_M)\Big[\frac{1}{\tau_M}e^{-2\alpha\tau_M}R + \frac{1}{\tau_M-\eta_M}e^{-2\alpha(\tau_M-\eta_M)}U\Big], \\
\Omega_{55} &= -\frac{\tau_M-\tau}{\tau_M}Re^{-2\alpha\tau_M}, \quad \Omega_{66} = -\frac{\eta_M}{\tau_M}Re^{-2\alpha\tau_M}, \\
\Sigma_{11} &= [Y_{11}\ Y_{12}\ 0_{n\times 3n}\ T_{11}], \quad \Sigma_{12} = [Z_{11}\ Z_{12}\ 0_{n\times 4n}], \quad \Sigma_{13} = [P_{21}\ P_{31}\ 0_{n\times 4n}], \\
\Sigma_{21} &= [Y_{21}\ Y_{22}\ 0_{n\times 4n}\ T_{21}\ 0], \quad \Sigma_{22} = [Z_{21}\ Z_{22}\ 0_{n\times 5n}\ T_{22}], \\
\Sigma_{23} &= [M_{21}\ M_{22}\ 0_{n\times 6n}], \quad \Sigma_{24} = [P_{22}\ P_{32}\ 0_{n\times 6n}].
\end{aligned} \tag{2.32}
$$

Then, system (2.16) is exponentially stable with the decay rate α for all delays (2.7) satisfying (2.5). If the above LMIs hold with $\alpha = 0$, then system (2.16) is exponentially stable with a small enough decay rate.
(ii) For a given $\gamma > 0$, assume that the following LMIs

$$
\left[\begin{array}{c|ccc}
 & P_{2i}^T A_2 & P_{2i}^T B_1 & C_1^T \\
 & P_{3i}^T A_2 & P_{3i}^T B_1 & 0 \\
\Psi_{ij}|_{\alpha=0} & 0 & 0 & 0 \\
 & 0 & 0 & D_1^T \\
\hline
* & -\gamma^2 I & 0 & D_2^T \\
* & * & -\gamma^2 I & 0 \\
* & * & * & -I
\end{array}\right] < 0, \quad i, j = 1, 2, \tag{2.33}
$$

are feasible, where the notations are given in (2.29) and (2.32). Then, system (2.8) is internally exponentially stable and has L_2-gain less than γ.

Proof (i) In order to apply V of (2.27), we consider two cases: $\tau \in [0, \eta_M)$ and $\tau \in [\eta_M, \tau_M]$.

Case 1: $\tau \in [0, \eta_M)$, where $\tau_1 = 0$ and thus, $V_1 = V_2 = 0$. Differentiating $\bar{V}$ along (2.16) yields that

$$
\begin{aligned}
[\dot{\bar{V}}(t)+2\alpha\bar{V}(t)]|_{\tau<\eta_M} \le\ & 2x^T(t)P\dot{x}(t)+\dot{x}^T(t)R\dot{x}(t)-\frac{1}{\tau_M}e^{-2\alpha\tau_M}\int_{t-\tau_M}^{t}\dot{x}^T(s)R\dot{x}(s)ds \\
& +x^T(t)(2\alpha P+S)x(t)-x^T(t-\tau_M)e^{-2\alpha\tau_M}Sx(t-\tau_M).
\end{aligned}
\tag{2.34}
$$

Following [80], we employ the representation

$$
-\int_{t-\tau_M}^{t}\dot{x}^T(s)R\dot{x}(s)ds=-\int_{t-\tau_M}^{t-\tau}\dot{x}^T(s)R\dot{x}(s)ds-\int_{t-\tau}^{t}\dot{x}^T(s)R\dot{x}(s)ds. \tag{2.35}
$$

Then the application of Jensen's inequality leads to

$$
\begin{aligned}
\int_{t-\tau}^{t}\dot{x}^T(s)R\dot{x}(s)ds &\ge \frac{1}{\tau}\int_{t-\tau}^{t}\dot{x}^T(s)dsR\int_{t-\tau}^{t}\dot{x}(s)ds, \\
\int_{t-\tau_M}^{t-\tau}\dot{x}^T(s)R\dot{x}(s)ds &\ge \frac{1}{\tau_M-\tau}\int_{t-\tau_M}^{t-\tau}\dot{x}^T(s)dsR\int_{t-\tau_M}^{t-\tau}\dot{x}(s)ds.
\end{aligned}
\tag{2.36}
$$

Here, for $\tau=0$ we understand by

$$
\frac{1}{\tau}\int_{t-\tau}^{t}\dot{x}(s)ds=\lim_{\tau\to 0}\frac{1}{\tau}\int_{t-\tau}^{t}\dot{x}(s)ds=\dot{x}(t).
$$

For $\tau_M-\tau=0$, the vector $\frac{1}{\tau_M-\tau}\int_{t-\tau_M}^{t-\tau}\dot{x}(s)ds$ is defined similarly as $\dot{x}(t-\tau_M)$. Then denoting

$$
\begin{aligned}
v_{11} &= \frac{1}{\tau}\int_{t-\tau}^{t}\dot{x}(s)ds, \\
v_{12} &= \frac{1}{\tau_M-\tau}\int_{t-\tau_M}^{t-\tau}\dot{x}(s)ds,
\end{aligned}
\tag{2.37}
$$

we obtain

$$
\begin{aligned}
[\dot{\bar{V}}(t)+2\alpha\bar{V}(t)]|_{\tau<\eta_M} \le\ & 2x^T(t)P\dot{x}(t)+\dot{x}^T(t)R\dot{x}(t)-e^{-2\alpha\tau_M}\frac{\tau}{\tau_M}v_{11}^TRv_{11} \\
& -e^{-2\alpha\tau_M}\frac{\tau_M-\tau}{\tau_M}v_{12}^TRv_{12}+x^T(t)(2\alpha P+S)x(t) \\
& -e^{-2\alpha\tau_M}x^T(t-\tau_M)Sx(t-\tau_M).
\end{aligned}
\tag{2.38}
$$

We further insert free-weighting $n\times n$ matrices by adding the following expressions to $\dot{\bar{V}}$ [80]:

$$
\begin{aligned}
0 &= 2[x^T(t)Y_{11}^T+\dot{x}^T(t)Y_{12}^T+x^T(t-\tau)T_{11}^T][-x(t)+x(t-\tau)+\tau v_{11}], \\
0 &= 2[x^T(t)Z_{11}^T+\dot{x}^T(t)Z_{12}^T][-x(t-\tau)+x(t-\tau_M)+(\tau_M-\tau)v_{12}].
\end{aligned}
\tag{2.39}
$$

Applying also the descriptor method, where the right-hand side of the expression

$$
0=2[x^T(t)P_{2j}^T+\dot{x}^T(t)P_{3j}^T][Ax(t)+A_1x(t-\tau)-\dot{x}(t)], \tag{2.40}
$$

with some $n \times n$ matrices $P_{2j}, P_{3j}, j = 1$, is added into the right-hand side of (2.38), and setting

$$\eta_1(t) = \mathrm{col}\{x(t), \dot{x}(t), v_{11}, v_{12}, x(t-\tau_M), x(t-\tau)\},$$

we have

$$[\dot{\bar{V}}(t) + 2\alpha\bar{V}(t)]|_{\tau<\eta_M} \le \eta_1^T(t)\Psi|_{\tau<\eta_M}(\tau)\eta_1(t) \tag{2.41}$$

with $\Psi|_{\tau<\eta_M}(\tau)$ given in (2.31).

Denote $\eta_{12}(t) = \mathrm{col}\{x(t), \dot{x}(t), v_{12}, x(t-\tau_M), x(t-\tau)\}$. It is noticed that the two LMIs $\Psi_{11} < 0$ and $\Psi_{12} < 0$ imply $\Psi|_{\tau<\eta_M}(\tau) < 0$ because

$$\frac{\eta_M - \tau}{\eta_M}\eta_{12}^T(t)\Psi_{11}\eta_{12}(t) + \frac{\tau}{\eta_M}\eta_1^T(t)\Psi_{12}\eta_1(t) = \eta_1^T(t)\Psi_{|\tau<\eta_M}(\tau)\eta_1(t) < 0.$$

This means that if $\Psi_{11} < 0$ and $\Psi_{12} < 0$, then $[\dot{\bar{V}}(t) + 2\alpha\bar{V}(t)]|_{\tau<\eta_M} < 0$.

Case 2: $\tau \in [\eta_M, \tau_M]$ and $\tau_1 = \tau - \eta_M$. Since

$$\frac{d}{dt}x(t-\tau_1) = (1-\dot{\tau}_1)\dot{x}(t-\tau_1) = 0,$$

it follows that

$$\begin{aligned}
\frac{d}{dt}V_1 &= -\frac{1}{\tau_M - \eta_M}[x(t) - x(t-\tau_1)]^T X[x(t) - x(t-\tau_1)] \\
&\quad +2\frac{\tau_M - \tau}{\tau_M - \eta_M}\dot{x}^T(t)X[x(t) - x(t-\tau_1)], \\
\frac{d}{dt}V_2 &= -\frac{1}{\tau_M - \eta_M}\int_{t-\tau_1}^{t} e^{2\alpha(s-t)}\dot{x}^T(s)U\dot{x}(s)ds + \frac{\tau_M - \tau}{\tau_M - \eta_M}\dot{x}^T(t)U\dot{x}(t) - 2\alpha V_2.
\end{aligned}$$

After differentiating $\bar{V}$ along (2.16), we employ the representation

$$\begin{aligned}
-\int_{t-\tau_M}^{t} \dot{x}^T(s)R\dot{x}(s)ds &= -\int_{t-\tau_M}^{t-\tau} \dot{x}^T(s)R\dot{x}(s)ds - \int_{t-\tau}^{t-\tau_1} \dot{x}^T(s)R\dot{x}(s)ds \\
&\quad -\int_{t-\tau_1}^{t} \dot{x}^T(s)R\dot{x}(s)ds.
\end{aligned} \tag{2.42}$$

Then following (2.36), applying the Jensen's inequality and denoting

$$\begin{aligned}
v_{21} &= \frac{1}{\tau - \eta_M}\int_{t-\tau_1}^{t} \dot{x}(s)ds, \\
v_{22} &= \frac{1}{\tau_M - \tau}\int_{t-\tau_M}^{t-\tau} \dot{x}(s)ds, \\
v_{23} &= \frac{1}{\eta_M}\int_{t-\tau}^{t-\tau_1} \dot{x}(s)ds,
\end{aligned} \tag{2.43}$$

we arrive at

$$\begin{aligned}
[\dot{\bar{V}}(t) + 2\alpha \bar{V}(t)]|_{\tau \geq \eta_M} \leq\ & 2x^T(t)P\dot{x}(t) + \dot{x}^T(t)\Big(R + \frac{\tau_M - \tau}{\tau_M - \eta_M}U\Big)\dot{x}(t) \\
& -(\tau - \eta_M)v_{21}^T\Big[\frac{1}{\tau_M}e^{-2\alpha\tau_M}R + \frac{1}{\tau_M - \eta_M}e^{-2\alpha(\tau_M - \eta_M)}U\Big]v_{21} \\
& -e^{-2\alpha\tau_M}\frac{\tau_M - \tau}{\tau_M}v_{22}^T R v_{22} - e^{-2\alpha\tau_M}\frac{\eta_M}{\tau_M}v_{23}^T R v_{23} \\
& +x^T(t)(2\alpha P + S)x(t) - x^T(t - \tau_M)e^{-2\alpha\tau_M}Sx(t - \tau_M) \\
& -\frac{1}{\tau_M - \eta_M}[x(t) - x(t - \tau_1)]^T X[x(t) - x(t - \tau_1)] \\
& +2\frac{\tau_M - \tau}{\tau_M - \eta_M}\dot{x}^T(t)X[x(t) - x(t - \tau_1)] \\
& +\frac{2\alpha(\tau_M - \tau)}{\tau_M - \eta_M}[x(t) - x(t - \tau_1)]^T X[x(t) - x(t - \tau_1)].
\end{aligned} \tag{2.44}$$

We also insert free-weighting $n \times n$ matrices by adding the following expressions

$$\begin{aligned}
0 &= 2[x^T(t)Y_{21}^T + \dot{x}^T(t)Y_{22}^T + x^T(t - \tau_1)T_{21}^T][-x(t) + x(t - \tau_1) + (\tau - \eta_M)v_{21}], \\
0 &= 2[x^T(t)Z_{21}^T + \dot{x}^T(t)Z_{22}^T + x^T(t - \tau)T_{22}^T][-x(t - \tau) + x(t - \tau_M) + (\tau_M - \tau)v_{22}], \\
0 &= 2[x^T(t)M_{21}^T + \dot{x}^T(t)M_{22}^T][-x(t - \tau_1) + x(t - \tau) + \eta_M v_{23}],
\end{aligned} \tag{2.45}$$

and (2.40) with $j = 2$, to the right-hand side of (2.44). Setting

$$\eta_2(t) = \text{col}\Big\{x(t), \dot{x}(t), x(t - \tau_M), v_{21}, v_{22}, v_{23}, x(t - \tau_1), x(t - \tau)\Big\},$$

we finally achieve that

$$[\dot{\bar{V}}(t) + 2\alpha \bar{V}(t)]|_{\tau \geq \eta_M} \leq \eta_2^T(t)\Psi|_{\tau \geq \eta_M}(\tau)\eta_2(t), \tag{2.46}$$

where $\Psi|_{\tau \geq \eta_M}(\tau)$ is given in (2.31).

Similar to the case 1, note that $\Psi_{21} < 0$ and $\Psi_{22} < 0$ imply $\Psi|_{\tau \geq \eta_M} < 0$. Hence, $[\dot{\bar{V}}(t) + 2\alpha \bar{V}(t)]|_{\tau \geq \eta_M} < 0$. Therefore, along (2.16) we have

$$\dot{V}(t, x_t, \dot{x}_t) + 2\alpha V(t, x_t, \dot{x}_t) \leq \max\{\eta_1^T(t)\Psi|_{\tau < \eta_M}\eta_1(t), \eta_2^T(t)\Psi|_{\tau \geq \eta_M}\eta_2(t)\} < 0, \tag{2.47}$$

and the proof of (i) is completed.

(ii) Consider the Lyapunov functional of (2.27) with $\alpha = 0$. By using the arguments similar to the proof of (i), we find that (2.21) holds if LMIs (2.33) are feasible.

When the network-induced delay is constant, i.e., $\eta_k \equiv \eta_M$, there is only one case of $\tau \in [\eta_M, \tau_M]$. Therefore, the following result is obtained from Theorem 2.1.

Corollary 2.1 *(i) Consider (2.8) with $\eta_k \equiv \eta_M$. For a given $\alpha > 0$, assume that there exist $n \times n$ matrices $P > 0$, $R > 0$, $U > 0$, $X > 0$, $S > 0$, P_{22}, P_{32}, T_{2i}, M_{2i}, Y_{2i} and Z_{2i}, $i = 1, 2$, such that the two LMIs $\Psi_{2i} < 0$, $i = 1, 2$, are feasible, where the notations are given in (2.29) and (2.32). Then (2.16) is exponentially stable with the* *decay rate α. If LMIs $\Psi_{2i} < 0$, $i = 1, 2$, hold with $\alpha = 0$, then (2.16) is exponentially stable with a small enough decay rate.*

(ii) Given $\gamma > 0$, if the two LMIs (2.33), where $i = 2$, $j = 1, 2$, are feasible, then (2.8) is internally exponentially stable and the cost function (2.11) achieves $J < 0$ for all nonzero $w \in L_2$, $v \in l_2$ and for the zero initial condition.

Remark 2.3 Note that the LMIs of Theorem 2.1 and Corollary 2.1 are affine in the system matrices. Therefore, in the case of polytopic-type uncertainty with A, B_1, B_2, C_1 and D_{12} from the uncertain time-varying polytope

$$\begin{aligned} \Omega &= \sum_{j=1}^{M} f_j(t)\Omega_j, \quad 0 \le f_j(t) \le 1, \quad \sum_{j=1}^{M} f_j(t) = 1, \\ \Omega_j &= \left[A^{(j)} \; B_1^{(j)} \; B_2^{(j)} \; C_1^{(j)} \; D_{12}^{(j)} \right], \end{aligned} \tag{2.48}$$

one has to solve these LMIs simultaneously for all the M vertices Ω_j, applying the same decision variables.

Remark 2.4 The method of this chapter can be extended to the case of $0 < \eta_m \le \eta_k \le \eta_M$ by modifying the functional (2.27) as follows:

$$V_{\eta_m}(t, x_t, \dot{x}_t) = \bar{V}_{\eta_m}(t) = V_{\eta_m}(x_t, \dot{x}_t) + \sum_{i=1}^{2} V_i(t, x_t, \dot{x}_t),$$

where V_1, V_2 are defined by (2.28) and

$$\begin{aligned} V_{\eta_m}(x_t, \dot{x}_t) = x^T(t)Px(t) &+ \int_{t-\eta_m}^{t} e^{2\alpha(s-t)} x^T(s) S_0 x(s) ds \\ &+ \int_{t-\tau_M}^{t-\eta_m} e^{2\alpha(s-t)} x^T(s) S_1 x(s) ds \\ &+ \eta_m \int_{-\eta_m}^{0} \int_{t+\theta}^{t} e^{2\alpha(s-t)} \dot{x}^T(s) R_0 \dot{x}(s) ds d\theta \\ &+ \frac{1}{\tau_M - \eta_m} \int_{-\tau_M}^{-\eta_m} \int_{t+\theta}^{t} e^{2\alpha(s-t)} \dot{x}^T(s) R_1 \dot{x}(s) ds d\theta, \end{aligned}$$

where $P > 0$, $S_i > 0$, $R_i > 0$, $i = 0, 1$. The latter functional leads to analysis in the following two cases: (1) $\eta_m + \eta_M \le \tau_M$ (e.g., if $\eta_k \le s_{k+1} - s_k$) and (2) $\eta_m + \eta_M > \tau_M$. In the first case, one can consider three intervals for τ:

$$\eta_m + \eta_M \le \tau_M \Longrightarrow \tau \in [\eta_m, \eta_M), \;\; \tau \in [\eta_M, \eta_m + \eta_M) \;\text{ and }\; \tau \in [\eta_m + \eta_M, \tau_M],$$

which will result in six LMIs (instead of four for $\eta_m = 0$). The second case leads to consider two intervals for τ:

$$\eta_m + \eta_M > \tau_M \Longrightarrow \tau \in [\eta_m, \eta_M) \;\text{ and }\; \tau \in [\eta_M, \tau_M]$$

and to four LMIs. It can be seen that such an extension essentially complicates the conditions and will not be given.

2.3 Application to Network-Based Design

This section is devoted to the application of the results derived in the previous section to design problems.

2.3.1 State-Feedback Design

In order to find the unknown gain K that exponentially stabilizes (2.13) with notations (2.14) and leads to $J_1 < 0$, we apply the matrix inequalities of Theorem 2.1 to (2.13). This leads to nonlinear matrix inequalities because of the terms $P_{2j}^T B_2 K$, $P_{3j}^T B_2 K$, $j = 1, 2$. Following [217], we assume that $P_{3i} = \varepsilon P_{21} = \varepsilon P_{22}$, $i = 1, 2$, where ε is a scalar. We arrive at the following result.

Corollary 2.2 *Given $\alpha > 0$, let there exist $n \times n$ matrices $\bar{P} > 0$, $\bar{R} > 0$, $\bar{U} > 0$, $\bar{X} > 0$, $\bar{S} > 0$, Q, $\bar{T}_{11}$, $\bar{T}_{2i}$, $\bar{M}_{2i}$, $\bar{Y}_{ij}$ and $\bar{Z}_{ij}$, $i, j = 1, 2$, an $n_u \times n$-matrix L and a tuning parameter $\varepsilon > 0$ such that four LMIs (2.30) are feasible, where P, R, U, X, S and T, M, Y, Z with subindexes are taken with bars and where the notations F_{24}, Σ_{13} and Σ_{24} are replaced by $\bar{F}_{13}$, $\bar{F}_{24}$, $\bar{\Sigma}_{13}$ and $\bar{\Sigma}_{24}$, respectively, with*

$$\begin{aligned} \bar{F}_{13} &= [AQ \quad -Q \quad 0_{n\times 3n} \quad B_2 L], \\ \bar{F}_{24} &= [AQ \quad -Q \quad 0_{n\times 5n} \quad B_2 L], \\ \bar{\Sigma}_{13} &= [I \quad \varepsilon I \quad 0_{n\times 4n}], \\ \bar{\Sigma}_{24} &= [I \quad \varepsilon I \quad 0_{n\times 6n}]. \end{aligned} \tag{2.49}$$

Then the stabilizing gain is given by $K = LQ^{-1}$. If in the above conditions only two LMIs $\Psi_{2,j} < 0$, $j = 1, 2$, are feasible, then the results are valid for (2.13) with constant delay $\eta_k \equiv \eta_M$.

Proof Consider (2.31) with the notations (2.29) and (2.32). Assuming $P_{3i} = \varepsilon P_{21} = \varepsilon P_{22}$, $i = 1, 2$, where ε is a scalar, we denote $Q = P_{21}^{-1}$, $\bar{P} = Q^T P Q$, $\bar{R} = Q^T R Q$, $\bar{U} = Q^T U Q$, $\bar{S} = Q^T S Q$, $\bar{X} = Q^T X Q$, $\bar{T}_{11} = Q^T T_{11} Q$, $\bar{T}_{2i} = Q^T T_{2i} Q$, $\bar{M}_{2i} = Q^T M_{2i} Q$, $\bar{Y}_{ij} = Q^T Y_{ij} Q$, $\bar{Z}_{ij} = Q^T Z_{ij} Q$, $i, j = 1, 2$ and $L = KQ$. Multiplication of $\Psi|_{\tau<\eta_M}(\tau)$ and $\Psi|_{\tau\geq\eta_M}(\tau)$ in (2.31) by $\mathrm{diag}\{Q^T, \cdots, Q^T\}$ and $\mathrm{diag}\{Q, \cdots, Q\}$, from the right and the left, completes the proof.

For the state-feedback H_∞ control, the resulting LMIs have the following form

$$\begin{bmatrix} & | & B_1 & C_1^T \\ & | & \varepsilon B_1 & 0 \\ \Psi_{ij}|_{\alpha=0} & | & 0 & 0 \\ & | & 0 & L^T D_{12}^T \\ - & - & - & - \\ * & | & -\gamma^2 I & 0 \\ * & | & * & -I \end{bmatrix} < 0, \quad i, j = 1, 2. \tag{2.50}$$

Remark 2.5 The results of Corollary 2.2 make use of the tuning scalar parameter ε. One way to address the tuning issue is to apply a numerical optimization algorithm, such as the program fminsearch in the optimization toolbox of MATLAB.

2.3.2 Static Output-Feedback H_∞ Control

It is well known that the static output-feedback stabilization is a non-convex problem. We suggest here some solution to this problem (which may be conservative). Assume that B_2 is of full rank. Then, there exists a mapping $x \mapsto \tilde{T}x$ with non-singular $n \times n$-matrix $\tilde{T}$, such that B_2 has the following partitioned form $B_2^T = [0 \ B^T]$, where $B \in \mathbb{R}^{n_u \times n_u}$ is non-singular. Hence, without loss of generality, we take B_2 in the above form.

Corollary 2.3 *For a given $\gamma > 0$ and tuning scalar parameters ε_i, $i = 2, 3$, and a constant matrix $G \in \mathbb{R}^{n_u \times (n-n_u)}$, assume that there exist $n \times n$ matrices $P > 0$, $R > 0$, $U > 0$, $X > 0$, $S > 0$, T_{11}, T_{2i}, M_{2i}, Y_{ij}, Z_{ij}, $i, j = 1, 2$ and matrices $K \in \mathbb{R}^{n_u \times n_y}$, $G_{k1} \in \mathbb{R}^{(n-n_u)\times(n-n_u)}$, $G_{k2} \in \mathbb{R}^{(n-n_u)\times n_u}$, $k = 2, 3$, such that the four LMIs (2.33) are feasible, where the notations are given in (2.9), (2.29) and (2.32), and where the slack variables P_{2i}, P_{3i}, $i = 1, 2$, are chosen of the following form:*

$$P_{2i} = \begin{bmatrix} G_{21} & G_{22} \\ G & \varepsilon_2 I \end{bmatrix}, \quad P_{3i} = \begin{bmatrix} G_{31} & G_{32} \\ G & \varepsilon_3 I \end{bmatrix}, \quad i = 1, 2. \tag{2.51}$$

Then (2.8) is internally exponentially stable, and the cost function (2.11) achieves $J < 0$ for all nonzero $w \in L_2$, $v \in l_2$ and for the zero initial condition. If in the above conditions only two LMIs, corresponding to $i = 2$, $j = 1, 2$, are feasible, then the results are valid for (2.8) with constant delay $\eta_k \equiv \eta_M$.

Proof Taking into account (2.51), we have

$$P_{ji}^T B_2 K C_2 = \mathrm{col}\{G^T B K C_2, \varepsilon_j B K C_2\}, \quad i = 1, 2, \quad j = 2, 3.$$

Substitution of (2.9) and (2.51) into matrix inequalities of Theorem 2.1 and Corollary 2.1 completes the proof.

The result for the static output-feedback exponential stabilization with a given decay rate can be formulated similarly.

2.4 Numerical Examples

Example 2.1 (*exponential stability and L_2-gain analysis*) Consider the system from [246, 259]:

$$\dot{x}(t) = \begin{bmatrix} 0 & 1 \\ 0 & -0.1 \end{bmatrix} x(t) + \begin{bmatrix} 0 \\ 0.1 \end{bmatrix} u(t) + \begin{bmatrix} 0.1 \\ 0.1 \end{bmatrix} w(t), \\ z(t) = [0 \ 1]x(t) + 0.1u(t), \tag{2.52}$$

where $u(t) = -[3.75 \ 11.5]x(t_k - \eta_k), \ t_k \le t < t_{k+1}$. For the closed-loop system with $w = 0$ and with pure (uniform) sampling, it was found in [162] that the system remains stable for all constant samplings less than 1.7295 and becomes unstable for samplings greater than 1.7295. The closed-loop system with $w = 0$ and with constant delay τ

$$\dot{x}(t) = \begin{bmatrix} 0 & 1 \\ 0 & -0.1 \end{bmatrix} x(t) + \begin{bmatrix} 0 & 0 \\ -0.375 & -1.15 \end{bmatrix} x(t - \tau) \tag{2.53}$$

is asymptotically stable for $\tau \le 1.16$ and becomes unstable for $\tau > 1.17$. The latter means that all the existing methods via time-independent Lyapunov functionals cannot guarantee the stability of (2.53) for the sampling intervals that may be greater than 1.17. When there is no network-induced delay, i.e., $\eta_k \equiv 0$, the resulting τ_M determines an upper bound on the variable sampling intervals $t_{k+1} - t_k$. The results (obtained by various methods in the literature and by Theorem 2.1 with $\alpha = 0$) for the admissible upper bounds on the sampling intervals, which preserve the stability, are listed in Table 2.1. From Table 2.1, we can see that the result by Theorem 2.1 almost coincides with the result of [46] and is close to the exact bound 1.72 for the constant sampling.

For the values of η_M given in Table 2.2, by applying various methods in the literature and by Theorem 2.1 with $\alpha = 0$, we obtain the maximum values of τ_M that preserve the stability (see Table 2.2). The LMIs of Theorem 2.1 with $X > 0$ and $U > 0$ (the nonzero X and U correspond to the discontinuous terms of Lyapunov functional) lead to less conservative results than the same LMIs, where $U = 0$ or $X = 0$ (see Table 2.2). It is noted that in this example the results of Corollary 2.1 for the constant $\eta_k \equiv \eta_M$ coincide with the results of Theorem 2.1 for the variable $0 \le \eta_k \le \eta_M$. Choosing next $\tau_M = 1.1137$, by applying Theorem 2.1, we obtain the maximum value of the decay rate α given in Table 2.3 for different bounds η_M.

Consider next the static output-feedback controller $u(t) = -0.1122y(t_k - \eta_k)$, $t_k \le t < t_{k+1}$, where

Table 2.1 Example 2.1: maximum upper bound on the variable sampling

Method	[172]	[161]	[157]	[62]	[46]	Theorem 2.1
τ_M	1.04	1.11	1.36	1.36	1.69	1.68

Table 2.2 Example 2.1: maximum value of τ_M for different η_M

$\tau_M \setminus \eta_M$	0	0.2	0.4	0.6	0.8
Reference [172]	1.04	1.04	1.04	1.04	1.04
Reference [161]	1.11	1.01	0.95	0.90	0.88
Theorem 2.1 ($U = 0$)	1.28	1.22	1.17	1.13	1.09
Theorem 2.1 ($X = 0$	1.61	1.17	1.10	1.08	1.07
Theorem 2.1	1.68	1.26	1.18	1.14	1.10

Table 2.3 Example 2.1: maximum value of α for different η_M

η_M	0	0.01	0.1	0.2
$\alpha(U > 0, X > 0)$	0.26	0.21	0.12	0.07
$\alpha(U > 0, X = 0)$	0.14	0.12	0.05	0.01

$$y(t_k - \eta_k) = [1\ 0]x(t_k - \eta_k) + 0.2v(t_k - \eta_k), \tag{2.54}$$

and where $\eta_M = 0.1$, $\tau_M = 1.5$. Applying LMIs of Theorem 2.1 (with the zero and with the nonzero X and U), we find that the resulting closed-loop system has an L_2-gain less than $\gamma = 1.27$ (for $X = U = 0$) and less than $\gamma = 1.17$ (for $X > 0$ and $U > 0$). Hence, the discontinuous terms of Lyapunov functional improve the performance (the exponential decay rate and the L_2-gain).

Example 2.2 (*asymptotic stability*) Consider the following simple and much-studied problem (see, e.g., [75, 170] and the references therein):

$$\dot{x}(t) = -x(s_k), \quad t_k \le t < t_{k+1}, \quad k \in \mathbb{Z}^+. \tag{2.55}$$

It is well known that the equation $\dot{x}(t) = -x(t - \tau)$ with constant delay τ is asymptotically stable for $\tau \le \pi/2$ and unstable for $\tau > \pi/2$, whereas for the fast-varying delay it is stable for $\tau < 1.5$ and there exists a destabilizing delay with an upper bound greater than 1.5. It is easy to check that in the case of pure sampling, the system remains stable for all constant samplings less than 2 and becomes unstable for samplings greater than 2. The conditions of [172] guarantee the asymptotic stability for all fast-varying delays from the interval $[0,\ 1.339]$.

By applying Theorem 2.1 with $\alpha = 0$ and $\eta_M = 0$, we find that for all variable samplings from the interval $[0, 1.99]$, the system remains stable. The latter interval is very close to the exact one $[0, 2)$.

By applying Theorem 2.1 and Corollary 2.1 in the cases of variable and constant delay η_k respectively, we obtain the maximum values of τ_M that preserve the stability (see Table 2.4).

Example 2.3 (*state-feedback stabilization*) Consider a two-axis example of a three-axis milling machine tool from [260]:

Table 2.4 Example 2.2: maximum value of τ_M for variable and constant delay η_k

$\tau_M \backslash \eta_M$	0	0.1	0.2	0.4	0.6	0.8	1	1.2
Theorem 2.1	1.99	1.64	1.57	1.50	1.46	1.43	1.40	1.36
Corollary 2.1	1.99	1.64	1.57	1.50	1.46	1.44	1.42	1.41

$$\dot{x}(t) = \begin{bmatrix} 0 & 1 & 0 & 0 \\ 0 & -18.18 & 0 & 0 \\ 0 & 0 & 0 & 1 \\ 0 & 0 & 0 & -17.86 \end{bmatrix} x(t) + \begin{bmatrix} 0 & 0 \\ 515.38 & 0 \\ 0 & 0 \\ 0 & 517.07 \end{bmatrix} u(t), \tag{2.56}$$

where the constant network-induced delays $\eta_k \equiv \eta_M = 50\,\text{ms}$ were considered, the sampling period was chosen to be 100 ms and it was assumed at most two successive data packet dropouts. Under the above assumptions, the resulting value of τ_M equals to 350 ms. Under some additional assumptions on the distribution of packet dropouts, a state-feedback controller has been found in [260] that exponentially stabilizes the system with the decay rate $\alpha = 1.0735$.

Without additional assumptions on the packet dropouts, we apply Corollary 2.2 with $\varepsilon = 0.4$ and we find that the state feedback with the gain

$$K = \begin{bmatrix} -0.0775 & -0.0043 & 0 & 0 \\ 0 & 0 & -0.0759 & -0.0042 \end{bmatrix},$$

stabilizes the system with a greater decay rate $\alpha = 1.2408$ for variable network-induced delays $0 \leq \eta_k \leq 50\,\text{ms}$.

Example 2.4 (*static output-feedback H_∞ control*) Consider system (2.52) with the measurement given by (2.54). It is assumed that the network-induced delay η_k satisfies $0 \leq \eta_k \leq \eta_M = 0.1$ and that $0 \leq t_{k+1} - s_k \leq \tau_M = 1.5$. Choosing $\varepsilon_2 = \varepsilon_3 = 10$, $G = 0.5$ and applying Corollary 2.3, we obtain a minimum performance level of $\gamma = 1.51$. The corresponding static output feedback control law is $u(t) = -0.1122y(t)$. As we have seen in Example 2.1, the above controller, in fact, leads to a smaller performance level of $\gamma = 1.17$ (which follows from the application of Theorem 2.1 to the resulting closed-loop system). The latter improvement of γ illustrates the conservatism of the design method.

2.5 Notes

A piecewise-continuous in time Lyapunov functional method was proposed for the analysis and design of linear NCSs. This method was developed in the framework of time-delay approach. The presented results depend on the upper bound η_M of the network-induced delays. The new analysis was applied to the state-feedback and to a novel static output-feedback H_∞ control. Different from the observer-based control, the static one is easy for implementation. The proposed method essentially reduces the conservatism. It gives insight for new constructions of Lyapunov functionals for systems with time-varying delays.

The conservatism was further reduced in [159] by a modified discontinuous LKF. The proposed method in this chapter was further extended in [97] to the stability and L_2-gain analysis of event-triggered NCSs.

Chapter 3
Wirtinger's Inequality and Sampled-Data Control

Extensions of the discontinuous Lyapunov functional constructions proposed in Chap. 2 to sampled-data systems in the presence of input delay η lead to complicated conditions. Moreover, these conditions become conservative if η is not small. In this chapter, we propose a direct Lyapunov approach via Wirtinger's inequality [148] to sampled-data stabilization in the presence of a constant input delay η. In this approach, novel discontinuous terms are added to "nominal" Lyapunov functionals for the stability analysis of systems with the delay η (either to simple or to complete ones). Being applied to sampled-data systems with $\eta = 0$, the new method recovers the result of [157], but it is more conservative than the one of [46]. However, the new analysis leads to simplified reduced-order LMIs and improves the existing results for $\eta > 0$. Comparatively to the standard time-independent Lyapunov functional terms for interval time-varying delays, the Wirtinger-based terms take advantage of the sawtooth evolution of the delays induced by sampled and hold and, thus, improve the results (both via simple and via discretized Lyapunov functionals).

The new method is applied to the state-feedback sampled-data stabilization. Moreover, a novel sampled-data static output-feedback problem is studied via discontinuous discretized Lyapunov functionals, where the delayed measurements are used for stabilization. This is a sampled-data counterpart of using an artificial delay for continuous-time stabilization studied in [111]. Note that the observer-based sampled-data control of systems with uncertain coefficients may become complicated and may lead to conservative results. From the other side, a simple static output feedback using the previous measurements can be easily designed and implemented.

3.1 Problem Formulation

Consider the following system

$$\dot{x}(t) = Ax(t) + Bu(t), \tag{3.1}$$

K. Liu et al., *Networked Control Under Communication Constraints*, Advances in Delays and Dynamics 11, https://doi.org/10.1007/978-981-15-4230-5_3

where $x(t) \in \mathbb{R}^n$ is the state vector, $u(t) \in \mathbb{R}^{n_u}$ is the control input, and A and B are system matrices with appropriate dimensions. Denote by t_k the updating instant time of the ZOH, and suppose that the updating signal (successfully transmitted signal from the sampler to the controller and to the ZOH) at the instant t_k has experienced a constant signal transmission delay η. We assume that the sampling intervals are bounded

$$t_{k+1} - t_k \leq h_s \ (i.e., \text{MATI}), \quad k \in \mathbb{Z}^+, \tag{3.2}$$

i.e.,

$$t_{k+1} - t_k + \eta \leq h_s + \eta \stackrel{\Delta}{=} \tau_M, \quad k \in \mathbb{Z}^+. \tag{3.3}$$

Here, τ_M denotes the maximum time span between the time $t_k - \eta$ at which the state is sampled and the time t_{k+1} at which the next update arrives at the destination.

The state-feedback controller has the form $u(t_k) = Kx(t_k - \eta)$, where K is the controller gain. Thus, considering the behavior of the ZOH, we have

$$u(t) = Kx(t_k - \eta), \quad t_k \leq t < t_{k+1}, \quad k \in \mathbb{Z}^+ \tag{3.4}$$

with t_{k+1} being the next updating instant time of the ZOH after t_k.

Defining $\tau(t) = t - t_k + \eta$, $t_k \leq t < t_{k+1}$, we obtain the following closed-loop system (3.1), (3.4):

$$\dot{x}(t) = Ax(t) + A_1 x(t - \tau(t)), \quad t_k \leq t < t_{k+1}, \quad k \in \mathbb{Z}^+, \tag{3.5}$$

where $A_1 = BK$. Under (3.3), we have $\eta \leq \tau(t) < t_{k+1} - t_k + \eta \leq \tau_M$ and $\dot{\tau}(t) = 1$ for $t \neq t_k$. For the sake of brevity, in the sequel, throughout the chapter, the notation τ stands for the time-varying delay $\tau(t)$.

The objective of the present chapter is to derive efficient LMI (asymptotic and exponential) stability conditions for system (3.5). Moreover, we will consider the static output-feedback stabilization of (3.1) under the sampled-data measured output $y(t_k) = Cx(t_k)$, $k \in \mathbb{Z}^+$, where $y(t_k) \in \mathbb{R}^{n_y}$ and C is a constant matrix. It is well known that using artificial delay in the (continuous-time) static output feedback can stabilize some systems, which are not stabilizable without delay [111]. For such systems, we will consider a sampled-data static output feedback that uses the previous measurements and we will derive LMI conditions for stabilization.

A novel Lyapunov functional construction is based on the extension of Wirtinger's inequality [77] to the vector case:

Lemma 3.1 ([148]) *Let $z(t) \in W[a, b)$ and $z(a) = 0$. Then for any $n \times n$-matrix $R > 0$, the following inequality holds:*

$$\int_a^b z^T(\xi) R z(\xi) d\xi \leq \frac{4(b-a)^2}{\pi^2} \int_a^b \dot{z}^T(\xi) R \dot{z}(\xi) d\xi. \tag{3.6}$$

3.2 Stabilization via Novel Lyapunov Functionals

The stability of system (3.5) can be analyzed via time-independent functionals of the form [44]:

$$V(x_t, \dot{x}_t) = V_n(x_t, \dot{x}_t) + V_Z(x_t, \dot{x}_t), \tag{3.7}$$

where V_n is a "nominal" functional for the "nominal" system with constant delay

$$\dot{x}(t) = Ax(t) + A_1 x(t - \eta) \tag{3.8}$$

and where (see, e.g., [80])

$$\begin{aligned} V_Z(x_t, \dot{x}_t) &= \int_{t-\tau_M}^{t-\eta} x^T(s) Z_1 x(s) ds + V_{Z_2}(\dot{x}_t), \\ V_{Z_2}(\dot{x}_t) &= (\tau_M - \eta) \int_{-\tau_M}^{-\eta} \int_{t+\theta}^{t} \dot{x}^T(s) Z_2 \dot{x}(s) ds d\theta, \quad Z_1 > 0, \quad Z_2 > 0. \end{aligned} \tag{3.9}$$

Remark 3.1 The time-dependent term introduced in [46] can be modified to the case of $\eta > 0$ as follows:

$$V_U(t, \dot{x}_t) = (t_{k+1} - t) \int_{t_k-\eta}^{t-\eta} \dot{x}^T(s) U \dot{x}(s) ds, \quad U > 0, \ t \in [t_k, t_{k+1}). \tag{3.10}$$

It is clear that V_U does not grow in the jumps since $V_{U|t=t_k} = 0$. Differentiation of V_U leads to

$$\frac{d}{dt} V_U(t, \dot{x}_t) = -\int_{t_k-\eta}^{t-\eta} \dot{x}^T(s) U \dot{x}(s) ds + (t_{k+1} - t) \dot{x}^T(t - \eta) U \dot{x}(t - \eta). \tag{3.11}$$

Hence, an additional term

$$V_{0U}(\dot{x}_t) = (\tau_M - \eta) \int_{t-\eta}^{t} \dot{x}^T(s) U \dot{x}(s) ds$$

is needed with

$$\frac{d}{dt} V_{0U} = (\tau_M - \eta) \dot{x}^T(t) U \dot{x}(t) - (\tau_M - \eta) \dot{x}^T(t - \eta) U \dot{x}(t - \eta).$$

This leads to the same positive term and the same negative integral term (for $U = (\tau_M - \eta) Z_2$) as in

$$\frac{d}{dt}V_{Z_2}(\dot{x}_t) = (\tau_M-\eta)^2\dot{x}^T(t)Z_2\dot{x}(t) - (\tau_M-\eta)\int_{t_k-\eta}^{t-\eta}\dot{x}^T(s)Z_2\dot{x}(s)ds \\ -(\tau_M-\eta)\int_{t-\tau_M}^{t_k-\eta}\dot{x}^T(s)Z_2\dot{x}(s)ds. \tag{3.12}$$

Therefore, $V_U + V_{0U}$ has no clear advantages over the standard double integral term V_{Z_2}.

In the present chapter, we suggest a discontinuous Lyapunov functional

$$V_d(t, x_t, \dot{x}_t) = \bar{V}_1(t) = V_n(x_t, \dot{x}_t) + V_W(t, x_t, \dot{x}_t) \tag{3.13}$$

with a novel discontinuous term

$$V_W(t, x_t, \dot{x}_t) = (\tau_M-\eta)^2\int_{t_k-\eta}^{t}\dot{x}^T(s)W\dot{x}(s)ds \\ -\frac{\pi^2}{4}\int_{t_k-\eta}^{t-\eta}[x(s)-x(t_k-\eta)]^T W[x(s)-x(t_k-\eta)]ds, \tag{3.14}$$

where $W > 0$, $t_k \le t < t_{k+1}$, $k \in \mathbb{Z}^+$. We note that V_W can be represented as a sum of the continuous in time term $(\tau_M-\eta)^2\int_{t-\eta}^{t}\dot{x}^T(s)W\dot{x}(s)ds \ge 0$ with the discontinuous one

$$V_{W1} \triangleq (\tau_M-\eta)^2\int_{t_k-\eta}^{t-\eta}\dot{x}^T(s)W\dot{x}(s)ds \\ -\frac{\pi^2}{4}\int_{t_k-\eta}^{t-\eta}[x(s)-x(t_k-\eta)]^T W[x(s)-x(t_k-\eta)]ds.$$

Since $[x(s)-x(t_k-\eta)]|_{s=t_k-\eta} = 0$, by the extended Wirtinger's inequality (3.6), we have $V_{W1} \ge 0$. Moreover, V_{W1} vanishes at $t{=}t_k$. Hence, the condition $\lim_{t\to t_k^-}\bar{V}_1(t) \ge \bar{V}_1(t_k)$ holds.

Differentiating V_W, we have

$$\frac{d}{dt}V_W = (\tau_M-\eta)^2\dot{x}^T(t)W\dot{x}(t) - \frac{\pi^2}{4}v^T(t)Wv(t), \\ v(t) = x(t_k-\eta) - x(t-\eta). \tag{3.15}$$

Remark 3.2 For $\eta = 0$, it is easily seen from (3.15) that the application of the functional $V_0 = x^T(t)Px(t) + V_W$ with $P > 0$ to (3.5) recovers the conditions of [157], which are based on the small-gain theorem. An advantage of the direct Lyapunov method considered in the present chapter over the small-gain theorem-based results is in its wider applications: to exponential bounds on the solutions of the initial value problems, to finding domains of attraction of some nonlinear systems.

3.2.1 Stabilization via Simple Lyapunov Functional

We start with the stability conditions via $V_d = V_{n1} + V_W$, where V_{n1} is a simple functional of the form

$$
\begin{aligned}
V_{n1}(x_t, \dot{x}_t) &= x^T(t)Px(t) + \int_{t-\eta}^{t} x^T(s)R_1x(s)ds \\
&+\eta \int_{-\eta}^{0}\int_{t+\theta}^{t} \dot{x}^T(s)R_2\dot{x}(s)dsd\theta,\ P > 0,\ R_1 > 0,\ R_2 > 0.
\end{aligned}
\tag{3.16}
$$

Theorem 3.1 *(i) Given $\eta \geq 0$, $h_s > 0$ and K, if there exist $n \times n$ matrices $P > 0$, $W > 0$, $R_i > 0$, $i = 1, 2$, such that the following LMI*

$$
\begin{bmatrix}
\Psi_1 & PA_1 + R_2 & PA_1 & A^T(h_s^2W + \eta^2R_2) \\
* & -R_1 - R_2 & 0 & A_1^T(h_s^2W + \eta^2R_2) \\
* & * & -\frac{\pi^2}{4}W & A_1^T(h_s^2W + \eta^2R_2) \\
* & * & * & -(h_s^2W + \eta^2R_2)
\end{bmatrix} < 0,
\tag{3.17}
$$

is feasible, where $\Psi_1 = \mathrm{He}(PA) + R_1 - R_2$. Then, system (3.5) is asymptotically stable.

(ii) Given $\eta \geq 0$, $h_s > 0$, if there exist $n \times n$ matrices $\bar{P} > 0$, Q, $\bar{W} > 0$, $\bar{R}_i > 0$, $i = 1, 2$, an $n_u \times n$-matrix L and a tuning parameter $\varepsilon > 0$ such that the following LMI

$$
\begin{bmatrix}
\mathrm{He}(AQ) + \bar{R}_1 - \bar{R}_2 & \bar{P} - Q + \varepsilon Q^TA^T & \bar{R}_2 + BL & BL \\
* & -\varepsilon \mathrm{He}(Q) + \eta^2\bar{R}_2 + h_s^2\bar{W} & \varepsilon BL & \varepsilon BL \\
* & * & -\bar{R}_1 - \bar{R}_2 & 0 \\
* & * & * & -\frac{\pi^2}{4}\bar{W}
\end{bmatrix} < 0,
\tag{3.18}
$$

is feasible. Then, the closed-loop system (3.1), (3.4) is asymptotically stable and the stabilizing gain is given by $K = LQ^{-1}$.

Proof (i) Differentiating $\bar{V}_1(t)$ along (3.5) and taking into account (3.15), we find

$$
\begin{aligned}
\dot{\bar{V}}_1(t) &= 2x^T(t)P\dot{x}(t) + x^T(t)R_1x(t) - x^T(t-\eta)R_1x(t-\eta) \\
&+\dot{x}^T(t)(\eta^2R_2 + h_s^2W)\dot{x}(t) - \frac{\pi^2}{4}v^T(t)Wv(t) - \eta\int_{t-\eta}^{t} \dot{x}^T(s)R_2\dot{x}(s)ds.
\end{aligned}
\tag{3.19}
$$

By Jensen's inequality, we have

$$
\begin{aligned}
\eta\int_{t-\eta}^{t} \dot{x}^T(s)R_2\dot{x}(s)ds &\geq \int_{t-\eta}^{t} \dot{x}^T(s)dsR_2\int_{t-\eta}^{t} \dot{x}(s)ds \\
&= [x(t) - x(t-\eta)]^TR_2[x(t) - x(t-\eta)].
\end{aligned}
\tag{3.20}
$$

Then, substitution of $Ax(t) + A_1x(t-\eta) + A_1v(t)$ for $\dot{x}(t)$ leads to

$$\begin{aligned}\dot{\bar{V}}_1(t) \le \zeta_1^T(t) &\begin{bmatrix} \Psi_1 & PA_1 + R_2 & PA_1 \\ * & -R_1 - R_2 & 0 \\ * & * & -\frac{\pi^2}{4}W \end{bmatrix} \zeta_1(t) \\ &+ [Ax(t) + A_1x(t-\eta) + A_1v(t)]^T(\eta^2 R_2 + h_s^2 W) \\ &\times [Ax(t) + A_1x(t-\eta) + A_1v(t)],\end{aligned}$$

where $\zeta_1(t) = \text{col}\{x(t), x(t-\eta), v(t)\}$. Hence, by Schur complements, (3.17) guarantees that $\dot{\bar{V}}_1(t) \le -\tilde{\beta}|x(t)|^2$ for some $\tilde{\beta} > 0$ which completes the proof of (i) (see (i) of Lemma 2.1).

(ii) For the state-feedback design, the descriptor method is used, where the right-hand side of the expression

$$2[x^T(t)P_2^T + \dot{x}^T(t)P_3^T][Ax(t) + A_1x(t-\eta) + A_1v(t) - \dot{x}(t)] = 0$$

with some $n \times n$-matrices P_2, P_3 is added to $\dot{\bar{V}}_1(t)$. Then, (3.19) and (3.20) lead to $\dot{\bar{V}}_1(t) \le \zeta_2^T(t)\Xi_s\zeta_2(t) \le -\tilde{\beta}|x(t)|^2$ for some $\tilde{\beta} > 0$, where $\zeta_2(t) = \text{col}\{x(t), \dot{x}(t), x(t-\eta), v(t)\}$, if

$$\Xi_s \triangleq \begin{bmatrix} \text{He}(P_2^T A) + R_1 - R_2 & P - P_2^T + A^T P_3 & R_2 + P_2^T A_1 & P_2^T A_1 \\ * & -\text{He}(P_3) + \eta^2 R_2 + h_s^2 W & P_3^T A_1 & P_3^T A_1 \\ * & * & -R_1 - R_2 & 0 \\ * & * & * & -\frac{\pi^2}{4}W \end{bmatrix} < 0. \quad (3.21)$$

Following [44, 217], we denote $P_3 = \varepsilon P_2$, where ε is a scalar, $Q = P_2^{-1}$, $\bar{P} = Q^T P Q$, $\bar{W} = Q^T W Q$, $\bar{R}_i = Q^T R_i Q$, $i = 1, 2$, and $L = KQ$. Multiplication of (3.21) by $\text{diag}\{Q^T, Q^T, Q^T, Q^T\}$ and $\text{diag}\{Q, Q, Q, Q\}$, from the left and the right, completes the proof of (ii).

Remark 3.3 The method of [173] for the stability of (3.5) (via functional (3.7) with $V_n = V_{n1}$, Jensen's inequality and the convexity arguments) leads to the following (affine in A and A_1) LMIs:

$$\begin{bmatrix} Z_2 & S_{12} \\ * & Z_2 \end{bmatrix} \ge 0, \quad (3.22)$$

$$\begin{bmatrix} \Psi_1 & R_2 & PA_1 & 0 & A^T(h_s^2 Z_2 + \eta^2 R_2) \\ * & \Psi_2 & Z_2 - S_{12} & S_{12} & 0 \\ * & * & -2Z_2 + \text{He}(S_{12}) & Z_2 - S_{12} & A_1^T(h_s^2 Z_2 + \eta^2 R_2) \\ * & * & * & -Z_1 - Z_2 & 0 \\ * & * & * & * & -(h_s^2 Z_2 + \eta^2 R_2) \end{bmatrix} < 0, \quad (3.23)$$

where S_{12} is $n \times n$ matrix and $\Psi_2 = -R_1 - R_2 + Z_1 - Z_2$.

Table 3.1 Numerical complexity of different methods

Method	Decision variables	No. of LMIs	The maximum order of LMI
Corollary 2.1 (i)	$12.5n^2 + 2.5n$	2	$7n$
Reference [173]	$3.5n^2 + 2.5n$	2	$5n$
Theorem 3.1 (i)	$2n^2 + 2n$	1	$4n$

Comparing LMI (3.17) with LMIs (3.22), (3.23), we can see that (3.17) is a lower-order single LMI with a fewer decision variables (W in (3.17) instead of Z_1, Z_2, S_{12} in (3.22), (3.23)). Note that the conditions in (i) of Corollary 2.1 are essentially more complicated than those of [173]. See Table 3.1 for numerical complexity of the above methods.

Remark 3.4 Consider now the LMI conditions via $V_{n1} + V_{Z_2}$ and Jensen's inequality, which contain the same number of decision variables and LMIs as Theorem 3.1. From (3.12) and Jensen's inequality, we have $\frac{d}{dt} V_{Z_2}(\dot{x}_t) \leq (\tau_M - \eta)^2 \dot{x}^T(t) Z_2 \dot{x}(t) - \nu^T(t) Z_2 \nu(t)$ ($\nu(t)$ is given in (3.15)), which leads to more restrictive LMI (3.17), where $W = Z_2$ and where the (3,3)-term $-\frac{\pi^2}{4} W$ is changed by the (more than twice) bigger term $-Z_2$.

More complicated LMI conditions via time-independent Lyapunov functionals and Jensen's inequality sometimes can be less restrictive than the Wirtinger-based conditions. Thus, in [46] for $\eta = 0$ in one example out of three the results by [157] (which are equivalent to Theorem 3.1) are more conservative than the results by [172]. In order to get less conservative LMI conditions, the functional $V_{n1} + V_Z + V_W$ can be applied by combining the arguments of [173] and of Theorem 3.1.

Remark 3.5 For the exponential stability analysis, we follow [195]. By changing the variable $\bar{x}(t) = x(t)e^{\lambda t}$, (3.5) can be rewritten as

$$\dot{\bar{x}}(t) = (A + \lambda I)\bar{x}(t) + e^{\lambda \tau} A_1 \bar{x}(t - \tau). \tag{3.24}$$

The asymptotic stability of (3.24) for some $\lambda > 0$ implies the exponential stability with the decay rate λ of (3.5). Since $e^{\lambda \tau} \in [\rho_1, \rho_2]$ with $\rho_1 = e^{\lambda \eta}$ and $\rho_2 = e^{\lambda \tau_M}$, (3.24) can be represented as the following polytopic form:

$$\dot{\bar{x}}(t) = \sum_{i=1}^{2} \mu_i(t)\{(A + \lambda I)\bar{x}(t) + \rho_i A_1 \bar{x}(t - \tau)\}, \tag{3.25}$$

where $\mu_1(t) = (\rho_2 - e^{\lambda \tau})/(\rho_2 - \rho_1)$ and $\mu_2(t) = (e^{\lambda \tau} - \rho_1)/(\rho_2 - \rho_1)$. The LMIs of Theorem 3.1 are affine in the system matrices. Therefore, one has to solve these LMIs simultaneously for the two vertices of system (3.25) given by $A_1{}^{(i)} = \rho_i A_1$, $i = 1, 2$, where the same decision matrices are applied.

Remark 3.6 The discrete-time counterpart of Wirtinger's inequality (3.6) was presented in [194] for stability analysis of discrete-time sampled-data systems: For a sequence of $N+1$ real n-dimensional vectors $\xi_0, \xi_1, \dots, \xi_N$ such that $\xi_0 = 0$, the following inequality holds:

$$\sum_{i=0}^{N-1} (\xi_{i+1} - \xi_i)^T R(\xi_{i+1} - \xi_i) \geq \varrho_N^2 \sum_{i=0}^{N-1} \xi_i^T R\xi_i, \tag{3.26}$$

where $0 < R \in \mathbb{R}^{n\times n}$ and $\varrho_N = 2\sin\frac{\pi}{2(2N+1)}$.

Note that the discrete-time Wirtinger's inequality (3.26) was employed in [141] to the security analysis of discrete-time multi-sensor NCSs.

Example 3.1 Consider the system ([259], Example 2.1)

$$\dot{x}(t) = \begin{bmatrix} 0 & 1 \\ 0 & -0.1 \end{bmatrix} x(t) + \begin{bmatrix} 0 \\ 0.1 \end{bmatrix} u(t). \tag{3.27}$$

We start with the analysis of the closed-loop system under the controller

$$u(t) = -[3.75\ 11.5]x(t_k - \eta), \quad t_k \leq t < t_{k+1}.$$

For the values of η given in Table 3.2, by applying (i) of Theorem 3.1 and Corollary 2.1, we obtain the maximum values of $\tau_M = h_s + \eta$ that preserve the stability (see Table 3.2). For $\eta = 0$, the results of [157] and of [62] lead to $\tau_M = 1.36$, which coincides with our results.

Choosing next $\tau_M = 1$, by applying Remark 3.5 and either (i) of Theorem 3.1 or [173] in the affine form (3.22), (3.23), we obtain the maximum value of the decay rate λ given in Table 3.3 for different values of η.

Table 3.2 Example 3.1: maximum value of τ_M for different η

$\tau_M \setminus \eta$	0.1	0.2	0.4	0.6
Reference [173]	1.05	1.06	1.07	1.07
Corollary 2.1(i)	1.33	1.26	1.18	1.14
Theorem 3.1 (i)	1.32	1.28	1.22	1.17

Table 3.3 Example 3.1: maximum value of λ for $\tau_M = 1$ and different η

$\lambda \setminus \eta$	0.1	0.2	0.4	0.6
Reference [173]	0.04	0.05	0.05	0.05
Corollary 2.1(i)	0.20	0.15	0.10	0.07
Theorem 3.1 (i)	0.26	0.23	0.17	0.12

We proceed next with the state-feedback design. Note that the poles of the open-loop system (3.27) have non-positive real parts. Therefore, by (ii) of Theorem 3.1 with $\varepsilon = 0.6$, $\eta = 0.9$, we obtain a low gain controller $u(t) = -10^{-15} \times [0.1482 \;\; 0.5412]x(t_k - \eta)$ which stabilizes (3.27) preserving the stability for $\tau_M \leq 10^8$. Choosing next $\eta = 0.2$, $\tau_M = 0.8$ and applying (ii) of Theorem 3.1 (as in Remark 3.5) with $\varepsilon = 0.9$, we find that the controller $u(t) = -[4.8260 \; 11.2343]x(t_k - \eta)$ exponentially stabilizes the system with the decay rate $\lambda = 0.50$. Next, applying the conditions of Theorem 3.1 (i), of Corollary 2.1(i) and of [173] (as in Remark 3.5) to the resulting closed-loop system, the maximum decay rate is found to be 0.52, 0.30 and 0.23, respectively. Hence, the method of Theorem 3.1 essentially simplifies the existing conditions and improves the results.

3.2.2 Stability via Discretized Lyapunov Functional

If system (3.8) with some constant delay $\bar{\eta} \in [0, \eta)$ is not stable (and thus, the simple Lyapunov functional V_{n_1} is not applicable), the nominal functional V_n can be chosen to be a complete one

$$\begin{aligned} V_{n2}(x_t, \dot{x}_t) &= x^T(t)Px(t) + 2x^T(t)\int_{-\eta}^{0} Q(s)x(t+s)ds \\ &+ \int_{-\eta}^{0}\int_{-\eta}^{0} x^T(t+s)R(s,\theta)dsx(t+\theta)d\theta \\ &+ \int_{-\eta}^{0} x^T(t+s)S(s)x(t+s)ds, \quad P > 0 \end{aligned} \tag{3.28}$$

with continuous and piecewise-linear functions $Q(s)$, $S(s)$ and $R(s,\theta)$ [71]. Following [71], we divide the delay interval $[-\eta, 0]$ into N segments $[\theta_p, \theta_{p-1}]$, $p = 1, ..., N$ of equal length $r = \eta/N$, where $\theta_p = -pr$. This divides the square $[-\eta, 0] \times [-\eta, 0]$ into $N \times N$ small squares $[\theta_p, \theta_{p-1}] \times [\theta_q, \theta_{q-1}]$. Each small square is further divided into two triangles. The continuous matrix functions $Q(s)$ and $S(s)$ are chosen to be linear within each segment, and the continuous matrix function $R(s,\theta)$ is chosen to be linear within each triangle:

$$\begin{aligned} &Q(\theta_p + \bar{\alpha}r) = (1-\bar{\alpha})Q_p + \bar{\alpha}Q_{p-1}, \\ &S(\theta_p + \bar{\alpha}r) = (1-\bar{\alpha})S_p + \bar{\alpha}S_{p-1}, \; \bar{\alpha} \in [0,1], \\ &R(\theta_p + \bar{\alpha}r, \theta_q + \bar{\beta}r) = \begin{cases} (1-\bar{\alpha})R_{pq} + \bar{\beta}R_{p-1,q-1} + (\bar{\alpha}-\bar{\beta})R_{p-1,q}, & \bar{\alpha} \geq \bar{\beta}, \\ (1-\bar{\beta})R_{pq} + \bar{\alpha}R_{p-1,q-1} + (\bar{\beta}-\bar{\alpha})R_{p,q-1}, & \bar{\alpha} < \bar{\beta}. \end{cases} \end{aligned}$$

We use $V_d = V_{n2} + V_W$. Then following the descriptor method and the arguments of Theorem 3.1, we arrive at

Corollary 3.1 *Given $\eta \geq 0$, $h_s > 0$ and K, the system (3.5) is asymptotically stable if there exist $n \times n$ matrices $P > 0$, P_2, P_3, $S_p = S_p^T$, Q_p, $R_{pq} = R_{qp}^T$, $p =$*

$0, 1, \ldots, N$, $q = 0, 1, \ldots, N$, $W > 0$, *such that the following LMIs*

$$\begin{bmatrix} P & \tilde{Q} \\ * & \tilde{R} + \tilde{S} \end{bmatrix} > 0, \tag{3.29}$$

$$\Xi_d \triangleq \begin{bmatrix} \Omega_d & \begin{bmatrix} D^s \\ 0 \end{bmatrix} & \begin{bmatrix} D^a \\ 0 \end{bmatrix} \\ * & -R_d - S_d & 0 \\ * & * & -3S_d \end{bmatrix} < 0, \tag{3.30}$$

hold, where $r = \frac{\eta}{N}$ *and*

$$\Omega_d = \begin{bmatrix} \Psi_{d11} & P - P_2^T + A^T P_3 & -Q_N + P_2^T A_1 & P_2^T A_1 \\ * & -\mathrm{He}(P_3) + h_s^2 W & P_3^T A_1 & P_3^T A_1 \\ * & * & -S_N & 0 \\ * & * & * & -\frac{\pi^2}{4} W \end{bmatrix} \tag{3.31}$$

and

$$\begin{aligned}
\Psi_{d11} &= \mathrm{He}(P_2^T A + Q_0) + S_0, \\
\tilde{Q} &= [Q_0\ Q_1\ \cdots\ Q_N], \\
\tilde{S} &= \mathrm{diag}\{1/r S_0, 1/r S_1, \ldots, 1/r S_N\}, \\
\tilde{R} &= \begin{bmatrix} R_{00} & R_{01} & \cdots & R_{0N} \\ R_{10} & R_{11} & \cdots & R_{1N} \\ \cdots & \cdots & \cdots\cdots \\ R_{N0} & R_{N1} & \cdots & R_{NN} \end{bmatrix}, \\
R_d &= \begin{bmatrix} R_{d11} & R_{d12} & \cdots & R_{d1N} \\ R_{d21} & R_{d22} & \cdots & R_{d2N} \\ \cdots & \cdots & \cdots\cdots \\ R_{dN1} & R_{dN2} & \cdots & R_{dNN} \end{bmatrix}, \\
R_{dpq} &= r(R_{p-1,q-1} - R_{pq}), \\
S_d &= \mathrm{diag}\{S_0 - S_1, S_1 - S_2, \ldots, S_{N-1} - S_N\}, \\
D^s &= [D_1^s\ D_2^s\ \cdots\ D_N^s],\ D^a = [D_1^a\ D_2^a\ \cdots\ D_N^a], \\
D_p^s &= \begin{bmatrix} r/2(R_{0,p-1} + R_{0p}) - (Q_{p-1} - Q_p) \\ r/2(Q_{p-1} + Q_p) \\ -r/2(R_{N,p-1} + R_{Np}) \end{bmatrix}, \\
D_p^a &= \begin{bmatrix} -r/2(R_{0,p-1} - R_{0p}) \\ -r/2(Q_{p-1} - Q_p) \\ r/2(R_{N,p-1} - R_{Np}) \end{bmatrix}.
\end{aligned} \tag{3.32}$$

Remark 3.7 Different from Corollary 3.1, the results of Theorem 3.1 are convex in η: If LMIs of Theorem 3.1 are feasible for some $\bar{\eta} > 0$, then they are feasible for all $\eta \in [0, \bar{\eta}]$. Therefore, Theorem 3.1 gives sufficient conditions for the stability of system (3.5) with the unknown but bounded constant delay $\eta \in [0, \bar{\eta}]$.

The conditions of Corollary 3.1 are derived via the descriptor method and, thus, can be easily applied to the state-feedback design by choosing, e.g., $P_3 = \varepsilon P_2$ [44].

Remark 3.8 Following the method of [173], the stability of system (3.5) via the time-independent functional $V_{n2} + V_Z$ leads to LMIs (3.22), (3.29), (3.30), where Ω_d is changed by

$$\tilde{\Omega}_d = \begin{bmatrix} \Psi_{d11} & P - P_2^T + A^T P_3 & -Q_N & P_2^T A_1 & 0 \\ * & -\mathrm{He}(P_3) + h_s^2 Z_2 & 0 & P_3^T A_1 & 0 \\ * & * & -S_N + Z_1 - Z_2 & Z_2 - S_{12} & S_{12} \\ * & * & * & -2Z_2 + \mathrm{He}(S_{12}) & Z_2 - S_{12} \\ * & * & * & * & -Z_1 - Z_2 \end{bmatrix}$$

with Ψ_{d11} given by (3.32). It is seen that also in the case of complete V_{n2}, the discontinuous Lyapunov functional leads to numerically simpler conditions than the time-independent one.

The results of Corollary 3.1 and Remark 3.8 can be applied to the exponential stability analysis by using the method of Remark 3.5.

Example 3.2 Consider the system from [73]:

$$\dot{x}(t) = \begin{bmatrix} 0 & 1 \\ -2 & 0.1 \end{bmatrix} x(t) + \begin{bmatrix} 0 \\ 1 \end{bmatrix} u(t), \tag{3.33}$$

where $u(t) = [1 \quad 0]x(t_k - \eta)$, $t_k \le t < t_{k+1}$. This system with $x(t_k - \eta)$ changed by $x(t - \eta)$ is stable for $0.1003 < \eta < 1.72$ and unstable if $\eta \in [0, 0.1]$. Thus, the simple Lyapunov functional-based results of [173], Corollary 2.1 and Theorem 3.1 are not applicable.

This is an example of the system that can be stabilized by using an artificial delay. For the values of $\eta > 0$ given in Table 3.4, by applying Corollary 3.1 and Remark 3.8 we obtain the maximum values of $\tau_M = h_s + \eta$ that preserve the stability.

Choosing next $\tau_M = 0.81$, by applying Corollary 3.1 and Remark 3.8 with $N = 2$ via Remark 3.5, we obtain the maximum value of the decay rate λ given in Table 3.5 for different values of η. Moreover, in this case the discontinuous discretized Lyapunov functional leads to reduced-order LMIs and improves the results via the time-independent one.

Table 3.4 Example 3.2: maximum value of τ_M for different η

$\tau_M \setminus \eta$		0.5	0.65	0.8
$N = 1$	$V_{n2} + V_W$	1.03	1.27	1.36
	$V_{n2} + V_Z$	0.84	1.05	1.16
$N = 2$	$V_{n2} + V_W$	1.07	1.39	1.65
	$V_{n2} + V_Z$	0.86	1.12	1.34

Table 3.5 Example 3.2: maximum value of λ for $\tau_M = 0.81$ and different η

	$\lambda \setminus \eta$	0.5	0.65	0.8
$N = 2$	$V_{n2} + V_W$	0.08	0.22	0.36
	$V_{n2} + V_Z$	0.02	0.18	0.35

3.3 Sampled-Data Stabilization by Using the Delayed Measurements

It is well known that using an artificial delay in the (continuous-time) static output feedback can stabilize some systems, which are not stabilizable without delay (see, e.g., [111, 165] and Example 3.2). Thus, the double integrator

$$\ddot{x}(t) = u(t), \quad y(t) = x(t) \tag{3.34}$$

can be stabilized by using a control action of the form $u(t) = -k_1 x(t - h_1) - k_2 x(t - h_2)$, where h_1 and h_2 are constant delays and $0 \leq h_1 < h_2$. The main criticism of the above method is that it has no advantages over the dynamic output feedback and that its implementation needs buffer for all the measurements $y(t + \theta)$, $\theta \in [-h_2, 0]$.

For the sampled-data control of systems with uncertain coefficients, the observer-based design is complicated and may lead to conservative results. From the other side, a simple static output feedback using the previous measurements can be easily designed and implemented. Thus, in the system of Example 3.2, one can insert an artificial delay η and apply the sampled-data controller with the sampling intervals satisfying $t_{k+1} - t_k \leq \tau_M - \eta$.

In this section, we will extend the sampled-data stabilization to the case, where (as in the double integrator) two sampled-data measurements are needed. Consider (3.1) and assume that the measured output $y(t_k) = Cx(t_k) \in \mathbb{R}^{n_y}$ is available at the discrete time instants $0 = t_0 < t_1 < \cdots < t_k < \cdots$ with the constant sampling interval $t_{k+1} - t_k = h$.

Consider the following static output-feedback controller, which uses the delayed measurement $y(t_{k-m})$:

$$\begin{aligned} u(t) &= K_1 y(t_k) + K_2 y(t_{k-m}) \\ &= K_1 C x(t_k) + K_2 C x(t_k - mh), \quad m \in \mathbb{N}, \quad t_k \leq t < t_{k+1}. \end{aligned} \tag{3.35}$$

The closed-loop system (3.1), (3.35) has the form

$$\dot{x}(t) = Ax(t) + A_{c1} x(t_k) + A_{c2} x(t_k - \eta), \tag{3.36}$$

where $\eta = mh$, $A_{c1} = BK_1 C$, $A_{c2} = BK_2 C$.

We extend the analysis of Sect. 3.2.2 to the system of (3.36) by adding the term introduced in [46]

$$V_U(t,\dot{x}_t) = (h - t + t_k)\int_{t_k}^{t} \dot{x}^T(s)U\dot{x}(s)ds, \quad U > 0,$$

to $V_d = V_{n2} + V_W$:

$$\begin{aligned} V_{sam}(t,x_t,\dot{x}_t) &= \bar{V}_2(t) \\ &= V_{n2}(x_t,\dot{x}_t) + V_W(t,x_t,\dot{x}_t) + V_U(t,\dot{x}_t), \quad t_k \le t < t_{k+1}, \end{aligned} \tag{3.37}$$

where $V_W(t,x_t,\dot{x}_t)$ is given by (3.14) with $\tau_M = (m+1)h$. The term V_U vanishes before t_k and after t_k. By using the arguments of Corollary 3.1 and [46], we arrive at the following result.

Corollary 3.2 *Given $h > 0$ and K_1, K_2, the system (3.36) is asymptotically stable if there exist $n \times n$ matrices $P > 0$, P_2, P_3, $S_p = S_p^T$, Q_p, $R_{pq} = R_{qp}^T$, $p = 0, 1, \ldots, N$, $q = 0, 1, \ldots, N$, and $U > 0$, $W > 0$, such that LMIs (3.29) and*

$$\bar{\Xi}_{di} \triangleq \begin{bmatrix} \bar{\Omega}_{di} & \begin{bmatrix} D^s \\ 0 \end{bmatrix} & \begin{bmatrix} D^a \\ 0 \end{bmatrix} \\ * & -R_d - S_d & 0 \\ * & * & -3S_d \end{bmatrix} < 0, \quad i = 1, 2, \tag{3.38}$$

hold, where $\tilde{Q}$, $\tilde{S}$, $\tilde{R}$, S_d, R_d, D^s and D^a are defined in (3.32). In (3.38),

$$\bar{\Omega}_{d1} = \Omega_d + \mathrm{diag}\{0_{n\times n}, hU, 0\},$$

$$\bar{\Omega}_{d2} = \begin{bmatrix} \Omega_d & \begin{bmatrix} -hP_2^T A_{c1} \\ -hP_3^T A_{c1} \\ 0 \end{bmatrix} \\ * & -hU \end{bmatrix},$$

with Ω_d given by (3.31), where A, A_1 and h_s are changed by $A + A_{c1}$, A_{c2} and h, respectively.

Remark 3.9 The LMIs of Corollary 3.2 are affine in A. Therefore, if A resides in the uncertain polytope

$$A = \sum_{j=1}^{M} \mu_j(t)A^{(j)}, \quad 0 \le \mu_j(t) \le 1, \quad \sum_{j=1}^{M} \mu_j(t) = 1,$$

one has to solve these LMIs simultaneously for all the M vertices $A^{(j)}$, applying the same decision matrices.

Example 3.3 Consider the following system:

$$\dot{x}(t) = \begin{bmatrix} 0 & 1 \\ g(t) & 0 \end{bmatrix} x(t) + \begin{bmatrix} 0 \\ 1 \end{bmatrix} u(t), \\ y(t_k) = [1 \ 0]x(t_k), \ t_k \le t < t_{k+1}, \ x(t) \in \mathbb{R}^2, \tag{3.39}$$

where $|g(t)| \le 0.1$. This system is not stabilizable by the non-delayed static output feedback $u(t) = Ky(t_k)$, $t_k \le t < t_{k+1}$. We take $m = 3$ and choose

$$u(t) = -0.35y(t_k) + 0.1y(t_{k-3}), \ t_k \le t < t_{k+1}, \ t_{k+1} - t_k = h. \tag{3.40}$$

We treat the closed-loop system (3.39)–(3.40) as a system with polytopic-type uncertainty defined by the two vertices corresponding to $g(t) = \pm 0.1$. By applying Remark 3.9 to the closed-loop system (3.39)–(3.40), we find the values of sampling period h that preserve the stability:

$$N = 1, \ h \in [10^{-5} \ 0.380], \ N = 2, \ h \in [10^{-5} \ 0.499].$$

3.4 Notes

In this chapter, novel discontinuous Lyapunov functionals were proposed for sampled-data systems in the presence of constant input delay. The construction of the functionals is based on the vector extension of Wirtinger's inequality. The new method leads to numerically simplified LMIs for the stability analysis, and it was applied to a novel problem of sampled-data stabilization by using the previous measurements.

Wirtinger's inequality of Lemma 3.1 was further widely adopted in the literature, e.g., in [263] for event-based H_∞ filtering of sampled-data systems, in [254] for synchronization analysis of Markovian neural networks. Based on Wirtinger's inequality, an integral inequality, which encompasses Jensen's inequality as a particular case, was derived in [197]. This inequality leads to a new stability criteria for linear time-delay and sampled-data systems. Furthermore, Wirtinger's inequality of Lemma 3.1 was extended in [189] for predictor-based exponential stabilization of NCSs with large uncertain delays.

Moreover, the extension of Corollary 3.2 to the case of asynchronous samplings was done in [193] by firstly reformulating the sampled-data system as an impulsive system and then by applying a looped-functional approach along with an integral inequality introduced in [197].

Chapter 4
Networked Control Under Round-Robin Protocol

In Chaps. 2 and 3 we have considered the case of one sensor node. In this chapter, we propose a time-delay approach to the stability and L_2-gain analysis of NCSs with multiple sensor nodes. For the sake of simplicity, we consider only two sensor nodes. Due to communication constraints, the scheduling of sensor information toward the controller is ruled by the classical Round-Robin protocol. The closed-loop system is modeled as a switched continuous-time system with multiple and ordered time-varying delays. The case of the ordered time-varying delays (where one delay is smaller than another) has not been studied yet in the literature. By developing appropriate Lyapunov–Krasovskii techniques in this case, we derive LMIs for the exponential stability and for the L_2-gain analysis. The efficiency and advantages of the proposed approach are illustrated by two benchmark examples. The numerical results essentially improve the hybrid system-based existing ones [84] and, for the stability analysis, are not far from those obtained via the discrete-time approach [29]. Note that the latter approach is not applicable to the performance analysis. Also, for the first time (under Round-Robin scheduling protocol), the network-induced delay is allowed to be greater than the sampling interval.

4.1 Stability and L_2-Gain Analysis of NCSs Under Round-Robin Protocol

4.1.1 Problem Formulation and a Switched System Model

Consider the following system controlled through a network (see Fig. 4.1):

$$\begin{aligned} \dot{x}(t) &= Ax(t) + Bu(t) + B_1 w(t), \\ z(t) &= C_0 x(t) + D_0 u(t), \end{aligned} \tag{4.1}$$

K. Liu et al., *Networked Control Under Communication Constraints*, Advances in Delays and Dynamics 11, https://doi.org/10.1007/978-981-15-4230-5_4

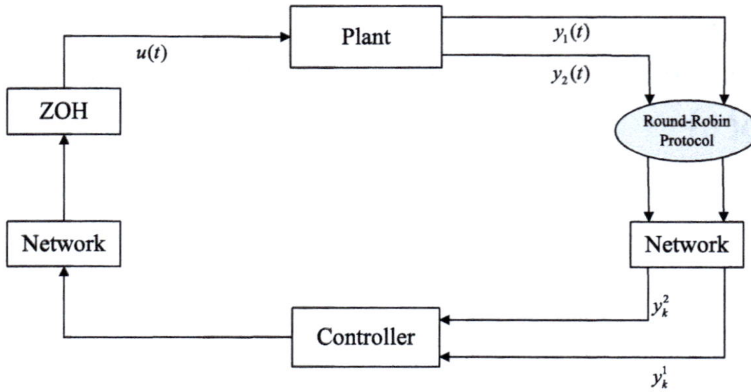

Fig. 4.1 NCSs with two sensors under Round-Robin protocol

where $x(t) \in \mathbb{R}^n$ is the state vector, $u(t) \in \mathbb{R}^{n_u}$ is the control input, $w(t) \in \mathbb{R}^{n_w}$ is the disturbance, $z(t) \in \mathbb{R}^{n_z}$ is controlled output, and A, B, B_1, C_0 and D_0 are system matrices with appropriate dimensions. These matrices can be uncertain with polytopic-type uncertainty. The system has several nodes (distributed sensors, a controller node and an actuator node) which are connected via two networks: a sensor network (relaying the sensors to the controller node) and a control network (from the controller node to the actuator). Consider the sensor nodes $y_i(t) = C_i x(t)$, $i = 1, 2$, and denote

$$C = \begin{bmatrix} C_1 \\ C_2 \end{bmatrix}, \quad y(t) = \begin{bmatrix} y_1(t) \\ y_2(t) \end{bmatrix} \in \mathbb{R}^{n_y}.$$

Let s_k denote the unbounded monotonously increasing sequence of sampling instants satisfying (2.2). At each sampling instant s_k, one of the outputs $y_i(t)$ is sampled and transmitted via the network. The choice of the active output node is ruled by a Round-Robin scheduling protocol: The outputs are transmitted one after another; i.e., $y_i(t)$ is transmitted only at the sampling instant $t = s_{2p+i-1}$, $p \in \mathbb{Z}^+$. After each transmission and reception, the values in $y_i(t)$ are updated with the newly received values, while the other values in $y(t)$ remain the same, as no additional information is received. This leads to the constrained data exchange expressed as

$$y_k^i = \begin{cases} y_i(s_k) = C_i x(s_k) + M_i v(s_k), & k = 2p + i - 1, \\ y_{k-1}^i, & k \neq 2p + i - 1, \end{cases} \quad p \in \mathbb{Z}^+, \tag{4.2}$$

where v is a measurement noise signal and M_i, $i = 1, 2$, are matrices with appropriate dimension.

We suppose that data loss is not possible and that the transmission of the information over the two networks (between the sensor and the actuator) is subject to a variable delay $\eta_k = \eta_k^{sc} + \eta_k^{ca}$, where η_k^{sc} and η_k^{ca} are the network-induced delays (from the sensor to the controller and from the controller to the actuator, respectively). Then, $t_k = s_k + \eta_k$ is the updating instant time of the ZOH.

Different from [29, 84], we do not restrict the network delays to be small with $t_k = s_k + \eta_k < s_{k+1}$, i.e., $\eta_k < s_{k+1} - s_k$. As in [161], we allow the delay to be non-small provided that the old sample cannot get to the destination (to the controller or to the actuator) after the most recent one

$$s_k + \eta_k^{sc} < s_{k+1} + \eta_{k+1}^{sc}, \quad s_k + \eta_k < s_{k+1} + \eta_{k+1}, \tag{4.3}$$

i.e., $\eta_k < t_{k+1} - s_k$. The assumption (4.3) is a necessary condition for making scheduling reasonable. A sufficient condition for (4.3) is that the delays are bounded $\eta_k^{sc} \in [\eta_m^{sc}, \eta_M^{sc}]$, $\eta_k \in [\eta_m, \eta_M]$, where η_m, η_m^{sc}, η_M (i.e., MAD), η_M^{sc} are known bounds with $\eta_M^{sc} - \eta_m^{sc} \leq \eta_M - \eta_m$, and the delay range is less than the sampling interval: $\eta_k - \eta_{k+1} \leq \eta_M - \eta_m < s_{k+1} - s_k$.

Assume that the network-induced delay η_k and the time span between the updating and the most recent sampling instants are bounded:

$$t_{k+1} - t_k + \eta_k \leq \tau_M, \; 0 \leq \eta_m \leq \eta_k \leq \eta_M, \; k \in \mathbb{Z}^+, \tag{4.4}$$

where τ_M, η_m and η_M are known bounds. Note that $\tau_M = \text{MATI} + \eta_M$. Then, we have

$$\begin{aligned} &t_{k+1} - t_k \leq \tau_M - \eta_m, \\ &t_{k+1} - t_{k-1} + \eta_{k-1} \leq 2\tau_M - \eta_m \stackrel{\Delta}{=} \bar{\tau}_M. \end{aligned} \tag{4.5}$$

Since $\text{MATI} = \tau_M - \eta_M \leq \tau_M - \eta_m$, the inequality $\eta_m > \frac{\tau_M}{2}$ implies that the communication delays are non-small due to $\eta_k \geq \eta_m > \tau_M - \eta_m$. Examples below in this chapter will show that when $\eta_m > \frac{\tau_M}{2}$, our method is still feasible.

We suppose that the controller and the actuator act in an event-driven manner. The general dynamic output-feedback controller is assumed to be given in the following form

$$\begin{aligned} &\dot{x}_c(t) = A_c x_c(t) + B_c y_k, \\ &u(t) = C_c x_c(t) + D_c y_k, \quad t \in [t_k, t_{k+1}), \quad k \in \mathbb{Z}^+, \end{aligned} \tag{4.6}$$

where $x_c(t) \in \mathbb{R}^{n_c}$ is the state of the controller and

$$y_k = \begin{bmatrix} y_k^1 \\ y_k^2 \end{bmatrix} \in \mathbb{R}^{n_y}, \quad k \in \mathbb{N},$$

is the most recently received output of the plant, which satisfies (4.2) and $y_0 = \begin{bmatrix} y_0^1 \\ 0 \end{bmatrix}$. The notations A_c, B_c, C_c and D_c are the matrices with appropriate dimensions.

4.1.1.1 Static Output-Feedback Controller

Consider first a particular case of (4.6) with $C_c = 0$, $D_c = K$, where we suppose that there exists a matrix $K = [K_1 \;\; K_2]$, $K_1 \in \mathbb{R}^{n_u \times n_1}$, $K_2 \in \mathbb{R}^{n_u \times (n_y - n_1)}$ such that

$A+BKC$ is Hurwitz. Consider the static output feedback of the form:

$$u_k = K_1 y_k^1 + K_2 y_k^2, \quad k \in \mathbb{N}$$

and $u_0 = K_1 y_0^1$. Then, the control law is piecewise-constant with

$$u(t) = u_k, \quad t \in [t_k, t_{k+1}).$$

The closed-loop system can be expressed in the form of the switched system

$$\begin{aligned}
\dot{x}(t) &= Ax(t) + A_1 x(t_k - \eta_k) + A_2 x(t_{k-1} - \eta_{k-1}) + B_1 w(t) \\
&\quad + D_{21} v(t_k - \eta_k) + D_{22} v(t_{k-1} - \eta_{k-1}), \\
z(t) &= C_0 x(t) + D_1 x(t_k - \eta_k) + D_2 x(t_{k-1} - \eta_{k-1}) + E_{21} v(t_k - \eta_k) \\
&\quad + E_{22} v(t_{k-1} - \eta_{k-1}), \quad t \in [t_k, t_{k+1}), \\
\dot{x}(t) &= Ax(t) + A_1 x(t_k - \eta_k) + A_2 x(t_{k+1} - \eta_{k+1}) + B_1 w(t) \\
&\quad + D_{21} v(t_k - \eta_k) + D_{22} v(t_{k+1} - \eta_{k+1}), \\
z(t) &= C_0 x(t) + D_1 x(t_k - \eta_k) + D_2 x(t_{k+1} - \eta_{k+1}) + E_{21} v(t_k - \eta_k) \\
&\quad + E_{22} v(t_{k+1} - \eta_{k+1}), \quad t \in [t_{k+1}, t_{k+2}),
\end{aligned} \tag{4.7}$$

where

$$\begin{aligned}
& k = 2p, \ p \in \mathbb{Z}^+, \ A_i = BK_iC_i, \ D_{2i} = BK_iM_i, \\
& D_i = D_0 K_i C_i, \ E_{2i} = D_0 K_i M_i, \ i = 1, 2, \\
& x(t_{-1} - \eta_{-1}) = 0, \ \ v(t_{-1} - \eta_{-1}) = 0.
\end{aligned} \tag{4.8}$$

From (4.4) and (4.5), it follows that

$$\begin{aligned}
t \in [t_k, t_{k+1}) \Rightarrow \quad & t - t_k + \eta_k \in [\eta_m, \tau_M], \\
& t - t_{k-1} + \eta_{k-1} \in [\eta_m, \ \bar{\tau}_M], \\
t \in [t_{k+1}, t_{k+2}) \Rightarrow \ & t - t_k + \eta_k \in [\eta_m, \ \bar{\tau}_M], \\
& t - t_{k+1} + \eta_{k+1} \in [\eta_m, \tau_M].
\end{aligned} \tag{4.9}$$

Moreover,

$$\begin{aligned}
& t - t_k + \eta_k < t - t_{k-1} + \eta_{k-1}, \quad t \in [t_k, t_{k+1}), \\
& t - t_k + \eta_k > t - t_{k+1} + \eta_{k+1}, \quad t \in [t_{k+1}, t_{k+2}).
\end{aligned}$$

For $t \in [t_k, t_{k+1})$, we can represent $t_k - \eta_k = t - \tau_1(t)$, $t_{k-1} - \eta_{k-1} = t - \tau_2(t)$, where

$$\begin{aligned}
& \tau_1(t) = t - t_k + \eta_k < \tau_2(t) = t - t_{k-1} + \eta_{k-1}, \\
& \tau_1(t) \in [\eta_m, \tau_M], \ \tau_2(t) \in [\eta_m, \ \bar{\tau}_M], \ \ t \in [t_k, t_{k+1}).
\end{aligned} \tag{4.10}$$

Therefore, (4.7) for $t \in [t_k, t_{k+1})$ can be considered as a system with two time-varying interval delays, where $\tau_1(t) < \tau_2(t)$. Similarly, for $t \in [t_{k+1}, t_{k+2})$ (4.7) is a system with two time-varying delays, one of which is less than another.

Assume that $v(t) = 0$, $t \le t_0$. For $k = 2p$, $p \in \mathbb{Z}^+$, denote

$$\begin{aligned} v_1(t) &= v(t_k - \eta_k), \qquad v_2(t) = v(t_{k-1} - \eta_{k-1}), \quad t \in [t_k, t_{k+1}), \\ v_1(t) &= v(t_{k+1} - \eta_{k+1}), \quad v_2(t) = v(t_k - \eta_k), \quad t \in [t_{k+1}, t_{k+2}). \end{aligned}$$

The switched continuous-time system (4.7) has three disturbances $w \in L_2[t_0, \infty)$, $v_i \in L_2[t_0, \infty)$, $i = 1, 2$, with

$$\begin{aligned} \|v_1\|_{L_2}^2 &= \int_{t_0}^{\infty} |v_1(t)|^2 dt = \sum_{p=0}^{\infty} (t_{2p+1} - t_{2p}) |v(t_{2p} - \eta_{2p})|^2 \\ &\quad + \sum_{p=0}^{\infty} (t_{2p+2} - t_{2p+1}) |v(t_{2p+1} - \eta_{2p+1})|^2 \\ &= \sum_{k=0}^{\infty} (t_{k+1} - t_k) |v(t_k - \eta_k)|^2, \\ \|v_2\|_{L_2}^2 &= \int_{t_0}^{\infty} |v_2(t)|^2 dt = \sum_{k=0}^{\infty} (t_{k+1} - t_k) |v(t_{k-1} - \eta_{k-1})|^2. \end{aligned}$$

For a given scalar $\gamma > 0$, we thus define the following performance index:

$$\begin{aligned} J &= \|z\|_{L_2}^2 - \gamma^2 [\|w\|_{L_2}^2 + \|v_1\|_{L_2}^2 + \|v_2\|_{L_2}^2] \\ &= \int_{t_0}^{\infty} [z^T(t) z(t) - \gamma^2 w^T(t) w(t)] dt \\ &\quad - \gamma^2 \sum_{k=0}^{\infty} (t_{k+1} - t_k) [v^T(t_k - \eta_k) v(t_k - \eta_k) + v^T(t_{k-1} - \eta_{k-1}) v(t_{k-1} - \eta_{k-1})]. \end{aligned} \tag{4.11}$$

The above H_∞-performance extends the indexes of [128, 217] to the case of Round-Robin scheduling protocol. It takes into account the updating rates of the measurement and is related to the energy of the measurement noise. The system (4.7) is said to have an L_2-gain less than γ if $J < 0$ along (4.7) for the zero initial function and for all nonzero $w \in L_2[t_0, \infty)$, $v \in l_2[t_0, \infty)$.

4.1.1.2 Dynamic Output-Feedback Controller

Consider now (4.1) under the dynamic output-feedback controller (4.6), where we assume that the controller is directly connected to the actuator, i.e., $\eta_k = \eta_k^{sc}$. The closed-loop system (4.1), (4.6) can be expressed in the form of (4.7), where the system state and the matrices are changed by the ones with the bars as follows:

$$\bar{x} = [x^T \; x_c^T]^T, \quad \bar{A} = \begin{bmatrix} A & BC_c \\ 0 & A_c \end{bmatrix}, \quad \bar{A}_1 = \begin{bmatrix} BD_c \begin{bmatrix} C_1 \\ 0 \end{bmatrix} & 0 \\ B_c \begin{bmatrix} C_1 \\ 0 \end{bmatrix} & 0 \end{bmatrix},$$

$$\bar{A}_2 = \begin{bmatrix} BD_c \begin{bmatrix} 0 \\ C_2 \end{bmatrix} & 0 \\ B_c \begin{bmatrix} 0 \\ C_2 \end{bmatrix} & 0 \end{bmatrix}, \quad \bar{B}_1 = \begin{bmatrix} B_1 \\ 0 \end{bmatrix}, \quad \bar{D}_{21} = \begin{bmatrix} BD_c \begin{bmatrix} M_1 \\ 0 \end{bmatrix} \\ B_c \begin{bmatrix} M_1 \\ 0 \end{bmatrix} \end{bmatrix},$$

$$\bar{D}_{22} = \begin{bmatrix} BD_c \begin{bmatrix} 0 \\ M_2 \end{bmatrix} \\ B_c \begin{bmatrix} 0 \\ M_2 \end{bmatrix} \end{bmatrix}, \quad \bar{C}_0 = [C_0 \;\; D_0 C_c], \quad \bar{D}_1 = \begin{bmatrix} D_0 D_c \begin{bmatrix} C_1 \\ 0 \end{bmatrix} & 0 \end{bmatrix},$$

$$\bar{D}_2 = \begin{bmatrix} D_0 D_c \begin{bmatrix} 0 \\ C_2 \end{bmatrix} & 0 \end{bmatrix}, \quad \bar{E}_{21} = D_0 D_c \begin{bmatrix} M_1 \\ 0 \end{bmatrix}, \quad \bar{E}_{22} = D_0 D_c \begin{bmatrix} 0 \\ M_2 \end{bmatrix}.$$

In this section, we will derive LMI conditions for the exponential stability of the disturbance-free switched model (4.7) and for the L_2-gain analysis of (4.7) via direct Lyapunov method. The results under variable network-induced delay are presented in Sect. 4.1.2. Note that in some NCSs (such as control area networks, where the transmission over the network is almost instantaneous compared to the plant dynamic) the network-induced transmission delay may be usually neglected (see, e.g., [164]). However, a constant delay may appear due to the measurement or due to the control computation. Therefore, Sect. 4.1.3 considers Round-Robin scheduling under variable sampling with constant delay. When there is no measurement delay, less conservative stability conditions for sampled-data systems under Round-Robin scheduling are derived in Sect. 4.1.4. Finally, the efficiency of the new criteria is illustrated via benchmark examples of batch reactor and cart–pendulum.

Remark 4.1 In the above reasoning, we assumed that the packet loss does not occur. However, for small delays $\eta_k < s_{k+1} - s_k$ we could accommodate the packet dropouts by modeling them as prolongations of the transmission interval [29, 84].

Lemma 4.1 *Let there exist positive numbers β, δ and a functional $V : \mathbb{R}^+ \times W[t_0 - \bar{\tau}_M, t_0] \times L_2[t_0 - \bar{\tau}_M, t_0] \to \mathbb{R}^+$ such that*

$$\beta|\phi(t_0)|^2 \le V(t, \phi, \dot{\phi}) \le \delta\|\phi\|_W^2. \tag{4.12}$$

Let the function $\bar{V}(t) = V(t, x_t, \dot{x}_t)$ be continuous from the right for $x(t)$ satisfying (4.7), absolutely continuous for $t \ne t_k$, and satisfy

$$\lim_{t \to t_k^-} \bar{V}(t) \ge \bar{V}(t_k). \tag{4.13}$$

(i) *If along (4.7) with $w = 0$ and $v = 0$*

$$\dot{\bar{V}}(t) \le -\tilde{\beta}|x(t)|^2 \; for \; t \ne t_k \; and \; for \; some \; scalar \; \tilde{\beta} > 0,$$

then (4.7) with $w = 0$ and $v = 0$ is asymptotically stable.
(ii) *If along (4.7) with $w = 0$ and $v = 0$ for some $\alpha > 0$*

$$\dot{\bar{V}}(t) + 2\alpha \bar{V}(t) \leq 0 \quad for\ t \neq t_k,$$

then (4.7) with $w = 0$ and $v = 0$ is exponentially stable with the decay rate α.
(iii) *For a given $\gamma > 0$, if along (4.7)*

$$\begin{array}{l}\dot{\bar{V}}(t) + z^T(t)z(t) - \gamma^2 w^T(t)w(t) \\ -\gamma^2[v^T(t_k - \eta_k)v(t_k - \eta_k) + v^T(t_{k-1} - \eta_{k-1})v(t_{k-1} - \eta_{k-1})] < 0, \ t \in [t_k, t_{k+1}), \ k \in \mathbb{Z}^+,\end{array} \tag{4.14}$$

then the performance index (4.11) achieves $J < 0$ for all nonzero $w \in L_2[t_0, \infty)$, $v \in l_2[t_0, \infty)$ and for the zero initial function.

Proof (i) and (ii) follow from the standard arguments for switched systems and time-delay systems (see, e.g., [46, 124, 215]).

(iii) Given $N >> 1$, we integrate inequality (4.14) from t_0 till t_N. Then for $w \neq 0$, $v \neq 0$, we have

$$\begin{array}{l}\bar{V}(t_N) - \bar{V}(t_{N-1}) + \bar{V}(t_{N-1}^-) - \bar{V}(t_{N-2}) \cdots + \bar{V}(t_1^-) - \bar{V}(t_0) \\ \quad + \int_{t_0}^{t_N} [z^T(s)z(s) - \gamma^2 w^T(s)w(s)]ds - \gamma^2 \sum_{k=0}^{N-1} (t_{k+1} - t_k)[v^T(t_k - \eta_k)v(t_k - \eta_k) \\ \qquad + v^T(t_{k-1} - \eta_{k-1})v(t_{k-1} - \eta_{k-1})] < 0.\end{array}$$

Taking into account that $\bar{V}(t_N) \geq 0$, $\bar{V}(t_k^-) - \bar{V}(t_k) \geq 0$ for $k = 1, \ldots, N-1$, and $\bar{V}(t_0) = 0$, for $N \to \infty$, we arrive at $J < 0$.

4.1.2 Stability and L_2-Gain Analysis of NCSs: Variable Delay

Consider the switched system (4.7) as a system with two ordered time-varying delays $\tau_1(t)$ and $\tau_2(t)$ either from $[\eta_m, \tau_M]$ or from $[\eta_m, \bar{\tau}_M]$ (cf., (4.10)). The stability and L_2-gain analysis will be based on the common (for both subsystems of the switched system) time-independent LKF for the exponential stability of systems with time-varying delay from the maximum delay interval $[\eta_m, \bar{\tau}_M]$ (see, e.g., [45, 80]):

$$V(x_t, \dot{x}_t) = \bar{V}(t) = V_0(x_t, \dot{x}_t) + V_1(x_t, \dot{x}_t), \tag{4.15}$$

where

$$
\begin{aligned}
V_0(x_t, \dot{x}_t) &= x^T(t)Px(t) + \int_{t-\eta_m}^{t} e^{2\alpha(s-t)} x^T(s) S_0 x(s) ds \\
&\quad + \eta_m \int_{-\eta_m}^{0} \int_{t+\theta}^{t} e^{2\alpha(s-t)} \dot{x}^T(s) R_0 \dot{x}(s) ds d\theta, \\
V_1(x_t, \dot{x}_t) &= \int_{t-\bar{\tau}_M}^{t-\eta_m} e^{2\alpha(s-t)} x^T(s) S_1 x(s) ds \\
&\quad + (\bar{\tau}_M - \eta_m) \int_{-\bar{\tau}_M}^{-\eta_m} \int_{t+\theta}^{t} e^{2\alpha(s-t)} \dot{x}^T(s) R_1 \dot{x}(s) ds d\theta
\end{aligned} \tag{4.16}
$$

with $P > 0,\ S_0 > 0,\ R_0 > 0,\ S_1 > 0,\ R_1 > 0,\ \alpha > 0$.

For the simplicity of presentation, we will use the notations:

$$
\begin{aligned}
F_{0i} &= [A\ 0\ A_i\ A_{3-i}\ 0],\ i = 1, 2, \\
F_1 &= [I\ \ 0_{n\times 4n}], \\
F_{12} &= [I\ \ -I\ 0\,0\,0], \\
F &= \begin{bmatrix} 0 & I & -I & 0 & 0 \\ 0 & 0 & I & -I & 0 \\ 0 & 0 & 0 & I & -I \end{bmatrix}.
\end{aligned} \tag{4.17}
$$

By taking into account the order of the delays $\tau_1(t)$, $\tau_2(t)$ and applying the convexity arguments of [173], we prove the following result.

Theorem 4.1 *(i) For given $0 \le \eta_m < \tau_M, \alpha > 0$, assume that there exist $n \times n$ matrices $P > 0,\ S_j > 0,\ R_j > 0,\ j = 0, 1,\ G_1^i,\ G_2^i,\ G_3^i,\ i = 1, 2$, such that the following four LMIs*

$$
\Omega_i = \begin{bmatrix} R_1 & G_1^i & G_2^i \\ * & R_1 & G_3^i \\ * & * & R_1 \end{bmatrix} \ge 0, \tag{4.18}
$$

$$
\Xi_i = \begin{bmatrix} \Phi_i & F_{0i}^T H \\ * & -H \end{bmatrix} < 0,\ i = 1, 2, \tag{4.19}
$$

are feasible, where

$$
\begin{aligned}
\Phi_i &= \mathrm{He}(F_1^T P F_{0i}) + \Upsilon - e^{-2\alpha\eta_m} F_{12}^T R_0 F_{12} - e^{-2\alpha\bar{\tau}_M} F^T \Omega_i F, \\
\Upsilon &= \mathrm{diag}\{S_0 + 2\alpha P, -e^{-2\alpha\eta_m}(S_0 - S_1), 0, 0, -e^{-2\alpha\bar{\tau}_M} S_1\}, \\
H &= \eta_m^2 R_0 + (\bar{\tau}_M - \eta_m)^2 R_1.
\end{aligned} \tag{4.20}
$$

Then, system (4.7) with $w = 0$ and $v = 0$ is exponentially stable with the decay rate α.

(ii) For a given $\gamma > 0$, assume that (4.18) and the following LMIs

$$\left[\begin{array}{c|cccc} & PB_1 & PD_{21} & PD_{22} & C_0^T \\ & 0 & 0 & 0 & 0 \\ \Xi_i|_{\alpha=0} & 0 & 0 & 0 & D_i^T \\ & 0 & 0 & 0 & D_{3-i}^T \\ & 0 & 0 & 0 & 0 \\ & HB_1 & HD_{21} & HD_{22} & 0 \\ \hline * & -\gamma^2 I & 0 & 0 & 0 \\ * & * & -\gamma^2 I & 0 & E_{21}^T \\ * & * & * & -\gamma^2 I & E_{22}^T \\ * & * & * & * & -I \end{array}\right] < 0, \quad i = 1, 2, \tag{4.21}$$

are feasible, where the notations are given in (4.17) and (4.20). Then, system (4.7) is internally exponentially stable and has L_2-gain less than γ.

Proof (i) Differentiating $\bar{V}(t)$ along (4.7) with $w = 0$ and $v = 0$, we have

$$\begin{aligned}\dot{\bar{V}}(t) + 2\alpha\bar{V}(t) \leq\ & 2x^T(t)P\dot{x}(t) + x^T(t)(S_0 + 2\alpha P)x(t) \\ & - x^T(t-\eta_m)(S_0 - S_1)e^{-2\alpha\eta_m}x(t-\eta_m) \\ & + \dot{x}^T(t)[\eta_m^2 R_0 + (\bar{\tau}_M - \eta_m)^2 R_1]\dot{x}(t) \\ & - x^T(t-\bar{\tau}_M)S_1 e^{-2\alpha\bar{\tau}_M}x(t-\bar{\tau}_M) \\ & - \eta_m e^{-2\alpha\eta_m}\int_{t-\eta_m}^{t}\dot{x}^T(s)R_0\dot{x}(s)ds \\ & - (\bar{\tau}_M - \eta_m)e^{-2\alpha\bar{\tau}_M}\int_{t-\bar{\tau}_M}^{t-\eta_m}\dot{x}^T(s)R_1\dot{x}(s)ds.\end{aligned}$$

Consider $t \in [t_k, t_{k+1})$. By Jensen's inequality, we have

$$\begin{aligned}\eta_m\int_{t-\eta_m}^{t}\dot{x}^T(s)R_0\dot{x}(s)ds &\geq \int_{t-\eta_m}^{t}\dot{x}^T(s)ds R_0\int_{t-\eta_m}^{t}\dot{x}(s)ds \\ &= [x(t) - x(t-\eta_m)]^T R_0[x(t) - x(t-\eta_m)].\end{aligned}$$

Taking into account that $t_{k-1} - \eta_{k-1} < t_k - \eta_k$ (i.e., the delays satisfy the relation (4.10)) and applying further Jensen's inequality, we obtain

$$\begin{aligned}& -(\bar{\tau}_M - \eta_m)\int_{t-\bar{\tau}_M}^{t-\eta_m}\dot{x}^T(s)R_1\dot{x}(s)ds \\ &= -(\bar{\tau}_M - \eta_m)\Big[\int_{t_k-\eta_k}^{t-\eta_m}\dot{x}^T(s)R_1\dot{x}(s)ds + \int_{t_{k-1}-\eta_{k-1}}^{t_k-\eta_k}\dot{x}^T(s)R_1\dot{x}(s)ds \\ &\quad + \int_{t-\bar{\tau}_M}^{t_{k-1}-\eta_{k-1}}\dot{x}^T(s)R_1\dot{x}(s)ds\Big] \\ &\leq -\frac{1}{\alpha_1}f_1(t) - \frac{1}{\alpha_2}f_2(t) - \frac{1}{\alpha_3}f_3(t),\end{aligned}$$

where

$$
\begin{aligned}
\alpha_1 &= \frac{t-\eta_m-t_k+\eta_k}{\bar{\tau}_M-\eta_m},\ \alpha_2 = \frac{t_k-\eta_k-t_{k-1}+\eta_{k-1}}{\bar{\tau}_M-\eta_m},\ \alpha_3 = \frac{\bar{\tau}_M-t+t_{k-1}-\eta_{k-1}}{\bar{\tau}_M-\eta_m},\\
f_1(t) &= [x(t-\eta_m)-x(t_k-\eta_k)]^T R_1[x(t-\eta_m)-x(t_k-\eta_k)],\\
f_2(t) &= [x(t_k-\eta_k)-x(t_{k-1}-\eta_{k-1})]^T R_1[x(t_k-\eta_k)-x(t_{k-1}-\eta_{k-1})],\\
f_3(t) &= [x(t_{k-1}-\eta_{k-1})-x(t-\bar{\tau}_M)]^T R_1[x(t_{k-1}-\eta_{k-1})-x(t-\bar{\tau}_M)].
\end{aligned}
$$

Denote

$$
\begin{aligned}
g_{1,2}(t) &= [x(t-\eta_m)-x(t_k-\eta_k)]^T G_1^1[x(t_k-\eta_k)-x(t_{k-1}-\eta_{k-1})],\\
g_{1,3}(t) &= [x(t-\eta_m)-x(t_k-\eta_k)]^T G_2^1[x(t_{k-1}-\eta_{k-1})-x(t-\bar{\tau}_M)],\\
g_{2,3}(t) &= [x(t_k-\eta_k)-x(t_{k-1}-\eta_{k-1})]^T G_3^1[x(t_{k-1}-\eta_{k-1})-x(t-\bar{\tau}_M)].
\end{aligned}
$$

Note that (4.18) with $i=1$ guarantees $\begin{bmatrix} R_1 & G_j^1 \\ * & R_1 \end{bmatrix} \geq 0,\ j=1,2,3$, and thus,

$$
\begin{bmatrix} f_i(t) & g_{i,j}(t) \\ g_{i,j}(t) & f_j(t) \end{bmatrix} \geq 0,\ i \neq j,\ i=1,2,\ j=2,3.
$$

Then setting $\xi(t) = \mathrm{col}\{x(t), x(t-\eta_m), x(t_k-\eta_k), x(t_{k-1}-\eta_{k-1}), x(t-\bar{\tau}_M)\}$ and applying the reciprocally convex combination lemma proposed in [173], we arrive at

$$
\begin{aligned}
&-(\bar{\tau}_M-\eta_m)\int_{t-\bar{\tau}_M}^{t-\eta_m} \dot{x}^T(s)R_1\dot{x}(s)ds\\
&\leq -\frac{1}{\alpha_1}f_1(t) - \frac{1}{\alpha_2}f_2(t) - \frac{1}{\alpha_3}f_3(t)\\
&\leq -f_1(t)-f_2(t)-f_3(t)-2g_{1,2}(t)-2g_{1,3}(t)-2g_{2,3}(t)\\
&= -\xi^T(t)F^T\Omega_1 F\xi(t),
\end{aligned}
$$

where F and Ω_1 are given by (4.17) and (4.18) with $i=1$, respectively.

Hence, it is found that

$$
\dot{\bar{V}}(t) + 2\alpha\bar{V}(t) \leq \xi^T(t)(\Phi_i + F_{0i}^T H F_{0i})\xi(t),\ i=1,
$$

where the notations are given by (4.17) and (4.20). Thus, by Schur complements, (4.18) and (4.19) with $i=1$ guarantee that $\dot{\bar{V}}(t) + 2\alpha\bar{V}(t) \leq 0$ for $t \in [t_k, t_{k+1})$.

Similarly, for $t \in [t_{k+1}, t_{k+2})$ (4.18) and (4.19) with $i=2$ guarantee $\dot{\bar{V}}(t) + 2\alpha\bar{V}(t) \leq 0$. Thus, by (ii) of Lemma 4.1, (4.7) with $w=0$ and $v=0$ is exponentially stable with the decay rate α.

(ii) By using the arguments of (i), we find that (4.14) holds along (4.7) if LMIs (4.18) and (4.21) are feasible, which completes the proof.

Remark 4.2 The switched system (4.7) can be alternatively analyzed by the standard arguments for systems with two (independent) time-varying delays. However, this leads to the overlapping delay intervals $\tau_1(t) \in [\eta_m, \tau_M]$ and $\tau_2(t) \in [\eta_m, \bar{\tau}_M]$, which may be conservative since the relation $\tau_1(t) < \tau_2(t)$ is ignored. We give now the standard LKF for two independent delays: $V(x_t, \dot{x}_t) = \sum_{i=0}^{2} V_i(x_t, \dot{x}_t)$, where

$V_i(x_t, \dot{x}_t)$, $i = 0, 1$, are given by (4.16) with $\bar{\tau}_M$ changed by τ_M and

$$V_2(x_t, \dot{x}_t) = \int_{t-\bar{\tau}_M}^{t-\eta_m} e^{2\alpha(s-t)} x^T(s) S_2 x(s) ds + (\bar{\tau}_M - \eta_m) \int_{-\bar{\tau}_M}^{-\eta_m} \int_{t+\theta}^{t} e^{2\alpha(s-t)} \dot{x}^T(s) R_2 \dot{x}(s) ds d\theta,$$

where $S_2 > 0,\ R_2 > 0,\ \alpha > 0$. Extending the arguments of [173] to the exponential stability of systems with two delays, we obtain the following LMIs:

$$\begin{bmatrix} R_i & G_i^1 & 0 & 0 \\ * & R_i & 0 & 0 \\ * & * & R_i & G_i^2 \\ * & * & * & R_i \end{bmatrix} \geq 0, \tag{4.22}$$

$$\tilde{\Xi}_i = \begin{bmatrix} \tilde{\Phi}_i & \tilde{F}_{0i}^T \tilde{H} \\ * & -\tilde{H} \end{bmatrix} < 0,\ i = 1, 2, \tag{4.23}$$

where

$$\begin{aligned}
\tilde{\Phi}_i &= \mathrm{He}(\tilde{F}_1^T P \tilde{F}_{0i}) + \tilde{\Upsilon} - e^{-2\alpha\eta_m} \tilde{F}_{12}^T R_0 \tilde{F}_{12} \\
&\quad - e^{-2\alpha\tau_M} \Big[[I\ \ I\ \ 0_{n\times 2n}]\tilde{F}\Big]^T \begin{bmatrix} R_1 & G_1^i \\ * & R_1 \end{bmatrix} [0_{n\times 2n}\ \ I\ \ I]\tilde{F}, \\
&\quad - e^{-2\alpha\bar{\tau}_M} \Big[[0_{n\times 2n}\ \ I\ \ I]\tilde{F}\Big]^T \begin{bmatrix} R_2 & G_2^i \\ * & R_2 \end{bmatrix} [I\ \ I\ \ 0_{n\times 2n}]\tilde{F},, \\
\tilde{\Upsilon} &= \mathrm{diag}\{S_0 + 2\alpha P, -e^{-2\alpha\eta_m}(S_0 - S_1 - S_2), 0, 0, -e^{-2\alpha\tau_M} S_1, -e^{-2\alpha\bar{\tau}_M} S_2\}, \\
\tilde{H} &= \eta_m^2 R_0 + (\tau_M - \eta_m)^2 R_1 + (\bar{\tau}_M - \eta_m)^2 R_2, \\
\tilde{F}_{0i} &= [A\ 0\ A_i\ A_{3-i}\ 0_{n\times 2n}],\ \tilde{F}_1 = [I\ \ 0_{n\times 5n}],\ \tilde{F}_{12} = [I\ \ -I\ \ 0_{n\times 4n}], \\
\tilde{F} &= \begin{bmatrix} 0 & I & -I & 0 & 0 & 0 \\ 0 & 0 & I & 0 & -I & 0 \\ 0 & I & 0 & -I & 0 & 0 \\ 0 & 0 & 0 & I & 0 & -I \end{bmatrix}.
\end{aligned} \tag{4.24}$$

The LMIs for L_2-gain analysis are given by (4.21) with Ξ and H changed by $\tilde{\Xi}$ and $\tilde{H}$, respectively. Note that LMIs (4.22), (4.23) for independent delays possess the same number of decision variables as LMIs (4.18), (4.19) of Theorem 4.1 (up to the symmetry of R_2, S_2), but have a higher-order:

Two LMIs of $7n \times 7n$ and two LMIs of $4n \times 4n$ in (4.22), (4.23),
Two LMIs of $6n \times 6n$ and two LMIs of $3n \times 3n$ in (4.18), (4.19).

The examples below in this chapter illustrate the improvement (in the maximum value of τ_M which preserves the stability and in the computational time) by taking into account the order of the delays.

Remark 4.3 Our method can be extended to the dynamic output feedback with two networks, where

$$u(t) = C_c x_c(t_k - \eta_k^{ca}) + D_c y_k, \ t \in [t_k, t_{k+1}).$$

In this case, the closed-loop system can be considered as the system with three delays: two ordered $\tau_1(t) = t - t_k + \eta_k < \tau_2(t) = t - t_{k-1} + \eta_{k-1}$ and the independent one $\tau_3(t) = t - t_k + \eta_k^{ca}$. Such a system can be analyzed by combining the standard Lyapunov-based method (for $\tau_3(t)$) with Theorem 4.1 (for the ordered delays $\tau_1(t)$ and $\tau_2(t)$).

4.1.3 *Stability and L_2-Gain Analysis of NCSs: Constant Delay*

In this section, we consider the case of negligible network-induced delay, where $\eta \geq 0$ is the constant measurement delay. It is noted that the conventional time-independent LKF of (4.15) (for systems with interval time-varying delays) was applied to NCSs (see, e.g., [53, 65]). However, these functionals did not take advantage of the sawtooth evolution of the delays induced by sample and hold. The latter drawback was removed in [46, 162], where time-dependent Lyapunov functionals for sampled-data systems were introduced.

We will adopt the time-dependent Lyapunov functional construction of Sect. 3.2, which is based on the extension of Wirtinger's inequality [77] to the vector case (see Lemma 3.1). We introduce the following discontinuous in time Lyapunov functional

$$\begin{aligned} V(t, x_t, \dot{x}_t) &= \bar{V}_1(t) \\ &= V_0(x_t, \dot{x}_t) + V_1(t, x_t, \dot{x}_t) + V_2(t, x_t, \dot{x}_t), \end{aligned} \tag{4.25}$$

where $V_0(x_t, \dot{x}_t)$ is given by (4.16) with $\eta_m = \eta$, $\alpha = 0$ and

$$\begin{aligned} V_1(t, x_t, \dot{x}_t) = {} & 4(\tau_M - \eta)^2 \int_{t_k-\eta}^{t} \dot{x}^T(s) W_1 \dot{x}(s) ds \\ & - \frac{\pi^2}{4} \int_{t_k-\eta}^{t-\eta} [x(s) - x(t_k - \eta)]^T W_1 [x(s) - x(t_k - \eta)] ds, \ t \in [t_k, t_{k+2}), \end{aligned}$$

$$
V_2(t, x_t, \dot{x}_t) = \begin{cases} 4(\tau_M - \eta)^2 \int_{t_{k-1}-\eta}^{t} \dot{x}^T(s) W_2 \dot{x}(s) ds \\ -\frac{\pi^2}{4} \int_{t_{k-1}-\eta}^{t-\eta} [x(s) - x(t_{k-1} - \eta)]^T W_2 [x(s) - x(t_{k-1} - \eta)] ds, \; t \in [t_k, t_{k+1}), \\ 4(\tau_M - \eta)^2 \int_{t_{k+1}-\eta}^{t} \dot{x}^T(s) W_2 \dot{x}(s) ds \\ -\frac{\pi^2}{4} \int_{t_{k+1}-\eta}^{t-\eta} [x(s) - x(t_{k+1} - \eta)]^T W_2 [x(s) - x(t_{k+1} - \eta)] ds, \; t \in [t_{k+1}, t_{k+2}) \end{cases}
$$

with $W_1 > 0,\ W_2 > 0,\ k = 2p$.

Note that V_0 is a "nominal" Lyapunov functional for the "nominal" system (4.26) with a constant delay [112, 171]:

$$
\dot{x}(t) = Ax(t) + (A_1 + A_2)x(t - \eta). \tag{4.26}
$$

The terms V_1, V_2 are extensions of the discontinuous Lyapunov functional constructions in Sect. 3.2. Note that V_1 can be represented as a sum of the continuous in time term $4(\tau_M - \eta)^2 \int_{t-\eta}^{t} \dot{x}^T(s) W_1 \dot{x}(s) ds \ge 0,\ t \in [t_k, t_{k+2})$, with the discontinuous (for $t = t_k$) one

$$
\begin{aligned} V_{W1} &\triangleq 4(\tau_M - \eta)^2 \int_{t_k-\eta}^{t-\eta} \dot{x}^T(s) W_1 \dot{x}(s) ds \\ &\quad - \frac{\pi^2}{4} \int_{t_k-\eta}^{t-\eta} [x(s) - x(t_k - \eta)]^T W_1 [x(s) - x(t_k - \eta)] ds, \quad t \in [t_k, t_{k+2}). \end{aligned}
$$

Note that $V_{W1|t=t_k} = 0$ and, by the extended Wirtinger's inequality (3.6), $V_{W1} \ge 0$ for all $t \ge t_0$. Therefore, V_1 does not grow in the jumps.

In a similar way, V_2 can be represented as a sum of the continuous in time term $4(\tau_M - \eta)^2 \int_{t-\eta}^{t} \dot{x}^T(s) W_2 \dot{x}(s) ds \ge 0$, with the discontinuous for $t = t_{k+1}$ term

$$
V_{W2} \triangleq \begin{cases} -\frac{\pi^2}{4} \int_{t_{k-1}-\eta}^{t-\eta} [x(s) - x(t_{k-1} - \eta)]^T W_2 [x(s) - x(t_{k-1} - \eta)] ds \\ \quad + 4(\tau_M - \eta)^2 \int_{t_{k-1}-\eta}^{t-\eta} \dot{x}^T(s) W_2 \dot{x}(s) ds, \quad t \in [t_k, t_{k+1}), \\ -\frac{\pi^2}{4} \int_{t_{k+1}-\eta}^{t-\eta} [x(s) - x(t_{k+1} - \eta)]^T W_2 [x(s) - x(t_{k+1} - \eta)] ds \\ \quad + 4(\tau_M - \eta)^2 \int_{t_{k+1}-\eta}^{t-\eta} \dot{x}^T(s) W_2 \dot{x}(s) ds, \quad t \in [t_{k+1}, t_{k+2}). \end{cases}
$$

We have $V_{W2|t=t_{k+1}} = 0$ and, by the extended Wirtinger's inequality (3.6), $V_{W2} \ge 0$ for all $t \ge t_0$; i.e., V_2 does not grow in the jumps. Therefore, $\bar{V}_1$ does not grow in the jumps: $\lim_{t \to t_k^-} \bar{V}_1(t) \ge \bar{V}_1(t_k)$ and $\lim_{t \to t_{k+1}^-} \bar{V}_1(t) \ge \bar{V}_1(t_{k+1})$ hold.

Theorem 4.2 *(i) Given $\tau_M > \eta \geq 0$, the system (4.7) with $\eta_k \equiv \eta$, $w = 0$ and $v = 0$ is asymptotically stable if there exist $n \times n$ matrices $P > 0$, $R_0 > 0$, $S_0 > 0$, $W_i > 0$, $i = 1, 2$, such that the following LMI*

$$\Xi = \begin{bmatrix} \Psi_1 & R_0 & PA_1 & PA_2 & A^T W \\ * & \Psi_2 & \frac{\pi^2}{4} W_1 & \frac{\pi^2}{4} W_2 & 0 \\ * & * & -\frac{\pi^2}{4} W_1 & 0 & A_1^T W \\ * & * & * & -\frac{\pi^2}{4} W_2 & A_2^T W \\ * & * & * & * & -W \end{bmatrix} < 0, \tag{4.27}$$

is feasible, where

$$\begin{aligned} \Psi_1 &= \mathrm{He}(PA) + S_0 - R_0, \\ \Psi_2 &= -\frac{\pi^2}{4} W_1 - \frac{\pi^2}{4} W_2 - S_0 - R_0, \\ W &= h^2 R_0 + 4(\tau_M - \eta)^2 (W_1 + W_2). \end{aligned} \tag{4.28}$$

(ii) For a given $\gamma > 0$, assume that the following LMI

$$\left[\begin{array}{c|cccc} & PB_1 & PD_{21} & PD_{22} & C_0^T \\ & 0 & 0 & 0 & 0 \\ \Xi & 0 & 0 & 0 & D_1^T \\ & 0 & 0 & 0 & D_2^T \\ & WB_1 & WD_{21} & WD_{22} & 0 \\ \hline * & -\gamma^2 I & 0 & 0 & 0 \\ * & * & -\gamma^2 I & 0 & E_{21}^T \\ * & * & 0 & -\gamma^2 I & E_{22}^T \\ * & * & * & * & -I \end{array}\right] < 0, \tag{4.29}$$

is feasible, where the notations are given by (4.28). Then, (4.7) with $\eta_k \equiv \eta$ is internally stable and has L_2-gain less than γ.

Proof (i) Differentiating $V_0(x_t, \dot{x}_t)$ and applying Jensen's inequality, we have

$$\begin{aligned} \frac{d}{dt} V_0(x_t, \dot{x}_t) \leq\ & 2x^T(t) P \dot{x}(t) + x^T(t) S_0 x(t) \\ & - x^T(t-\eta) S_0 x(t-\eta) + \eta^2 \dot{x}^T(t) R_0 \dot{x}(t) \\ & - [x(t) - x(t-\eta)]^T R_0 [x(t) - x(t-\eta)]. \end{aligned}$$

Consider $t \in [t_k, t_{k+1})$. Along (4.7) with $\eta_k \equiv \eta$, $w = 0$ and $v = 0$, we have

$$\begin{aligned}\frac{d}{dt}\sum_{i=1}^{2}V_i(t,x_t,\dot{x}_t) = \dot{x}^T(t)[4(\tau_M-\eta)^2(W_1+W_2)]\dot{x}(t)\\ -\frac{\pi^2}{4}[x(t-\eta)-x(t_k-\eta)]^T W_1[x(t-\eta)-x(t_k-\eta)]\\ -\frac{\pi^2}{4}[x(t-\eta)-x(t_{k-1}-\eta)]^T W_2[x(t-\eta)-x(t_{k-1}-\eta)].\end{aligned}$$

Then, substitution of $Ax(t)+A_1x(t_k-\eta)+A_2x(t_{k-1}-\eta)$ for $\dot{x}(t)$ leads to

$$\begin{aligned}\dot{\bar{V}}_1(t) \le \zeta^T(t)\begin{bmatrix}\Psi_1 & R_0 & PA_1 & PA_2\\ * & \Psi_2 & \frac{\pi^2}{4}W_1 & \frac{\pi^2}{4}W_2\\ * & * & -\frac{\pi^2}{4}W_1 & 0\\ * & * & * & -\frac{\pi^2}{4}W_2\end{bmatrix}\zeta(t)\\ +[Ax(t)+A_1x(t_k-\eta)+A_2x(t_{k-1}-\eta)]^T W\\ \times[Ax(t)+A_1x(t_k-\eta)+A_2x(t_{k-1}-\eta)],\end{aligned}$$

where $\zeta(t)=\text{col}\{x(t),x(t-\eta),x(t_k-\eta),x(t_{k-1}-\eta)\}$ and W is given in (4.28). Hence, by Schur complements, (4.27) guarantees that $\dot{\bar{V}}_1(t)\le-\tilde{\beta}|x(t)|^2$ for some $\tilde{\beta}>0$.

Similarly, when $t\in[t_{k+1},t_{k+2})$, $k=2p$, we prove that (4.27) guarantees $\dot{\bar{V}}_1(t)\le -\tilde{\beta}|x(t)|^2$ for some $\tilde{\beta}>0$, which proves the asymptotic stability (see Lemma 4.1(i)).

(ii) By using the arguments of (i), we find that (4.14) with $\dot{\bar{V}}(t)=\dot{\bar{V}}_1(t)$ holds along (4.7) with $\eta_k\equiv\eta$ if LMI (4.29) is feasible, which completes the proof.

Remark 4.4 Compared to the stability LMI conditions of Theorem 4.1, of Remark 4.2 and of [135], the LMI of Theorem 4.2 is essentially simpler (single LMI of $5n\times 5n$ with fewer decision variables) and is less conservative (see examples below in this chapter).

Remark 4.5 Following Remark 3.24, we can find the decay rate of the exponential stability for (4.7) with $\eta_k\equiv\eta$, $w=0$ and $v=0$ by changing the variable $\bar{x}(t)=x(t)e^{\alpha t}$ and by applying LMI (4.27) to the resulting system with polytopic-type uncertainty.

For the case of constant delay η, we can combine the methods of Theorems 4.1 and 4.2. Thus, the following corollary is obtained.

Corollary 4.1 *Given $\tau_M>\eta\ge 0$, the system (4.7) with $\eta_k\equiv\eta$, $w=0$ and $v=0$ is asymptotically stable if there exist $n\times n$ matrices $P>0$, $S_i>0$, $R_i>0$, $i=0,1$, G_1^i, G_2^i, G_3^i, $W_i>0$, $i=1,2$, such that (4.18) and $\Xi_i|_{\alpha=0}$, $i=1,2$, hold, where Ξ_i, $i=1,2$, are defined in (4.19) with η_m changed by η and Φ_i, and H changed by*

$$
\begin{aligned}
\Phi_i &= \mathrm{He}(F_1^T P F_{0i}) + \Upsilon - F_{12}^T R_0 F_{12} - F^T \Omega_i F, \\
&- \frac{\pi^2}{4} F_{23}^T W_i F_{23} - \frac{\pi^2}{4} F_{24}^T W_{3-i} F_{24}, \\
H &= \eta^2 R_0 + 4(\tau_M - \eta)^2 (R_1 + W_1 + W_2), \\
F_{23} &= [0 \;\; I \;\; -I \;\; 0 \;\; 0], \quad F_{24} = [0 \;\; I \;\; 0 \;\; -I \;\; 0].
\end{aligned} \tag{4.30}
$$

Note that the LMIs of Corollary 4.1 with $W_i = 0$, $i = 1, 2$, or with $R_1 = S_1 = G_j^i = 0$, $i = 1, 2$, $j = 1, 2, 3$, are reduced to the ones of Theorem 4.1 or of Theorem 4.2, respectively.

4.1.4 Stability and L₂-Gain Analysis of Sampled-Data Systems

When there is no measurement delay, i.e., $\eta_k \equiv 0$, the problem for NCSs is reduced to the one for sampled-data systems with scheduling (see, e.g., [164]), where the closed-loop system has the form of (4.7) with $\eta_k \equiv 0$, $k \in \mathbb{Z}^+$. As we will see in the example below, for $\eta \to 0$ the conditions of Theorem 4.2 become conservative. Less conservative conditions can be derived in this case via different from (4.25) continuous in time Lyapunov functionals.

For the constant sampling, where $t_{k+1} - t_k = \tau_M$, $k \in \mathbb{Z}^+$, choose the Lyapunov functional of the form

$$
\begin{aligned}
V(t, x_t, \dot{x}_t) = \bar{V}_2(t) = x^T(t) P x(t) + \sum_{i=1}^{2} V_i(t, \dot{x}_t) + \sum_{i=1}^{2} V_{Xi}(t, x_t), \\
P > 0, \; t \in [t_k, t_{k+2}), \; k = 2p, \; p \in \mathbb{Z}^+,
\end{aligned} \tag{4.31}
$$

where

$$
\begin{aligned}
V_1(t, \dot{x}_t) &= (t_{k+2} - t) \int_{t_k}^{t} e^{2\alpha(s-t)} \dot{x}^T(s) U_1 \dot{x}(s) ds, \\
V_2(t, \dot{x}_t) &= \begin{cases} (t_{k+1} - t) \int_{t_{k-1}}^{t} e^{2\alpha(s-t)} \dot{x}^T(s) U_2 \dot{x}(s) ds, & t \in [t_k, t_{k+1}), \\ (t_{k+3} - t) \int_{t_{k+1}}^{t} e^{2\alpha(s-t)} \dot{x}^T(s) U_2 \dot{x}(s) ds, & t \in [t_{k+1}, t_{k+2}), \end{cases} \\
V_{X1}(t, x_t) &= (t_{k+2} - t) \xi_0^T(t) \begin{bmatrix} \frac{\mathrm{He}(X)}{2} & -X + X_1 \\ * & -\bar{X}_1 \end{bmatrix} \xi_0(t), \\
V_{X2}(t, x_t) &= \begin{cases} (t_{k+1} - t) \xi_{-1}^T(t) \begin{bmatrix} \frac{\mathrm{He}(X_2)}{2} & -X_2 + X_3 \\ * & -\bar{X}_3 \end{bmatrix} \xi_{-1}(t), & t \in [t_k, t_{k+1}), \\ (t_{k+3} - t) \xi_1^T(t) \begin{bmatrix} \frac{\mathrm{He}(X_2)}{2} & -X_2 + X_3 \\ * & -\bar{X}_3 \end{bmatrix} \xi_1(t), & t \in [t_{k+1}, t_{k+2}) \end{cases}
\end{aligned}
$$

with $U_1 > 0$, $U_2 > 0$, $k = 2p$, and

$$\xi_i(t) = \mathrm{col}\{x(t), x(t_{k+i})\}(i = 0, \pm 1),$$
$$\bar{X}_1 = \mathrm{He}(X_1) - \frac{X + X^T}{2},$$
$$\bar{X}_3 = \mathrm{He}(X_3) - \frac{X_2 + X_2^T}{2}.$$

The terms V_i and V_{Xi}, $i = 1, 2$, extend the constructions of [46] to the case of multiple sampling intervals. These terms are continuous in time along (4.7) with $\eta_k = 0$ since

$$\begin{aligned} V_{1|t=t_k^-} &= V_{1|t=t_k} = 0, \; V_{1|t=t_{k+1}^-} = V_{1|t=t_{k+1}} \geq 0, \\ V_{2|t=t_k^-} &= V_{2|t=t_k} \geq 0, \; V_{2|t=t_{k+1}^-} = V_{2|t=t_{k+1}} = 0, \end{aligned}$$

$$\begin{aligned} V_{X1|t=t_k^-} &= V_{X1|t=t_k} = 0, \quad V_{X1|t=t_{k+1}^-} = V_{X1|t=t_{k+1}} \geq 0, \\ V_{X2|t=t_k^-} &= V_{X2|t=t_k} \geq 0, \quad V_{X2|t=t_{k+1}^-} = V_{X2|t=t_{k+1}} = 0. \end{aligned}$$

The condition $V(t, x_t, \dot{x}_t) \geq \bar{\beta}|x(t)|^2$ holds for $t \in [t_k, t_{k+1})$, $k = 2p$, if

$$\begin{bmatrix} P + \tau_M \frac{\mathrm{He}(X)}{2} & \tau_M(-X + X_1) \\ * & -\tau_M \bar{X}_1 \end{bmatrix} > 0,$$

$$\begin{bmatrix} P + \tau_M \mathrm{He}(X) + \tau_M \frac{\mathrm{He}(X_2)}{2} & 2\tau_M(-X + X_1) & \tau_M(-X_2 + X_3) \\ * & -2\tau_M \bar{X}_1 & 0 \\ * & * & -\tau_M \bar{X}_3 \end{bmatrix} > 0,$$

and for $t \in [t_{k+1}, t_{k+2})$, $k = 2p$, if

$$\begin{bmatrix} P + \tau_M \frac{\mathrm{He}(X_2)}{2} & \tau_M(-X_2 + X_3) \\ * & -\tau_M \bar{X}_3 \end{bmatrix} > 0.$$

$$\begin{bmatrix} P + \tau_M \mathrm{He}(X_2) + \tau_M \frac{\mathrm{He}(X)}{2} & \tau_M(-X + X_1) & 2\tau_M(-X_2 + X_3) \\ * & -\tau_M \bar{X}_1 & 0 \\ * & * & -2\tau_M \bar{X}_3 \end{bmatrix} > 0,$$

Remark 4.6 The resulting LMI conditions can be found in [135]. Moreover, the Lyapunov functional V of (4.31) with $X = X_i = 0$, $i = 1, 2, 3$, is applicable to systems with variable sampling $t_{k+1} - t_k \leq \tau_M$.

Remark 4.7 LMIs of Theorems 4.1 and 4.2, of Remarks 4.2 and 4.6, of Corollary 4.1 and of [135] are affine in the system matrices. Therefore, in the case of system matrices from the uncertain time-varying polytope

$$
\begin{aligned}
\Theta &= \sum_{j=1}^{N} g_j(t)\Theta_j, \ 0 \le g_j(t) \le 1, \ \sum_{j=1}^{N} g_j(t) = 1, \\
\Theta_j &= \left[A^{(j)} \ B^{(j)} \ B_1^{(j)} \ C_0^{(j)} \ D_0^{(j)} \right],
\end{aligned}
$$

one has to solve these LMIs simultaneously for all the N vertices Θ_j, applying the same decision matrices.

4.2 Numerical Examples

Example 4.1 (*batch reactor*) We illustrate the efficiency of the given conditions on the benchmark example of a batch reactor under the dynamic output feedback with $\eta_k = \eta_k^{sc}$ (see, e.g., [17, 29, 84]), where

$$
A = \begin{bmatrix} 1.380 & -0.208 & 6.715 & -5.676 \\ -0.581 & -4.2902 & 0 & 0.675 \\ 1.067 & 4.273 & -6.654 & 5.893 \\ 0.048 & 4.273 & 1.343 & -2.104 \end{bmatrix},
$$

$$
\begin{aligned}
&B = \begin{bmatrix} 0 & 0 \\ 5.679 & 0 \\ 1.136 & -3.146 \\ 1.136 & 0 \end{bmatrix}, \quad B_1 = \begin{bmatrix} 10 & 0 \\ 0 & 5 \\ 10 & 0 \\ 0 & 5 \end{bmatrix}, \\
&C_0 = \begin{bmatrix} 1 & 0 & 1 & -1 \\ 0 & 1 & 0 & 0 \end{bmatrix}, \quad D_0 = \begin{bmatrix} 0 & 0 \\ 0 & 0 \end{bmatrix}, \\
&\left[\begin{array}{c|c} A_c & B_c \\ \hline C_c & D_c \end{array}\right] = \left[\begin{array}{cc|cc} 0 & 0 & 0 & 1 \\ 0 & 0 & 1 & 0 \\ \hline -2 & 0 & 0 & -2 \\ 0 & 8 & 5 & 0 \end{array}\right].
\end{aligned}
\tag{4.32}
$$

As in [84], the controlled output $z(t)$ is chosen to be equal to the measured output $y(t) = \begin{bmatrix} y_1(t) \\ y_2(t) \end{bmatrix}$. Thus,

$$
C = \left[\begin{array}{c} C_1 \\ \hline C_2 \end{array}\right] = \left[\begin{array}{cccc} 1 & 0 & 1 & -1 \\ \hline 0 & 1 & 0 & 0 \end{array}\right], \quad M_1 = M_2 = 0. \tag{4.33}
$$

We start with the stability analysis in the disturbance-free case, where $w = 0$. When there is no communication delay, i.e., $\eta_k \equiv 0$, by applying the method of Remark 4.6 with $\alpha = 0$ we find the maximum values of $\tau_M = \mathrm{MATI} + \eta_M$ that preserve the asymptotic stability (see Table 4.1). Our results are close to the discretization-based results of [29], whereas the latter results are complicated to the performance analysis.

Table 4.1 Example 4.1: maximum value of τ_M for $\eta_k \equiv 0$

Method	τ_M = MATI
Reference [17]	0.0082
Reference [84]	0.0088
Reference [29]	0.0645
Remark 4.6: $t_{k+1} - t_k \equiv \tau_M$	0.0655
Remark 4.6 ($X = X_i = 0,\ i = 1, 2, 3$): $t_{k+1} - t_k \leq \tau_M$	0.0622
Theorem 4.2 with $\eta = 0$: $t_{k+1} - t_k \leq \tau_M$	0.049

Table 4.2 Example 4.1: maximum value of τ_M for different η_m

$\tau_M \backslash \eta_m$	0	0.004	0.02	0.03	0.04
Reference [84] ($\eta_M = 0.004$)	0.0088	0.0088	–	–	–
Reference [29] ($\eta_M = 0.03$)	0.068	0.068	0.068	0.068	–
Remark 4.2 (variable η_k)	0.036	0.038	0.047	0.053	0.059
Theorem 4.1 (variable η_k)	0.042	0.044	0.053	0.058	0.063
Theorem 4.2 (constant η_m)	0.049	0.051	0.057	0.061	0.065
Corollary 4.1 (constant η_m)	0.051	0.054	0.060	0.063	0.067

Table 4.3 Example 4.1: the computational time for maximum τ_M

$\eta_m = 0.04$	Remark 4.2	Theorem 4.1	Theorem 4.2	Corollary 4.1
τ_M	0.059	0.063	0.065	0.067
Time	8.77	6.55	0.97	13.48

Further, for the values of η_m given in Table 4.2, by applying Theorem 4.1 and Remark 4.2 with $\alpha = 0$, Theorem 4.2 and Corollary 4.1 with constant delay $\eta = \eta_m$, we obtain the maximum values of τ_M that preserve the stability (see Table 4.2). From Table 4.2, it is seen that the results of Theorems 4.1 and 4.2 essentially improve the hybrid system-based result in [84]. Moreover, our results are applicable when the delay is larger than the sampling interval (e.g., when $\eta_m = 0.04$). For $\eta_m = 0.04$ and the corresponding maximum τ_M, we give also the computational time (in seconds) for different methods (see Table 4.3). It is seen that the improvement (till 15% increase of the maximum τ_M and till 25% decrease of computational time) is achieved by taking into account the order of the delays in Theorem 4.1 (for variable delay η_k). Theorem 4.2 essentially decreases the computational time (for constant delay $\eta_k \equiv \eta_m$).

Consider next the perturbed model of the batch reactor, i.e., $w \neq 0$. As in [84], we assume that $y_i(s_k) = C_i x(s_k)$ where C_i, $i = 1, 2$, are given in (4.33) and consider $J = \int_{t_0}^{\infty} [z^T(t)z(t) - \gamma^2 w^T(t)w(t)]dt$. For the values of η_m given in Table 4.4, by applying Theorem 4.1 we find the minimum values of γ for different values of η_m (see Table 4.4). From Table 4.4, it is seen that our results are favorably compared

Table 4.4 Example 4.1: minimum γ for different τ_M and η_m

η_m	0	0	0.02	0.03	0.04
τ_M	0.0056	0.0149	0.03	0.04	0.05
Reference [84] ($\eta_M = \frac{\tau_M}{2}$)	2.50	200	–	–	–
Remark 4.2 (var delay)	2.07	2.32	2.51	2.90	3.97
Theorem 4.1 (var delay)	2.06	2.25	2.43	2.74	3.48
Theorem 4.2 (con delay)	2.02	2.13	2.30	2.52	2.97

with [84]. As previously, Theorem 4.1 is applicable when the delay is larger than the sampling interval.

Example 4.2 (*inverted pendulum*) Consider the following linearized model of the inverted pendulum on a cart

$$\begin{bmatrix} \dot{x} \\ \ddot{x} \\ \dot{\theta} \\ \ddot{\theta} \end{bmatrix} = \begin{bmatrix} 0 & 1 & 0 & 0 \\ 0 & 0 & \frac{-mg}{M} & 0 \\ 0 & 0 & 0 & 1 \\ 0 & 0 & \frac{(M+m)g}{Ml} & 0 \end{bmatrix} \begin{bmatrix} x \\ \dot{x} \\ \theta \\ \dot{\theta} \end{bmatrix} + \begin{bmatrix} 0 \\ \frac{a}{M} \\ 0 \\ \frac{-a}{Ml} \end{bmatrix} u + \begin{bmatrix} 1 \\ 1 \\ 1 \\ 1 \end{bmatrix} w, \quad (4.34)$$
$$z(t) = \begin{bmatrix} 1 & 1 & 1 & 1 \end{bmatrix} \begin{bmatrix} x & \dot{x} & \theta & \dot{\theta} \end{bmatrix}^T + 0.1u$$

with $M = 3.9249\,\text{kg}, m = 0.2047\,\text{kg}, l = 0.2302\,\text{m}, g = 9.81\,\text{N/kg}, a = 25.3\,\text{N/V}$. In the model, x and θ represent the cart position coordinate and the pendulum angle from vertical, respectively.

We start with the disturbance-free case, where $w = 0$. The pendulum can be stabilized by a state feedback $u(t) = K \begin{bmatrix} x & \dot{x} & \theta & \dot{\theta} \end{bmatrix}^T$ with the gain

$$K = [5.825 \;\; 5.883 \quad -24.941 \;\; 5.140],$$

which leads to the closed-loop system eigenvalues $\{-100, \; -2 + 2i, \; -2 - 2i, \; -2\}$. In practice, the variables $\theta, \; \dot{\theta}$ and $x, \; \dot{x}$ are not accessible simultaneously. We consider

$$C_1 = \begin{bmatrix} 1 & 0 & 0 & 0 \\ 0 & 1 & 0 & 0 \end{bmatrix}, \quad C_2 = \begin{bmatrix} 0 & 0 & 1 & 0 \\ 0 & 0 & 0 & 1 \end{bmatrix}. \quad (4.35)$$

The applied control is obtained from the following blocks of K

$$K_1 = \begin{bmatrix} 5.825 & 5.883 \end{bmatrix}, \quad K_2 = \begin{bmatrix} 24.941 & 5.140 \end{bmatrix}.$$

Using the classical discretization-based model for the case of constant sampling and for the values of constant delay η, Fig. 4.2 shows the stability domain for the inverted pendulum example with Round-Robin scheduling protocol (in constant sampling interval/delay plane). By applying further Theorem 4.1 with $\alpha = 0$ for the val-

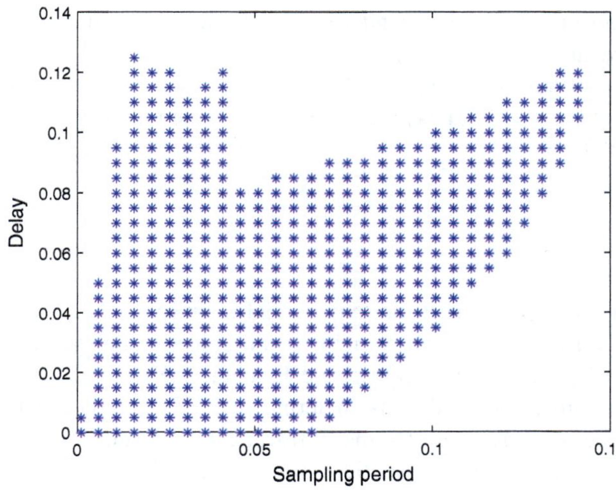

Fig. 4.2 Estimation of stability domain for Round-Robin protocol with constant sampling and constant delay based on discretization

Table 4.5 Example 4.2: maximum value of τ_M for different η_m

$\eta_m \backslash \tau_M$	Theorem 4.1		Reference [135]	Theorem 4.2	Analytical
1.0×10^{-3}	4.7×10^{-3}		3.7×10^{-3}	4.9×10^{-3}	8.5×10^{-3}
2.0×10^{-3}	5.4×10^{-3}	$\eta \equiv \eta_m$	4.5×10^{-3}	5.5×10^{-3}	1.05×10^{-2}
3.0×10^{-3}	6.1×10^{-3}		5.2×10^{-3}	6.1×10^{-3}	1.25×10^{-2}
4.0×10^{-3}	6.8×10^{-3}		6.0×10^{-3}	6.8×10^{-3}	1.45×10^{-2}

Table 4.6 Example 4.2: maximum value of τ_M for $\eta_k \equiv 0$

Method	τ_M
Analytical: $t_{k+1} - t_k \equiv \tau_M$	6.8×10^{-3}
Remark 4.6: $t_{k+1} - t_k \equiv \tau_M$	6.4×10^{-3}
Remark 4.6 ($X = X_i = 0$, $i = 1, 2, 3$): $t_{k+1} - t_k \leq \tau_M$	5.3×10^{-3}
Theorem 4.2 with $\eta = 0$: $t_{k+1} - t_k \leq \tau_M$	4.3×10^{-3}

ues of η_m given in Table 4.5 and Theorem 4.2 with $\eta \equiv \eta_m$, we find the maximum values of τ_M that preserve the asymptotic stability (see Table 4.5).

When there is no communication delay, i.e., $\eta_k \equiv 0$, by applying Remark 4.6 with $\alpha = 0$, we find the maximum values of τ_M that preserve the asymptotic stability (see Table 4.6). It is seen from Table 4.6 that our results for the constant sampling are close to the analytical one and are less conservative than for the variable sampling.

Choose further $\tau_M = 4.0 \times 10^{-3}$, $\eta_m = 2.0 \times 10^{-3}$. By applying Theorem 4.1, we find that the system is exponentially stable with the decay rate $\alpha = 1.94$. Consider next the perturbed model of pendulum and the noisy measurements, i.e., $w \neq 0$ and $v \neq 0$. We assume that $y_i(s_k) = C_i x(s_k) + 0.1v(s_k),\ i = 1, 2$. By applying Theorem 4.1, we find that for $\tau_M = 4.0 \times 10^{-3}$, $\eta_m = 2.0 \times 10^{-3}$, the system has an L_2-gain less than $\gamma = 1.24$, whereas for $\tau_M \to 0$ the resulting $\gamma = 1.19$.

4.3 Notes

In this chapter, we investigated the exponential stability and L_2-gain analysis of NCSs with Round-Robin protocol, variable communication delay and variable sampling intervals. By developing appropriate Lyapunov–Krasovskii-based methods, we derived the sufficient conditions in terms of LMIs for the switched closed-loop system. The batch reactor example illustrates the advantages of our proposed method over the existing ones: essential improvement of the results comparatively to the hybrid system approach, performance analysis comparatively to the discrete-time approach, non-small network-induced delay comparatively to both existing approaches.

The Round-Robin protocol was further considered in different control problems in the literature, e.g., in [226] for distributed estimation with H_∞ consensus, in [7] to design network-based H_∞ filter for a parabolic system, in [266] for set-membership filtering problem of discrete time-varying system and in [121, 122, 236] to achieve master–salve synchronization.

Chapter 5
NCSs in the Presence of TOD and Round-Robin Protocols

In Chap. 4 we have developed a time-delay approach to NCSs with two sensor nodes under Round-Robin protocol in the presence of non-small communication delays. The closed-loop system has been presented as a switched system with two and ordered time-varying delays. Some preliminary results on time-delay approach to NCSs with two sensor nodes and TOD protocol were presented in [132], where the closed-loop system was modeled as a hybrid-delayed system. A time-dependent Lyapunov functional was introduced in [132] to derive stability conditions for the hybrid-delayed system. Note that the extension from two to multiple sensor nodes is far from being straightforward. On the one hand, the switched system model with multiple ordered delays for Round-Robin protocol may lead to complicated conditions. On the other hand, the time-dependent Lyapunov functional of [132] is not applicable any more.

In this chapter, we consider linear (probably, uncertain) NCSs with additive essentially bounded disturbances in the presence of multiple sensor nodes, TOD or Round-Robin protocol, variable transmission delays and sampling intervals. We give a unified hybrid system model under both TOD and Round-Robin protocols for the closed-loop system that contains time-varying delays in the continuous dynamics and in the reset conditions. A novel Lyapunov–Krasovskii method is introduced for the stability analysis of the hybrid-delayed systems, which is based on discontinuous in time Lyapunov functionals.

5.1 Problem Formulation and the Novel Hybrid Model

5.1.1 The Description of NCSs

Consider the system architecture in Fig. 5.1 with plant

K. Liu et al., *Networked Control Under Communication Constraints*, Advances in Delays and Dynamics 11, https://doi.org/10.1007/978-981-15-4230-5_5

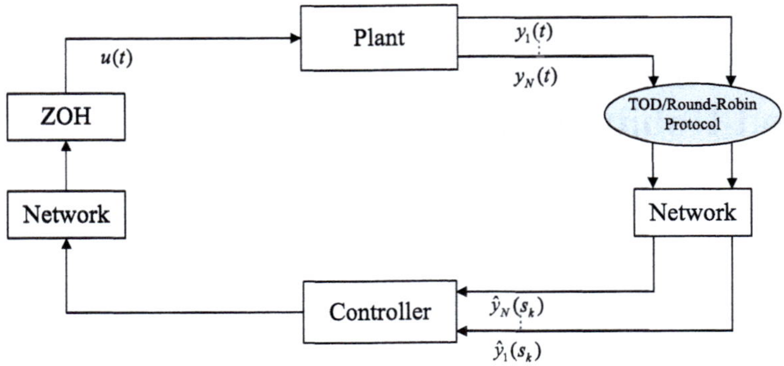

Fig. 5.1 System architecture with multiple sensors

$$\dot{x}(t) = Ax(t) + Bu(t) + D\omega(t), \ t \geq 0, \tag{5.1}$$

where $x(t) \in \mathbb{R}^n$ and $u(t) \in \mathbb{R}^m$ denote the state and the control input, respectively; $\omega(t) \in \mathbb{R}^q$ is the essentially bounded disturbance. Assume that there exists a real number $\Delta > 0$ such that $\|\omega[0, t]\|_\infty \leq \Delta$ for all $t \geq 0$, where $\|w[0, t]\|_\infty$ denotes the essential supremum of the Euclidean norm $\|w[t_0, t]\|$. The system matrices A, B and D can be uncertain with polytopic-type uncertainties.

The system is equipped with N distributed sensors, a controller and an actuator, which are connected via the network. The measurements are given by $y_i(t) = C_i x(t) \in \mathbb{R}^{n_i}$ $(i = 1, \dots, N, \ \sum_{i=1}^{N} n_i = n_y)$, which are sampled at s_k, satisfying (2.2). Then, we denote $C = [C_1^T \ \dots \ C_N^T]^T$, $y(t) = [y_1^T(t) \ \dots \ y_N^T(t)]^T \in \mathbb{R}^{n_y}$.

At each sampling instant s_k, one of the outputs $y_i(s_k) \in \mathbb{R}^{n_i}$ is transmitted via the sensor network. We suppose that the data loss does not occur and that the transmission of the information over the network experiences an uncertain, time-varying delay η_k. Then, $t_k = s_k + \eta_k$ is the updating time instant of the ZOH device.

Assume that the maximum sampling interval and the maximum delay between the sampling instant s_k and its updating instant t_k are bounded by MATI and MAD, respectively. Following [161] and Chap. 4, we allow the transmission delays to be non-small provided that the transmission order of the data packets is maintained for reception. Assume that the network-induced delay η_k and the time span between the updating and the most recent sampling instants are bounded:

$$t_{k+1} - t_k + \eta_k \leq \tau_M, \ 0 \leq \eta_m \leq \eta_k \leq \text{MAD}, \ k \in \mathbb{Z}^+, \tag{5.2}$$

where τ_M denotes the maximum time span between the time

$$s_k = t_k - \eta_k \tag{5.3}$$

at which the state is sampled and the time t_{k+1} at which the next update arrives at the destination. Here, η_m and MAD are known bounds and $\tau_M = \text{MATI} + \text{MAD}$.

Note that MATI $= \tau_M - \text{MAD} \leq \tau_M - \eta_m$, $\eta_m > \frac{\tau_M}{2}$, i.e., $\eta_m > \tau_M - \eta_m$ leads to MATI $\leq \tau_M - \eta_m < \eta_m \leq \eta_k$, which implies that the network delays are non-small. In the examples of Sect. 5.5, we will show that our method is applicable to $\eta_m > \frac{\tau_M}{2}$.

5.1.2 A Hybrid Model via Time-Delay Approach

We will consider the TOD and Round-Robin protocols that orchestrate the sensor data transmission. Denote by $\hat{y}(s_k) = [\hat{y}_1^T(s_k) \ \dots \ \hat{y}_N^T(s_k)]^T \in \mathbb{R}^{n_y}$ the output information submitted to the scheduling protocol. At each sampling instant s_k, one of the system nodes $i \in \{1, \dots, N\}$ is active; that is, only one of $\hat{y}_i(s_k)$ values is updated with the recent output $y_i(s_k)$. Let $i_k^* \in \{1, \dots, N\}$ denote the active output node at the sampling instant s_k, which will be chosen due to scheduling protocols. Then, we have

$$\hat{y}_i(s_k) = \begin{cases} y_i(s_k), & i = i_k^*, \\ \hat{y}_i(s_{k-1}), & i \neq i_k^*. \end{cases} \tag{5.4}$$

The choice of i_k^* either will be periodic (in Round-Robin protocol) or will depend on the transmission error (in TOD protocol) which is defined below. Consider the error between the system output $y_i(s_k)$ and the last available information $\hat{y}_i(s_{k-1})$:

$$\begin{aligned} &e_i(t) = \hat{y}_i(s_{k-1}) - y_i(s_k), \ \hat{y}_i(s_{-1}) \triangleq 0, \ i = 1, \dots, N, \\ &e(t) = \text{col}\{e_1(t), \dots, e_N(t)\}, \ t \in [t_k, t_{k+1}), \ k \in \mathbb{Z}^+, \ e(t) \in \mathbb{R}^{n_y}. \end{aligned} \tag{5.5}$$

It is supposed that the controller and the actuator are event-driven. The most recent output information on the controller side is denoted by $\hat{y}(s_k)$.

5.1.2.1 Static Output-Feedback Controller

Assume that there exists a matrix $K = [K_1 \ \dots \ K_N]$, $K_i \in \mathbb{R}^{m \times n_i}$ such that $A + BKC$ is Hurwitz. Then, the static output-feedback controller has the form

$$u(t) = K\hat{y}(s_k) = \sum_{i=1}^{N} K_i \hat{y}_i(s_k), \ t \in [t_k, t_{k+1}).$$

Therefore, due to (5.4), the controller can be written as

$$u(t) = K_{i_k^*} y_{i_k^*}(t_k - \eta_k) + \sum_{i=1, i \neq i_k^*}^{N} K_i \hat{y}_i(t_{k-1} - \eta_{k-1}), \ t \in [t_k, t_{k+1}), \tag{5.6}$$

where i_k^* is the index of the active node at s_k and η_k is the communication delay.

Denote

$$\tau(t) = t - t_k + \eta_k, \ t \in [t_k, t_{k+1}). \tag{5.7}$$

Then, $\eta_m \leq \tau(t) \leq \tau_M$. We thus obtain the impulsive closed-loop model with the following continuous dynamics:

$$\begin{cases} \dot{x}(t) = Ax(t) + A_1 x(t - \tau(t)) + \sum_{i=1, i \neq i_k^*}^{N} B_i e_i(t) + D\omega(t), \\ \dot{e}(t) = 0, \ t \in [t_k, t_{k+1}), \end{cases} \tag{5.8}$$

where $A_1 = BKC$, $B_i = BK_i$, $i = 1, \dots, N$.

Taking into account (5.5), we obtain

$$\begin{aligned} e_i(t_{k+1}) &= \hat{y}_i(s_k) - y_i(s_{k+1}) = y_i(s_k) - y_i(s_{k+1}) \\ &= C_i x(s_k) - C_i x(s_{k+1}), \ i = i_k^*, \end{aligned}$$

and

$$\begin{aligned} e_i(t_{k+1}) &= \hat{y}_i(s_k) - y_i(s_{k+1}) = \hat{y}_i(s_{k-1}) - y_i(s_{k+1}) \\ &= \hat{y}_i(s_{k-1}) - y_i(s_k) + y_i(s_k) - y_i(s_{k+1}) \\ &= e_i(t_k) + C_i[x(s_k) - x(s_{k+1})], \ i \neq i_k^*, \ i \in \mathbb{N}. \end{aligned}$$

Thus, the delayed reset system is given by

$$\begin{cases} x(t_{k+1}) &= x(t_{k+1}^-), \\ e_i(t_{k+1}) &= C_i[x(t_k - \eta_k) - x(t_{k+1} - \eta_{k+1})], \ i = i_k^*, \\ e_i(t_{k+1}) &= e_i(t_k) + C_i[x(t_k - \eta_k) - x(t_{k+1} - \eta_{k+1})], \ i \neq i_k^*, \ i \in \mathbb{N}. \end{cases} \tag{5.9}$$

Therefore, (5.8)–(5.9) is a novel hybrid model of the NCS. It contains the piecewise-continuous delay $\tau(t)$ in the continuous-time dynamics (5.8). Even for $\eta_k = 0$, we have the delayed state $x(t_k) = x(t - \tau(t))$ with $\tau(t) = t - t_k$.

Remark 5.1 In [84] (in the framework of impulsive/hybrid system approach), a piecewise-continuous error $e(t) = \hat{y}(t_k) - y(t)$, $t \in [s_k, s_{k+1}]$ is defined. It leads to the non-delayed continuous dynamics. The derivation of reset equations is based on the assumption of small communication delays that is avoided in our approach. In our approach, $e(t)$ is different: It is given by (5.5) and is piecewise-constant. As a consequence, our hybrid model is different with the delayed continuous dynamics. Moreover, in the absence of scheduling protocols, the closed-loop system is given by non-hybrid system (5.8), where $e(t) \equiv 0$. The latter is consistent with the time-delay model considered, e.g., in [64, 66].

Note that in our model, the first updating time t_0 corresponds to the time instant when the first data is received by the actuator. Assume that the initial conditions for (5.8)–(5.9) are given by

$$x_{t_0} \in W[-\tau_M, 0], \quad e(t_0) = -Cx(t_0 - \eta_0) = -Cx_0.$$

5.1.2.2 Dynamic Output-Feedback Controller

Assume that the controller is directly connected to the actuator. Consider a dynamic output-feedback controller of the form

$$\begin{aligned} \dot{x}_c(t) &= A_c x_c(t) + B_c \hat{y}(s_k), \\ u(t) &= C_c x_c(t) + D_c \hat{y}(s_k), \ t \in [t_k, t_{k+1}), \ k \in \mathbb{Z}^+, \end{aligned}$$

where $x_c(t) \in \mathbb{R}^{n_c}$, A_c, B_c, C_c and D_c are the matrices with appropriate dimensions. Let $e_i(t), i = 1, \ldots, N$, be defined by (5.5). The closed-loop system can be expressed in the form of (5.8)–(5.9), where x, e_i and the matrices are changed by the ones with the bars as follows:

$$\begin{aligned} \bar{x} &= \begin{bmatrix} x \\ x_c \end{bmatrix}, \ \bar{A} = \begin{bmatrix} A & BC_c \\ 0_{n_c \times n} & A_c \end{bmatrix}, \ \bar{B}_i = \begin{bmatrix} BD_c \\ B_c \end{bmatrix}, \ \bar{D} = \begin{bmatrix} D \\ 0_{n_c \times q} \end{bmatrix}, \\ \bar{A}_1 &= \begin{bmatrix} BD_cC \ 0_{n \times n_c} \\ B_cC \ \ 0_{n_c \times n_c} \end{bmatrix}, \ \bar{C}_1 = \begin{bmatrix} C_1 \ 0 \\ 0 \ \ 0 \end{bmatrix}, \ \bar{C}_i \in \mathbb{R}^{n_y \times (n+n_c)}, \\ \bar{C}_2 &= \begin{bmatrix} 0_{n \times n_1} & C_2^T \ 0 \\ 0_{n_c \times n_1} & 0 \ \ 0 \end{bmatrix}^T, \ldots, \bar{C}_N = \begin{bmatrix} 0 \ C_N^T \\ 0 \ \ 0 \end{bmatrix}^T, \ \bar{e}_1(t) = [e_1^T(t) \ 0]^T, \\ \bar{e}_2(t) &= [0_{1 \times n_1} \ e_2^T(t) \ 0]^T, \ldots \ \bar{e}_N(t) = [0 \ e_N^T(t)]^T, \ \bar{e}_i(t) \in \mathbb{R}^{n_y}, \ i = 1, \ldots, N. \end{aligned}$$

5.1.3 Scheduling Protocols

5.1.3.1 TOD Protocol

In the TOD protocol, the output node $i \in \{1, \ldots, N\}$ with the greatest weighted error will have the highest priority to be accessed to the network.

Definition 5.1 (Weighted TOD protocol) Let $Q_i > 0, \ i = 1, \ldots, N,$ be some weighting matrices. At the sampling instant s_k, the weighted TOD protocol is a protocol for which the active output node with the index i_k^* is defined as any index that satisfies

$$\left|\sqrt{Q_{i_k^*}} e_{i_k^*}(t)\right|^2 \geq \left|\sqrt{Q_i} e_i(t)\right|^2, \ t \in [t_k, t_{k+1}), \ k \in \mathbb{Z}^+, \ i = 1, \ldots, N. \tag{5.10}$$

A possible choice of i_k^* is given by

$$i_k^* = \min\left\{\arg\max_{i\in\{1,\ldots,N\}} |\sqrt{Q_i}\left(\hat{y}_i(s_{k-1}) - y_i(s_k)\right)|^2\right\},$$

i.e., if several errors are the same, we transmit the node i with a minimum index. The conditions for computing the weighting matrices $Q_1, \ldots, Q_N$ will be given in Theorem 5.1.

Remark 5.2 For implementation of TOD protocol in wireless networks, we refer to [21].

5.1.3.2 Round-Robin Protocol

The active output node is chosen in a periodic order:

$$\begin{array}{l} i_k^* = i_{k+N}^*, \text{ for all } k \in \mathbb{Z}^+, \\ i_j^* \neq i_l^*, \text{ for } 0 \le j < l \le N-1. \end{array} \tag{5.11}$$

Remark 5.3 Note that the closed-loop system under Round-Robin protocol in Sect. 4.1 of Chap. 4 is a switched system with ordered delays $\tau_1(t) < \cdots < \tau_N(t)$, where $\tau_i(t) = t - t_{k-i+1} + \eta_{k-i+1}$, $i = 1, \ldots, N$. A Lyapunov–Krasovskii analysis of the latter model is based on the standard time-independent Lyapunov functional for interval delay.

Remark 5.4 The inclusion of packet dropouts under scheduling protocols is relatively easy if one assumes that there is an additional perfect (without packet dropouts) feedback channel to send a reception/dropout acknowledgment to the active sensor and if this acknowledgment is completed within one sampling period. Then as in [84], the packet dropouts can be modeled as prolongations of the transmission interval.

Definition 5.2 The hybrid system (5.8)–(5.9) with essentially bounded disturbance ω is said to be partially input-to-state stable (ISS) with respect to x (or x-ISS) if there exist constants $b > 0$, $\delta > 0$ and $c > 0$ such that for $t \ge t_0$ the following

$$|x(t)|^2 \le be^{-\delta(t-t_0)}\left[\|x_{t_0}\|_W^2 + |e(t_0)|^2\right] + c\|\omega[t_0, t]\|_\infty^2$$

holds for the solutions of the hybrid system initialized with $x_{t_0} = \phi \in W[-\tau_M, 0]$ and $e(t_0) \in \mathbb{R}^{n_y}$. The hybrid system (5.8)–(5.9) is ISS if additionally the following bound

$$|e(t)|^2 \le be^{-\delta(t-t_0)}\left[\|x_{t_0}\|_W^2 + |e(t_0)|^2\right] + c\|\omega[t_0, t]\|_\infty^2$$

is valid for $t \ge t_0$.

The objective is to derive conditions for the partial ISS of the hybrid system (5.8)–(5.9) with respect to the variable of interest x. In [26], the notion of partial stability was also used. In Sect. 5.2, the ISS of (5.8)–(5.9) under TOD protocol with N sensor nodes will be studied. For $N = 2$, less restrictive conditions will be derived in Sect. 5.3, and it will be shown that the same conditions guarantee x-ISS of (5.8)–(5.9) under Round-Robin protocol. In Sect. 5.4, the latter conditions will be extended to Round-Robin protocol with $N \geq 2$.

5.2 ISS Under TOD Protocol: General N

Note that the differential equation for x given by (5.8) depends on $e_i(t) = e_i(t_k)$, $t \in [t_k, t_{k+1})$ with $i \neq i_k^*$ only. Consider the following Lyapunov functional:

$$\begin{aligned}
V_e(t) &= V(t, x_t, \dot{x}_t) + \sum_{i=1}^{N} e_i^T(t) Q_i e_i(t), \\
V(t, x_t, \dot{x}_t) &= \tilde{V}(t, x_t, \dot{x}_t) + V_G, \\
V_G &= \sum_{i=1}^{N} (\tau_M - \eta_m) \int_{s_k}^{t} e^{2\alpha(s-t)} |\sqrt{G_i} C_i \dot{x}(s)|^2 ds, \\
\tilde{V}(t, x_t, \dot{x}_t) &= x^T(t) P x(t) + \int_{t-\eta_m}^{t} e^{2\alpha(s-t)} x^T(s) S_0 x(s) ds \\
&\quad + \int_{t-\tau_M}^{t-\eta_m} e^{2\alpha(s-t)} x^T(s) S_1 x(s) ds \\
&\quad + \eta_m \int_{-\eta_m}^{0} \int_{t+\theta}^{t} e^{2\alpha(s-t)} \dot{x}^T(s) R_0 \dot{x}(s) ds d\theta \\
&\quad + (\tau_M - \eta_m) \int_{-\tau_M}^{-\eta_m} \int_{t+\theta}^{t} e^{2\alpha(s-t)} \dot{x}^T(s) R_1 \dot{x}(s) ds d\theta,
\end{aligned} \tag{5.12}$$

where $P > 0,\ S_j > 0,\ R_j > 0,\ G_i > 0,\ Q_i > 0,\ \alpha > 0,\ j = 0, 1, i = 1, \ldots, N$, $t \in [t_k, t_{k+1})$, $k \in \mathbb{Z}^+$ and where we define $x(t) = x_0$ for $t < 0$.

Here, the terms

$$e_i^T(t) Q_i e_i(t) \equiv e_i^T(t_k) Q_i e_i(t_k),\ t \in [t_k, t_{k+1})$$

are piecewise-constant, and $\tilde{V}(t, x_t, \dot{x}_t)$ presents the standard Lyapunov functional for systems with interval delays $\tau(t) \in [\eta_m, \tau_M]$. The novel piecewise-continuous in time term V_G is inserted to cope with the delays in the reset conditions. It is continuous on $[t_k, t_{k+1})$ and does not grow in the jumps (when $t = t_{k+1}$), since

$$
\begin{aligned}
V_{G|t=t_{k+1}} - V_{G|t=t_{k+1}^-} &= \sum_{i=1}^{N}(\tau_M - \eta_m)\int_{s_{k+1}}^{t_{k+1}} e^{2\alpha(s-t_{k+1})}|\sqrt{G_i}C_i\dot{x}(s)|^2 ds \\
&\quad - \sum_{i=1}^{N}(\tau_M - \eta_m)\int_{s_k}^{t_{k+1}^-} e^{2\alpha(s-t_{k+1})}|\sqrt{G_i}C_i\dot{x}(s)|^2 ds \\
&\leq - \sum_{i=1}^{N}(\tau_M - \eta_m)e^{-2\alpha\tau_M}\int_{s_k}^{s_{k+1}} |\sqrt{G_i}C_i\dot{x}(s)|^2 ds \\
&\leq - \sum_{i=1}^{N} e^{-2\alpha\tau_M}\left|\sqrt{G_i}C_i[x(s_k) - x(s_{k+1})]\right|^2,
\end{aligned} \tag{5.13}
$$

where we applied Jensen's inequality. The function $V_e(t)$ is thus continuous and differentiable over $[t_k, t_{k+1})$. The following lemma gives the sufficient conditions for the ISS of (5.8)–(5.10).

Lemma 5.1 *Suppose that there exist positive constants α, b, $0 < Q_i \in \mathbb{R}^{n_i \times n_i}$, $0 < U_i \in \mathbb{R}^{n_i \times n_i}$, $0 < G_i \in \mathbb{R}^{n_i \times n_i}$, $i = 1, \ldots, N$, and $V_e(t)$ of (5.12) such that along (5.8) the following inequality*

$$
\dot{V}_e(t) + 2\alpha V_e(t) - \frac{1}{\tau_M - \eta_m}\sum_{i=1, i\neq i_k^*}^{N} |\sqrt{U_i}e_i(t)|^2 - 2\alpha\left|\sqrt{Q_{i_k^*}}e_{i_k^*}(t)\right|^2 - b|\omega(t)|^2 \leq 0 \tag{5.14}
$$

holds for $t \in [t_k, t_{k+1})$. Assume additionally that

$$
\Omega_i \triangleq \begin{bmatrix} -\frac{1-2\alpha(\tau_M - \eta_m)}{N-1}Q_i + U_i & Q_i \\ * & Q_i - G_i e^{-2\alpha\tau_M} \end{bmatrix} < 0, \ i = 1, \ldots, N. \tag{5.15}
$$

Then, $V_e(t)$ does not grow in the jumps along (5.8)–(5.10):

$$
\Theta \triangleq V_e(t_{k+1}) - V_e(t_{k+1}^-) + \sum_{i=1, i\neq i_k^*}^{N} |\sqrt{U_i}e_i(t_k)|^2 + 2\alpha(\tau_M - \eta_m)\left|\sqrt{Q_{i_k^*}}e_{i_k^*}(t_k)\right|^2 \leq 0. \tag{5.16}
$$

Moreover, the following bounds hold for the solutions of (5.8)–(5.10) initialized by $x_{t_0} \in W[-\tau_M, 0]$, $e(t_0) \in \mathbb{R}^{n_y}$:

$$
\begin{aligned}
V(t, x_t, \dot{x}_t) &\leq e^{-2\alpha(t-t_0)}V_e(t_0) + \frac{b}{2\alpha}\Delta^2, \ t \geq t_0, \\
V_e(t_0) &= V(t_0, x_{t_0}, \dot{x}_{t_0}) + \sum_{i=1}^{N}|\sqrt{Q_i}e_i(t_0)|^2,
\end{aligned} \tag{5.17}
$$

and

$$\sum_{i=1}^{N} |\sqrt{Q_i} e_i(t)|^2 \leq \tilde{c} e^{-2\alpha(t-t_0)} V_e(t_0) + \frac{b}{2\alpha}\Delta^2, \tag{5.18}$$

where $\tilde{c} = e^{2\alpha(\tau_M - \eta_m)}$, *implying ISS of (5.8)–(5.10).*

Proof Since $\int_{t_k}^{t} e^{-2\alpha(t-s)} ds \leq \tau_M - \eta_m$, $t \in [t_k, t_{k+1})$ and $|\omega(t)| \leq \Delta$, by the comparison principle, (5.14) implies

$$\begin{aligned} V_e(t) \leq e^{-2\alpha(t-t_k)} V_e(t_k) &+ \sum_{i=1, i\neq i_k^*}^{N} \{|\sqrt{U_i} e_i(t_k)|^2\} \\ &+ 2\alpha(\tau_M - \eta_m)\left|\sqrt{Q_{i_k^*}} e_{i_k^*}(t_k)\right|^2 + b\Delta^2 \int_{t_k}^{t} e^{-2\alpha(t-s)} ds, \; t \in [t_k, t_{k+1}). \end{aligned} \tag{5.19}$$

Note that (5.15) yields $0 < 2\alpha(\tau_M - \eta_m) < 1$ and $U_i \leq \frac{1-2\alpha(\tau_M-\eta_m)}{N-1} Q_i \leq Q_i$, $i = 1, \ldots, N$. Hence,

$$V(t, x_t, \dot{x}_t) \leq e^{-2\alpha(t-t_k)} V_e(t_k) + b\Delta^2 \int_{t_k}^{t} e^{-2\alpha(t-s)} ds, \; t \in [t_k, t_{k+1}). \tag{5.20}$$

Since $\tilde{V}_{|t=t_{k+1}} = \tilde{V}_{|t=t_{k+1}^-}$ and $e(t_{k+1}^-) = e(t_k)$, we obtain

$$\begin{aligned} \Theta = &\sum_{i=1}^{N}\left[|\sqrt{Q_i} e_i(t_{k+1})|^2 - |\sqrt{Q_i} e_i(t_k)|^2\right] + \sum_{i=1, i\neq i_k^*}^{N} |\sqrt{U_i} e_i(t_k)|^2 \\ &+ 2\alpha(\tau_M - \eta_m)\left|\sqrt{Q_{i_k^*}} e_{i_k^*}(t_k)\right|^2 + V_{G|t=t_{k+1}} - V_{G|t=t_{k+1}^-}. \end{aligned}$$

Then taking into account (5.13), we find

$$\begin{aligned} \Theta \leq &\left|\sqrt{Q_{i_k^*}} e_{i_k^*}(t_{k+1})\right|^2 + \sum_{i=1, i\neq i_k^*}^{N} |\sqrt{Q_i} e_i(t_{k+1})|^2 - [1 - 2\alpha(\tau_M - \eta_m)]\left|\sqrt{Q_{i_k^*}} e_{i_k^*}(t_k)\right|^2 \\ &- \sum_{i=1, i\neq i_k^*}^{N}\left[e_i^T(t_k)(Q_i - U_i) e_i(t_k)\right] - \sum_{i=1}^{N} e^{-2\alpha\tau_M}\left|\sqrt{G_i} C_i [x(s_k) - x(s_{k+1})]\right|^2. \end{aligned}$$

Noting that under TOD protocol, we have

$$-\left|\sqrt{Q_{i_k^*}} e_{i_k^*}(t_k)\right|^2 \leq -\frac{1}{N-1}\sum_{i=1, i\neq i_k^*}^{N} |\sqrt{Q_i} e_i(t_k)|^2.$$

Denote $\zeta_i = \mathrm{col}\{e_i(t_k), C_i[x(s_k) - x(s_{k+1})]\}$. Then employing (5.3) and (5.9), we arrive at

$$\Theta \leq -\left|\sqrt{G_{i_k^*}e^{-2\alpha\tau_M} - Q_{i_k^*}}C_{i_k^*}[x(s_k) - x(s_{k+1})]\right|^2 + \sum_{i=1, i\neq i_k^*}^{N} \zeta_i^T \Omega_i \zeta_i \leq 0,$$

that yields (5.16).

The inequalities (5.16) and (5.19) with $t = t_{k+1}^-$ imply

$$V_e(t_{k+1}) \leq e^{-2\alpha(t_{k+1}-t_k)}V_e(t_k) + b\Delta^2 \int_{t_k}^{t_{k+1}} e^{-2\alpha(t_{k+1}-s)}ds.$$

Then, we obtain

$$\begin{aligned} V_e(t_{k+1}) &\leq e^{-2\alpha(t_{k+1}-t_{k-1})}V_e(t_{k-1}) + b\Delta^2 \int_{t_{k-1}}^{t_{k+1}} e^{-2\alpha(t_{k+1}-s)}ds \\ &\leq e^{-2\alpha(t_{k+1}-t_0)}V_e(t_0) + b\Delta^2 \int_{t_0}^{t_{k+1}} e^{-2\alpha(t_{k+1}-s)}ds. \end{aligned} \tag{5.21}$$

Replacing in (5.21) $k+1$ by k and using (5.20), we arrive at (5.17), which yields x-ISS of (5.8)–(5.10) because

$$\lambda_{\min}(P)|x(t)|^2 \leq V(t, x_t, \dot{x}_t), \ \ V(t_0, x_{t_0}, \dot{x}_{t_0}) \leq \delta \|x_{t_0}\|_W^2$$

for some scalar $\delta > 0$. Moreover, (5.21) with $k+1$ replaced by k implies (5.18), since for $t \in [t_k, t_{k+1})$

$$e^{-2\alpha(t_k-t_0)} = e^{-2\alpha(t-t_0)}e^{-2\alpha(t_k-t)} \leq \tilde{c}e^{-2\alpha(t-t_0)}.$$

By using Lemma 5.1 and the standard arguments for delay-dependent analysis, we derive LMI conditions for ISS of (5.8)–(5.10) (see appendix for the proof).

Theorem 5.1 *Given $0 \leq \eta_m < \tau_M$, $\alpha > 0$, assume that there exist positive scalar b, $n \times n$ matrices $P > 0$, $S_0 > 0$, $R_0 > 0$, $S_1 > 0$, $R_1 > 0$, S_{12}, $n_i \times n_i$ matrices $Q_i > 0$, $U_i > 0$, $G_i > 0$, $i = 1, \ldots, N$, such that (5.15) and the following LMIs*

$$\Phi = \begin{bmatrix} R_1 & S_{12} \\ * & R_1 \end{bmatrix} \geq 0, \tag{5.22}$$

$$\begin{bmatrix} \Sigma_i - (F^i)^T \Phi F^i e^{-2\alpha\tau_M} & \Xi_i^T H \\ * & -H \end{bmatrix} < 0, \ i = 1, \ldots, N, \tag{5.23}$$

are feasible, where

$$
\begin{aligned}
H &= \eta_m^2 R_0 + (\tau_M - \eta_m)^2 R_1 + (\tau_M - \eta_m) \sum_{l=1}^{N} C_l^T G_l C_l, \\
\Sigma_i &= \mathrm{He}((F_1^i)^T P \Xi_i) + \Upsilon_i - (F_2^i)^T R_0 F_2^i e^{-2\alpha\eta_m}, \\
F_1^i &= [I_n \; 0_{n\times(3n+n_y-n_i+q)}], \\
F_2^i &= [I_n \; -I_n \; 0_{n\times(2n+n_y-n_i+q)}], \\
F^i &= \begin{bmatrix} 0_{n\times n} & I_n & -I_n & 0_{n\times n} & 0_{n\times(n_y-n_i+q)} \\ 0_{n\times n} & 0_{n\times n} & I_n & -I_n & 0_{n\times(n_y-n_i+q)} \end{bmatrix}, \\
\Xi_i &= [A \; 0_{n\times n} \; A_1 \; 0_{n\times n} \; \tilde{\Xi}_i \; D], \\
\tilde{\Xi}_i &= [B_2 \; \dots \; B_N], \; i = 1, \\
\tilde{\Xi}_i &= [B_1 \; \dots \; B_{N-1}], \; i = N, \\
\tilde{\Xi}_i &= [B_1 \; \dots \; B_{j|j\neq i} \dots B_N], \; i = 2, \dots, N-1, \\
\Upsilon_i &= \mathrm{diag}\{S_0 + 2\alpha P, -(S_0 - S_1)e^{-2\alpha\eta_m}, 0_{n\times n}, -S_1 e^{-2\alpha\tau_M}, \; \phi_i, \; -bI_q\}, \\
\phi_i &= \mathrm{diag}\{\psi_2, \dots, \psi_N\}, \; i = 1, \\
\phi_i &= \mathrm{diag}\{\psi_1, \dots, \psi_{N-1}\}, \; i = N, \\
\phi_i &= \mathrm{diag}\{\psi_1, \dots, \psi_{j|j\neq i}, \dots, \psi_N\}, \; i = 2, \dots, N-1, \\
\psi_j &= -\frac{1}{\tau_M - \eta_m} U_j + 2\alpha Q_j, \; j = 1, \dots, N.
\end{aligned}
\tag{5.24}
$$

Then, the solutions of the hybrid system (5.8)–(5.10) satisfy the bound (5.17), where $V(t, x_t, \dot{x}_t)$ is given by (5.12), implying ISS of (5.8)–(5.10). If the above LMIs are feasible with $\alpha = 0$, then the bound (5.17) holds with a small enough $\alpha_0 > 0$.

Remark 5.5 The LMIs of Theorem 5.1 are always feasible for small enough delay bound τ_M. For simplicity, we will explain this in the case of $\alpha = 0$. Choose $G_i = (N-1)Q_i$. The application of Schur complements leads the LMI (5.15) with $\alpha = 0$ to a solution $U_i = \frac{Q_i}{(N-2)(N-1)}$, $N > 2$, $N \in \mathbb{N}$.

Consider next the LMIs (5.22) and (5.23). Since $A + A_1$ is Hurwitz, there always exists matrix $P > 0$ such that $P(A + A_1) + (A + A_1)^T P < 0$. The functional $\tilde{V}(t, x_t, \dot{x}_t)$ is a standard Lyapunov functional for delay-dependent analysis. The matrices $G_i > 0$ and $U_i > 0$, $i = 1, \dots, N$, appear only in H-term and on the main diagonal of the LMI (5.23), respectively. The latter LMI is independent of Q_i. Then given $G_i > 0$ and $U_i > 0$, the LMIs (5.22) and (5.23) are feasible for small enough $\tau_M > 0$ [56]. Hence, the LMIs (5.15), (5.22) and (5.23) are feasible for small enough τ_M.

Remark 5.6 From (5.2)–(5.3), it follows that $t_{k+1} - t_k$ is lower bounded by 0. Under additional assumption that there is a positive lower bound T on $t_{k+1} - t_k$, it is possible to relax the condition (5.16). Thus, for the stability analysis of (5.8)–(5.10) with $\omega(t) \equiv 0$, the inequality (5.16) can be replaced by a less restrictive one

$$
V_e(t_{k+1}) - \mu V_e(t_{k+1}^-) + \sum_{i=1, i\neq i_k^*}^{N} |\sqrt{U_i} e_i(t_k)|^2 + 2\alpha(\tau_M - \eta_m)\left|\sqrt{Q_{i_k^*}} e_{i_k^*}(t_k)\right|^2 \leq 0
\tag{5.25}
$$

with some $\mu > 1$. Then by the arguments of Lemma 5.1, the condition (5.25) holds if there exist $0 < 2\alpha(\tau_M - \eta_m) < 1$ and $U_i < Q_i$, $i = 1, \dots, N$, such that

$$\begin{bmatrix} -\frac{\mu - 2\alpha(\tau_M - \eta_m)}{N-1} Q_i + U_i + (1-\mu)Q_i & Q_i \\ * & Q_i - \mu G_i e^{-2\alpha\tau_M} \end{bmatrix} < 0, \ i = 1, \dots, N. \quad (5.26)$$

Moreover, the following bound is achieved

$$V(t, x_t, \dot{x}_t) \le \mu^k e^{-2\alpha(t-t_0)} V_e(t_0) \le e^{-(2\alpha - \ln\mu/T)(t-t_0)} V_e(t_0)$$

for $t \ge t_0$. The latter inequality holds due to $k \le \frac{t-t_0}{T}$. Therefore, the exponential stability is guaranteed under the dwell time condition $T > \frac{\ln\mu}{2\alpha}$.

The LMI (5.26) is less restrictive than (5.15): Given U_i, (5.26) allows smaller Q_i and, hence, smaller G_i, which may enlarge τ_M that solve (5.26), (5.22) and (5.23). Note that in the examples below (under assumption of constant sampling and constant network-induced delays η, where $T = \tau_M - \eta$) the relaxed condition (5.25) does not improve the results (does not enlarge τ_M).

5.3 ISS Under TOD/Round-Robin Protocol: $N = 2$

For $N = 2$, less restrictive conditions than those of Theorem 5.1 for the x-ISS of (5.8)–(5.9) will be derived via a different from (5.12) Lyapunov functional:

$$V_e(t) = V(t, x_t, \dot{x}_t) + \frac{t_{k+1} - t}{\tau_M - \eta_m} \{e_i^T(t) Q_i e_i(t)\}_{|i \ne i_k^*}, \quad (5.27)$$

where $Q_1 > 0$, $Q_2 > 0$, $\alpha > 0$, $t \in [t_k, t_{k+1})$, $k \in \mathbb{Z}^+$, $i_k^* \in \{1, 2\}$ and $V(t, x_t, \dot{x}_t)$ is given by (5.12) with $G_i = Q_i e^{2\alpha\tau_M}$. The term $\frac{t_{k+1}-t}{\tau_M - \eta_m}[e_i^T(t_k) Q_i e_i(t_k)]$ is inspired by the similar construction of Lyapunov functionals for the sampled-data systems [46, 162, 192]. The following statement holds.

Lemma 5.2 *Given $N = 2$, if there exist positive constants α, b and $V_e(t)$ of (5.27) such that along (5.8), the following inequality*

$$\dot{V}_e(t) + 2\alpha V_e(t) - b|\omega(t)|^2 \le 0, \ t \in [t_k, t_{k+1}) \quad (5.28)$$

holds. Then, $V_e(t)$ does not grow in the jumps along (5.8)–(5.10) ((5.8), (5.9), (5.11)), where

$$\Theta \triangleq V_e(t_{k+1}) - V_e(t_{k+1}^-) \le 0. \quad (5.29)$$

The bound (5.17) is valid for the solutions of (5.8)–(5.10) ((5.8), (5.9), (5.11)) with the initial condition $x_{t_0} \in W[-\tau_M, 0]$, $e(t_0) \in \mathbb{R}^{n_y}$, implying the x-ISS of (5.8)–(5.10) ((5.8), (5.9), (5.11)).

Proof Since $|\omega(t)| \leq \Delta$, (5.28) implies

$$V_e(t) \leq e^{-2\alpha(t-t_k)} V_e(t_k) + b\Delta^2 \int_{t_k}^{t} e^{-2\alpha(t-s)} ds, \; t \in [t_k, t_{k+1}). \tag{5.30}$$

Noting that

$$\begin{aligned} V_e(t_{k+1}) \leq \; & \tilde{V}_{|t=t_{k+1}} + |\sqrt{Q}_i e_i(t_{k+1})|^2_{|i \neq i^*_{k+1}} \\ & + \sum_{i=1}^{2} (\tau_M - \eta_m) \int_{t_{k+1}-\eta_{k+1}}^{t_{k+1}} e^{2\alpha(s-t_{k+1})} |\sqrt{G_i} C_i \dot{x}(s)|^2 ds, \end{aligned}$$

and employing (5.13), we obtain

$$\begin{aligned} \Theta & \leq e_i^T(t_{k+1}) Q_i e_i(t_{k+1})_{|i \neq i^*_{k+1}} + V_{G|t=t_{k+1}} - V_{G|t=t^-_{k+1}} \\ & \leq e_i^T(t_{k+1}) Q_i e_i(t_{k+1})_{|i \neq i^*_{k+1}} - \sum_{i=1}^{2} \left| \sqrt{Q_i} C_i [x(t_k - \eta_k) - x(t_{k+1} - \eta_{k+1})] \right|^2. \end{aligned}$$

We will prove that $\Theta \leq 0$ under TOD and Round-Robin protocols, respectively. Under TOD protocol, we have

$$e_i^T(t_{k+1}) Q_i e_i(t_{k+1})_{|i \neq i^*_{k+1}} \leq e^T_{i^*_k}(t_{k+1}) Q_{i^*_k} e_{i^*_k}(t_{k+1})$$

for $i^*_{k+1} = i^*_k$, whereas

$$e_i^T(t_{k+1}) Q_i e_i(t_{k+1})_{|i \neq i^*_{k+1}} = e^T_{i^*_k}(t_{k+1}) Q_{i^*_k} e_{i^*_k}(t_{k+1}) \tag{5.31}$$

for $i^*_{k+1} \neq i^*_k$. Then taking into account (5.9), we obtain

$$\begin{aligned} \Theta \leq & \left| \sqrt{Q_i} C_i [x(t_k - \eta_k) - x(t_{k+1} - \eta_{k+1})] \right|^2_{|i=i^*_k} \\ & - \sum_{i=1}^{2} \left| \sqrt{Q_i} C_i [x(t_k - \eta_k) - x(t_{k+1} - \eta_{k+1})] \right|^2 \leq 0. \end{aligned}$$

Under Round-Robin protocol, we have $i^*_{k+1} \neq i^*_k$ meaning that (5.31) holds and that $\Theta \leq 0$. Then, the result follows by the arguments of Lemma 5.1.

Remark 5.7 Different from Lemma 5.1, Lemma 5.2 guarantees (5.21), which does not give a bound on $e_{i^*_k}(t_k)$ since $V_e(t)$ for $t \in [t_k, t_{k+1})$ does not depend on $e_{i^*_k}(t_k)$. That is why Lemma 5.2 guarantees only x-ISS. However, as explained in Remark 5.9, under Round-Robin protocol x-ISS implies the boundedness of e.

In the next section, we will extend the result of Lemma 5.2 under Round-Robin protocol to the case of $N \geq 2$. Theorem 5.2 (in the particular case of $N = 2$) will provide LMIs for the x-ISS of (5.8)–(5.10) ((5.8), (5.9), (5.11)).

5.4 ISS Under Round-Robin Protocol: $N \geq 2$

Under Round-Robin protocol (5.11), the reset system (5.9) can be rewritten as

$$\begin{cases} x(t_{k+1}) &= x(t_{k+1}^-), \\ e_{i^*_{k-j}}(t_{k+1}) &= C_{i^*_{k-j}}[x(s_{k-j}) - x(s_{k+1})], \ j = 0, \dots, N-1 \text{ if } k \geq N-1, \end{cases} \tag{5.32}$$

where the index $k - j$ corresponds to the last updated measurement in the node i^*_{k-j}.

Consider the following Lyapunov functional:

$$\begin{aligned} V_e(t) &= V(t, x_t, \dot{x}_t) + V_Q, \ t \geq t_{N-1}, \\ V(t, x_t, \dot{x}_t) &= \tilde{V}(t, x_t, \dot{x}_t) + V_G, \end{aligned} \tag{5.33}$$

where $\tilde{V}(t, x_t, \dot{x}_t)$ is given by (5.12). The discontinuous in time terms V_Q and V_G are defined as follows:

$$\begin{aligned} V_Q &= \sum_{j=1}^{N-1} \frac{t_{k+1} - t}{j(\tau_M - \eta_m)} \left|\sqrt{Q_{i^*_{k-j}}} e_{i^*_{k-j}}(t)\right|^2, \ k \geq N-1, \ t \in [t_k, t_{k+1}), \\ V_G &= \begin{cases} \sum_{i=1}^{N} (\tau_M - \eta_m) \int_{s_k}^{t} e^{2\alpha(s-t)} |\sqrt{G_i} C_i \dot{x}(s)|^2 ds, \ k \geq N, \ t \in [t_k, t_{k+1}), \\ \sum_{i=1}^{N} (\tau_M - \eta_m) \int_{s_0}^{t} e^{2\alpha(s-t)} |\sqrt{G_i} C_i \dot{x}(s)|^2 ds, \ t \in [t_{N-1}, t_N), \end{cases} \end{aligned} \tag{5.34}$$

where for $i = 1, \dots, N$

$$G_i = (N-1) Q_i e^{2\alpha[\tau_M + (N-2)(\tau_M - \eta_m)]} > 0. \tag{5.35}$$

Here, V_e does not depend on $e_{i^*_k}(t_k)$. Note that given $i = 1, \dots, N$, e_i-term appears $N-1$ times in V_Q for every N intervals $[t_{k+j}, t_{k+j+1})$, $j = 0, \dots, N-1$ (except the interval with $i^*_{k+j} = i$). This motivates $N-1$ in (5.35) because V_G is supposed to compensate V_Q term.

As in the previous sections, the term V_G is inserted to cope with the delays in the reset conditions. It is continuous on $[t_k, t_{k+1})$ and does not grow in the jumps (when $t = t_{k+1}$), since for $k > N-1$ (cf., (5.13))

$$V_{G|t=t_{k+1}} - V_{G|t=t_{k+1}^-} \leq -\sum_{i=1}^{N} (\tau_M - \eta_m) \int_{s_k}^{s_{k+1}} e^{2\alpha(s - t_{k+1})} |\sqrt{G_i} C_i \dot{x}(s)|^2 ds \tag{5.36}$$

and for $k = N - 1$

$$V_{G|t=t_N} - V_{G|t=t_N^-} \leq -\sum_{i=1}^{N}(\tau_M - \eta_m)\int_{s_0}^{s_N} e^{2\alpha(s-t_N)}|\sqrt{G_i}C_i\dot{x}(s)|^2 ds. \tag{5.37}$$

The term V_Q grows in the jumps as follows:

$$\begin{aligned} V_{Q|t=t_{k+1}} - V_{Q|t=t_{k+1}^-} &= \sum_{j=1}^{N-1}\frac{t_{k+2}-t_{k+1}}{j(\tau_M-\eta_m)}\left|\sqrt{Q_{i^*_{k+1-j}}}e_{i^*_{k+1-j}}(t_{k+1})\right|^2 \\ &\leq \sum_{j=0}^{N-2}\frac{1}{j+1}\left|\sqrt{Q_{i^*_{k-j}}}C_{i^*_{k-j}}[x(s_{k-j})-x(s_{k+1})]\right|^2 \\ &\leq \sum_{j=0}^{N-2}(\tau_M-\eta_m)\int_{s_{k-j}}^{s_{k+1}}\left|\sqrt{Q_{i^*_{k-j}}}C_{i^*_{k-j}}\dot{x}(s)\right|^2 ds, \end{aligned}$$

where we used Jensen's inequality and the bound

$$\begin{aligned} s_{k+1}-s_{k-j} &= s_{k+1}-s_k+s_k-\cdots+s_{k-j+1}-s_{k-j} \\ &\leq (j+1)(\tau_M-\eta_m). \end{aligned} \tag{5.38}$$

Since $1 \leq e^{2\alpha[\tau_M+(N-2)(\tau_M-\eta_m)]}e^{2\alpha(s_{k-j}-t_{k+1})}$ for $j = 0, \ldots, N-2$, we obtain

$$\begin{aligned} V_{Q|t=t_{k+1}} - V_{Q|t=t_{k+1}^-} \leq &\sum_{j=0}^{N-2}(\tau_M-\eta_m)e^{2\alpha[\tau_M+(N-2)(\tau_M-\eta_m)]} \\ &\times\int_{s_{k-j}}^{s_{k+1}} e^{2\alpha(s-t_{k+1})}\left|\sqrt{Q_{i^*_{k-j}}}C_{i^*_{k-j}}\dot{x}(s)\right|^2 ds. \end{aligned} \tag{5.39}$$

The following lemma gives the sufficient conditions for the x-ISS of (5.8), (5.11), (5.32) (see appendix for the proof).

Lemma 5.3 *If there exist positive constants α, b and $V_e(t)$ of (5.33) such that along (5.8) the inequality (5.28) is satisfied for $k \geq N-1$, then the following bound holds along the solutions of (5.8), (5.11), (5.32):*

$$V_e(t_{k+1}) \leq e^{-2\alpha(t_{k+1}-t_{N-1})}V_e(t_{N-1}) + \Psi_{k+1} + b\Delta^2\int_{t_{N-1}}^{t_{k+1}} e^{-2\alpha(t_{k+1}-s)}ds,\ k \geq N-1, \tag{5.40}$$

where

$$\begin{aligned} \Psi_{k+1} = &-(\tau_M-\eta_m)e^{2\alpha[\tau_M+(N-2)(\tau_M-\eta_m)]} \\ &\times\Big[\sum_{j=0}^{N-3}(N-2-j)\int_{s_{k-j-1}}^{s_{k+1}} e^{2\alpha(s-t_{k+1})}\left|\sqrt{Q_{i^*_{k-j}}}C_{i^*_{k-j}}\dot{x}(s)\right|^2 ds \\ &+(N-1)\int_{s_k}^{s_{k+1}} e^{2\alpha(s-t_{k+1})}\left|\sqrt{Q_{i^*_{k+1}}}C_{i^*_{k+1}}\dot{x}(s)\right|^2 ds\Big] \leq 0. \end{aligned} \tag{5.41}$$

Table 5.1 Numerical complexity of stability conditions under different protocols (for $y_1,\ y_2 \in \mathbb{R}^{n/2}$)

Method	Decision variables	Number and order of LMIs
Theorem 4.1 (Round-Robin)	$8.5n^2 + 2.5n$	Two of $6n \times 6n$, two of $3n \times 3n$
Theorem 5.1 (TOD)	$4.25n^2 + 4n$	Two of $5.5n \times 5.5n$, two of $2n \times 2n$
Theorem 5.2 (TOD/Round-Robin)	$3.75n^2 + 3n$	Two of $5.5n \times 5.5n$, one of $2n \times 2n$

Moreover, for all $t \geq t_{N-1}$

$$\begin{aligned} V(t, x_t, \dot{x}_t) &\leq e^{-2\alpha(t-t_{N-1})} V_e(t_{N-1}) + \frac{b}{2\alpha}\Delta^2, \\ V_e(t_{N-1}) &= V(t_{N-1}, x_{t_{N-1}}, \dot{x}_{t_{N-1}}) + \sum_{i=1}^{N} |\sqrt{Q_i} e_i(t_{N-1})|^2. \end{aligned} \tag{5.42}$$

The latter inequality guarantees the x-ISS of (5.8), (5.11), (5.32) for $t \geq t_{N-1}$.

By using Lemma 5.3, the arguments of Theorem 5.1 and the fact that for $j = 1, \ldots, N-1$,

$$\frac{d}{dt}\frac{t_{k+1} - t}{j(\tau_M - \eta_m)} = -\frac{1}{j(\tau_M - \eta_m)} \leq -\frac{1}{(N-1)(\tau_M - \eta_m)},$$

we arrive at the following result.

Theorem 5.2 *Given $0 \leq \eta_m < \tau_M$ and $\alpha > 0$, assume that there exist positive scalar b, $n \times n$ matrices $P > 0$, $S_0 > 0$, $R_0 > 0$, $S_1 > 0$, $R_1 > 0$, S_{12} and $n_i \times n_i$ matrices $Q_i > 0, i = 1, \ldots, N$, such that (5.22) and (5.23) are feasible with $U_i = \frac{Q_i}{N-1}$, where G_i is given by (5.35). Then for $N > 2$, the solutions of the hybrid system (5.8), (5.11), (5.32) satisfy the bound (5.42) with $V(t, x_t, \dot{x}_t)$ given by (5.33), meaning x-ISS for $t \geq t_{N-1}$. For $N = 2$, the solutions of the hybrid system (5.8)–(5.10) ((5.8), (5.9), (5.11)) satisfy the bound (5.17) meaning x-ISS (for $t \geq t_0$). Moreover, if the above LMIs are feasible with $\alpha = 0$, then the solution bounds hold with a small enough $\alpha_0 > 0$.*

Remark 5.8 Let $N = 2$ and compare the number of scalar decision variables of the resulting LMIs under different protocols. The conditions of Theorem 5.2 are essentially more simpler than those of Theorem 4.1 and lead to complementary results compared to the conditions of Theorem 4.1 (see examples below). See Table 5.1 for the complexity of the LMI conditions under different protocols.

Remark 5.9 For $N = 2$ and $\alpha = 0$, the LMIs of Theorem 5.1 are more restrictive than those of Theorem 5.2: (5.15) of Theorem 5.1 yields $(N-1)U_i < Q_i < G_i$, whereas in Theorem 5.2 we have $(N-1)^2 U_i = (N-1)Q_i = G_i$ that leads to larger U_i for the same G_i. The latter helps for the feasibility of (5.23), where $U_i > 0$ appears on the main diagonal only (with minus). However, Theorem 5.1 achieves ISS with respect to the full state col$\{x, e\}$ and provides the solution bounds for $t \geq t_0$, while Theorem 5.2 only guarantees x-ISS.

Note that Theorem 5.2 under Round-Robin protocol guarantees the boundedness of e as well. From (5.38) and $s_0 = 0$, it follows that $s_{N-1} - s_0 \leq (N-1)(\tau_M - \eta_m)$. Due to (5.2), we have $t_N \leq s_{N-1} + \tau_M \leq N\tau_M$. Moreover, $e(t_N)$ in (5.32) depends on $x(0), \ldots, x(t_N - \eta_N)$. Therefore, the relations (5.32) yield

$$|e_i(t)|^2 \leq c' \sup_{\theta \in [-N\tau_M, 0]} |x(t+\theta)|^2, \; t \geq t_N$$

with some $c' > 0$, which together with (5.42) imply

$$|e_i(t)|^2 \leq c''[e^{-2\alpha(t-t_{N-1})} V_e(t_{N-1}) + \Delta^2]$$

for some $c'' > 0$ and all $t \geq t_N + N\tau_M$.

Remark 5.10 The LMIs of Theorems 5.1 and 5.2 are affine in the system matrices. Therefore, in the case of system matrices from an uncertain time-varying polytope

$$\begin{aligned}
&\Omega = \sum_{j=1}^{M} g_j(t)\Omega_j, \; 0 \leq g_j(t) \leq 1, \\
&\sum_{j=1}^{M} g_j(t) = 1, \; \Omega_j = \left[A^{(j)} \; B^{(j)} \; D^{(j)} \right],
\end{aligned}$$

where g_j, $j = 1, \ldots, M$, are uncertain time-varying parameters, one has to solve these LMIs simultaneously for all the M vertices Ω_j, applying the same decision matrices.

5.5 Numerical Examples

Example 5.1 (*Uncertain inverted pendulum*) Consider an inverted pendulum mounted on a car. We focus on the stability analysis in the absence of disturbance. Following [68], we assume that the friction coefficient between the air and the car, f_c, and the air and the bar, f_b, is not exactly known and time-varying: $f_c(t) \in [0.15, 0.25]$ and $f_b(t) \in [0.15, 0.25]$. The linearized model can be written as (5.1), where the matrices

$$A = E^{-1}A_f \text{ and } B = E^{-1}B_0 \tag{5.43}$$

are determined from

$$E = \begin{bmatrix} 1 & 0 & 0 & 0 \\ 0 & 1 & 0 & 0 \\ 0 & 0 & 3/2 & -1/4 \\ 0 & 0 & -1/4 & 1/6 \end{bmatrix},$$
$$A_f = \begin{bmatrix} 0 & 0 & 1 & 0 \\ 0 & 0 & 0 & 1 \\ 0 & 0 & -(f_c + f_b) & f_b/2 \\ 0 & 5/2 & f_b/2 & -f_b/3 \end{bmatrix} \quad \text{and} \quad B_0 = \begin{bmatrix} 0 \\ 0 \\ 1 \\ 0 \end{bmatrix}. \tag{5.44}$$

Here, A belongs to uncertain polytope, defined by four vertices corresponding to $f_c/f_b = 0.15$ and $f_c/f_b = 0.25$. The pendulum can be stabilized by a state feedback $u(t) = Kx(t)$, where $x = [x_1, x_2, x_3, x_4]^T$, with the gain

$$K = [11.2062 \quad -128.8597 \quad 10.7823 \quad -22.2629]. \tag{5.45}$$

In this model, x_1 and x_2 represent the cart position and the velocity, whereas x_3, x_4 represent the pendulum angle from vertical and its angular velocity, respectively. In practice, x_1, x_2 and x_3, x_4 (presenting spatially distributed components of the state of the pendulum–cart system) are not accessible simultaneously. Suppose that the state variables are not accessible simultaneously. Consider first $N = 2$ and

$$C_1 = \begin{bmatrix} 1 & 0 & 0 & 0 \\ 0 & 1 & 0 & 0 \end{bmatrix}, \quad C_2 = \begin{bmatrix} 0 & 0 & 1 & 0 \\ 0 & 0 & 0 & 1 \end{bmatrix}.$$

The applied controller gain K has the following blocks:

$$K_1 = [11.2062 \quad -128.8597], \quad K_2 = [10.7823 \quad -22.2629].$$

For the values of η_m given in Table 5.2, we apply Theorems 5.1 and 5.2 with $\alpha = 0$, $b = 0$ via Remark 5.10 and find the maximum values of $\tau_M = \text{MATI} + \text{MAD}$ that preserve the stability of the hybrid system (5.8)–(5.9) with $\omega(t) = 0$ with respect to x. From Table 5.2, it is observed that under TOD or Round-Robin protocol the conditions of Theorem 5.2 possess less decision variables and stabilize the system for larger τ_M than the results in Theorem 4.1 under Round-Robin protocol. Moreover, when $\eta_m > \frac{\tau_M}{2}$ ($\eta_m = 0.02, 0.04$), the method is still feasible (communication delays are larger than the sampling intervals). The computational time for solving the LMIs (in seconds) under TOD protocol is essentially less than that under Round-Robin protocol in Theorem 4.1 (till 36% decrease).

Consider next $N = 4$, where $C_1, \ldots, C_4$ are the rows of I_4 and $K_1, \ldots, K_4$ are the entries of K given by (5.45). Here, the maximum values of τ_M that preserve the stability of (5.8)–(5.9) with $\omega(t) = 0$ with respect to x are given in Table 5.3. Also here, Theorem 5.2 leads to less conservative results than Theorem 5.1.

Example 5.2 (*Batch reactor*) The effectiveness of the derived conditions is demonstrated by the example of a batch reactor under the dynamic output-feedback con-

Table 5.2 Example 5.1 ($N = 2$): maximum value of $\tau_M = \text{MATI} + \text{MAD}$

$\tau_M \backslash \eta_m$	0	0.005	0.01	0.02	0.04	Decision variables
Theorem 4.1 (Round-Robin)	0.023	0.026	0.029	0.035	0.046	146
Theorem 5.1 (TOD)	0.014	0.018	0.022	0.029	0.044	84
Theorem 5.2 (TOD/Round-Robin)	0.025	0.028	0.031	0.036	0.047	72

Table 5.3 Example 5.1 ($N = 4$): maximum value of $\tau_M = \text{MATI} + \text{MAD}$

$\tau_M \backslash \eta_m$	0	0.01
Theorem 5.1 (TOD)	0.003	0.012
Theorem 5.2 (Round-Robin)	0.006	0.015

Table 5.4 Example 5.2: maximum value of $\tau_M = \text{MATI} + \text{MAD}$ for different η_m

$\tau_M \backslash \eta_m$	0	0.004	0.01	0.02	0.03	0.04
Reference [84] (MAD = 0.004)	0.0108	0.0133	–	–	–	–
Reference [29] (MAD = 0.03)	0.069	0.069	0.069	0.069	0.069	–
Theorem 5.1 (TOD)	0.019	0.022	0.027	0.034	0.042	0.050
Theorem 5.2 (TOD/Round-Robin)	0.035	0.037	0.041	0.047	0.053	0.059
Theorem 4.1 (Round-Robin)	0.042	0.044	0.048	0.053	0.058	0.063

troller, where $N = 2$, the matrices A, B, A_c, B_c, C_c and D_c are given in (4.32) and

$$C = \left[\frac{C_1}{C_2}\right] = \left[\frac{1\ 0\ 1\ -1}{0\ 1\ 0\ \ 0}\right].$$

For the values of η_m given in Table 5.4, we apply Theorems 5.1 and 5.2 with $\alpha = 0$, $b = 0$ and find the maximum values of $\tau_M = \text{MATI} + \text{MAD}$ that preserve the stability of the hybrid system (5.8)–(5.9) with $\omega(t) = 0$ with respect to x. From Table 5.4, it is seen that the results of our method essentially improve the results in [84] and are more conservative than those obtained via the discrete-time approach. In [10], the sum of squares method was developed in the framework of hybrid system approach, and the same result $\tau_M = 0.035$ as ours in Theorem 5.2 for $\eta_m = 0$, MAD $= 0.01$ was achieved. We note that the sum of squares method has not been applied yet to ISS. Moreover, our conditions are simple LMIs with a fewer decision variables. When $\eta_m > \frac{\tau_M}{2}$ ($\eta_m = 0.03, 0.04$), our method is still feasible (communication delays are larger than the sampling intervals). The computational time under TOD protocol is essentially less than that under Round-Robin protocol in Theorem 4.1 (till 32% decrease).

5.6 Notes

In this chapter, a time-delay approach was developed for the ISS of NCSs with scheduling protocols, variable transmission delays and variable sampling intervals. A unified hybrid system model with time-varying delays in the continuous dynamics and in the reset equations was introduced for the closed-loop system under both TOD and Round-Robin protocols, and a new Lyapunov–Krasovskii method was developed. The ISS conditions of the delayed hybrid system were derived in terms of LMIs. Numerical examples were provided to demonstrate the effectiveness of the proposed methods.

Chapter 6
Networked Control Under Stochastic Protocol

In Chaps. 4 and 5 a crucial point is that the data packet dropouts are not allowed for large communication delays under scheduling protocols. In the framework of hybrid systems, a stochastic protocol was introduced and analyzed in [220] for the input-output stability of NCSs in the presence of data packet dropouts or collisions. An iid sequence of Bernoulli random variables is applied to describe the stochastic protocol. However, the communication delays are not included in the analysis. The stability of NCSs under a stochastic protocol, where the activated node is modeled by a Markov chain, was studied in [28] by applying the discrete-time modeling framework. In [28], the data packet dropouts can be regarded as prolongations of the sampling interval for small delays.

In this chapter, to overcome the lack of stability analysis of NCSs under scheduling protocols with large communication delays and data packet dropouts, we develop a time-delay approach considering multiple sensors under a stochastic scheduling protocol. The resulting closed-loop system is a stochastic impulsive system with delays both in the continuous dynamics and in the reset equations. We treat two classes of stochastic protocols. The first one is defined by an iid stochastic process. The activation probability of each node for this protocol is a given constant, whereas it is assumed that the collisions occur with a certain probability. The second protocol is defined by a discrete-time Markov chain with a known transition probability matrix taking into account collisions. By developing appropriate Lyapunov–Krasovskii techniques, we derive conditions for the exponential mean-square stability of the closed-loop system. As in Chaps. 4 and 5, different from the hybrid and discrete-time approaches, we allow the transmission delays to be larger than the sampling intervals in the presence of scheduling protocols. The efficiency of the proposed approach is illustrated by a batch reactor example. Preliminary results on the stabilization of NCSs with two sensor nodes under iid stochastic protocol have been presented in [136].

K. Liu et al., *Networked Control Under Communication Constraints*, Advances in Delays and Dynamics 11, https://doi.org/10.1007/978-981-15-4230-5_6

6.1 Problem Formulation

6.1.1 The Description of NCSs

Consider the system architecture in Fig. 6.1 with plant

$$\dot{x}(t) = Ax(t) + Bu(t), \tag{6.1}$$

where $x(t) \in \mathbb{R}^n$ is the state vector, $u(t) \in \mathbb{R}^m$ is the control input and A, B are system matrices of appropriate dimensions. The initial condition is given by $x(0) = x_0$.

The NCS has N distributed sensors, a controller and an actuator connected via two wireless networks. Their measurements are given by $y_i(t) = C_i x(t)$, $i = 1, \ldots, N$. Let $C = [C_1^T \ \cdots \ C_N^T]^T$, $y(t) = [y_1^T(t) \ \cdots \ y_N^T(t)]^T \in \mathbb{R}^{n_y}$. Denote by s_k the unbounded and monotonously increasing sequence of sampling instants satisfying (2.2). At each sampling instant s_k, at most one of the outputs $y_i(s_k) \in \mathbb{R}^{n_i}$, $\sum_{i=1}^N n_i = n_y$, is transmitted over the network.

We suppose that the transmission of the information (between the sensor and the actuator) is subject to a variable delay $\eta_k = \eta_k^{sc} + \eta_k^{ca} + \eta_k^c$, where η_k^{sc} and η_k^{ca} are the network-induced delays (from the sensor to the controller and from the controller to the actuator, respectively) and where η_k^c is the computational delay in the controller node. Denote $s_k + \eta_k$ by t_k. Different from [29, 84], we do not restrict the network delays to be small with $\eta_k < s_{k+1} - s_k$. Following [161], and Chaps. 4 and 5, we allow the delay to be large provided that packet ordering is maintained. Assume that the network-induced delay η_k and the time span between the instant t_{k+1} and the current sampling instant s_k are bounded:

$$t_{k+1} - t_k + \eta_k \le \tau_M, \ 0 \le \eta_m \le \eta_k \le \mathrm{MAD}, \ k \in \mathbb{Z}^+, \tag{6.2}$$

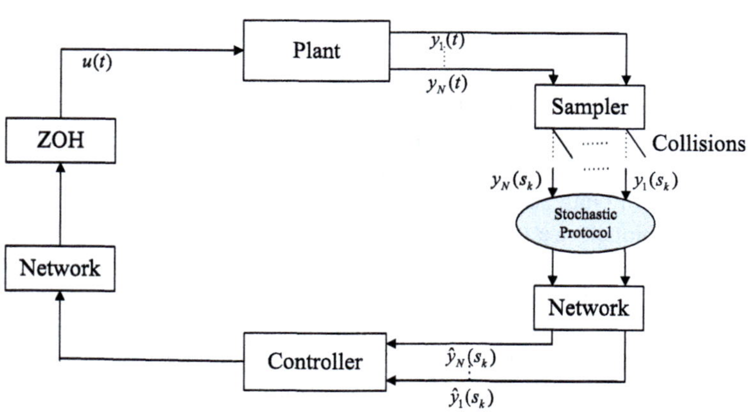

Fig. 6.1 NCSs under stochastic protocol

Here, η_m and MAD are known bounds and $\tau_M = \text{MATI} + \text{MAD}$. The inequality $\eta_m > \tau_M/2$ implies the case of large delay. For the given example in Sect. 6.4, we show that our method is applicable also to $\eta_m > \tau_M/2$.

Remark 6.1 Different from [161], where subscript k in t_k corresponds to the measurements that are not lost, in this chapter k corresponds to the sampling time. This is because we consider the probability of collisions or data packet dropouts (see further details below). Therefore, t_k is the actual or the fictitious (when collisions occur or the data packet is lost) updating time instant of the ZOH device.

Remark 6.2 We follow a commonly used assumption on the boundedness of the network-induced delays, e.g., [29, 63]. Another possibility is the Markov chain model of the network-induced delays, e.g., [166].

6.1.2 *The Impulsive System Model*

At each sampling instant s_k, at most one of the system nodes $i \in \{1, \ldots, N\}$ is active. In some cases, the collisions may occur when the nodes access the network [220]. If this happens, then the packet with sensor data will be dropped. At the sampling instant s_k, let $\sigma_k \in \mathcal{I} = \{0, 1, \ldots, N\}$ denote the active output node, which will be chosen according to the stochastic protocol. Here, $\sigma_k = 0$ means that either the collisions occur when the nodes access the network or the data packet is lost during the transmission over the network from the sensor to the controller. We suppose the data loss is not possible during the transmission from the controller to the actuator.

Denote by $\hat{y}(s_k) = [\hat{y}_1^T(s_k) \; \cdots \; \hat{y}_N^T(s_k)]^T \in \mathbb{R}^{n_y}$ the most recently received output information on the controller side. We consider the error between the system output $y(s_k)$ and the last available information $\hat{y}(s_{k-1})$:

$$\begin{aligned} &e(t) = \text{col}\{e_1(t), \cdots, e_N(t)\} \equiv \hat{y}(s_{k-1}) - y(s_k), \\ &t \in [t_k, t_{k+1}), \; k \in \mathbb{Z}^+, \; \hat{y}(s_{-1}) \stackrel{\Delta}{=} 0, \; e(t) \in \mathbb{R}^{n_y}. \end{aligned} \tag{6.3}$$

Suppose that the controller and the actuator are event-driven.

6.1.2.1 Static Output-Feedback Controller

Assume that there exists a matrix $K = [K_1 \; \cdots \; K_N]$, $K_i \in \mathbb{R}^{m \times n_i}$ such that $A + BKC$ is Hurwitz. Then, the static output-feedback controller has the form

$$u(t) = K_{\sigma_k} y_{\sigma_k}(s_k) + \sum_{i=1, i \neq \sigma_k}^{N} K_i \hat{y}_i(s_{k-1}), \; t \in [t_k, t_{k+1}), \; k \in \mathbb{Z}^+, \tag{6.4}$$

where $K_{\sigma_k} y_{\sigma_k}(s_k) = 0$ when $\sigma_k = 0$. Therefore, we obtain the following continuous dynamics:

$$\begin{cases} \dot{x}(t) = Ax(t) + A_1 x(t_k - \eta_k) + \sum\limits_{i=1, i\neq\sigma_k}^{N} B_i e_i(t), \\ \dot{e}(t) = 0, \quad t \in [t_k, t_{k+1}), \end{cases} \tag{6.5}$$

where $A_1 = BKC,\ B_i = BK_i,\ i = 1, \ldots, N$.

From (6.3), it follows that

$$\begin{aligned} e_i(t_{k+1}) &= \hat{y}_i(s_k) - y_i(s_{k+1}) \\ &= y_i(s_k) - y_i(s_{k+1}),\ \ i = \sigma_k \in \mathscr{I}\backslash\{0\}, \\ e_i(t_{k+1}) &= \hat{y}_i(s_k) - y_i(s_{k+1}) \\ &= \hat{y}_i(s_{k-1}) - y_i(s_{k+1}),\ i \neq \sigma_k,\ i \in \mathscr{I}\backslash\{0\}. \end{aligned}$$

Thus, the delayed reset system is given by

$$\begin{cases} x(t_{k+1}) \ = x(t_{k+1}^-), \\ e_i(t_{k+1}) = C_i[x(s_k) - x(s_{k+1})],\ i = \sigma_k \in \mathscr{I}\backslash\{0\}, \\ e_i(t_{k+1}) = e_i(t_{k+1}^-) + C_i[x(s_k) - x(s_{k+1})],\ i \neq \sigma_k,\ i \in \mathscr{I}\backslash\{0\}. \end{cases} \tag{6.6}$$

Applying the time-delay approach to sampled-data control, we denote $\tau(t) = t - t_k + \eta_k$. Then, $\tau(t) \in [\eta_m, \tau_M]$ (cf., (6.2)) and $x(t_k - \eta_k) = x(t - \tau(t))$ for $t \in [t_k, t_{k+1})$. Therefore, the impulsive system model (6.5)–(6.6) contains the piecewise-continuous delay $\tau(t)$ in the continuous-time dynamics (6.5). Even for $\eta_k = 0$, we have the delayed state $x(t_k) = x(t - \tau(t))$ with $\tau(t) = t - t_k$. The initial condition for (6.5)–(6.6) has the form of $x(t) = \phi(t)$, $t \in [t_0 - \tau_M,\ t_0]$, $\phi(0) = x_0$ and $e(t_0) = -Cx(t_0 - \eta_0) = -Cx_0$, where $\phi(t)$ is a continuous function on $[t_0 - \tau_M,\ t_0]$.

6.1.2.2 Dynamic Output-Feedback Controller

Assume that the controller is directly connected to the actuator. Consider a dynamic output-feedback controller of the form

$$\begin{aligned} \dot{x}_c(t) &= A_c x_c(t) + B_c \hat{y}(s_k), \\ u(t) &= C_c x_c(t) + D_c \hat{y}(s_k),\ t \in [t_k, t_{k+1}),\ k \in \mathbb{Z}^+, \end{aligned}$$

where $x_c(t) \in \mathbb{R}^{n_c}$ is the state of the controller, and A_c, B_c, C_c and D_c are matrices of appropriate dimensions. Let $e_i(t)$, $i = 1, \ldots, N$, be defined by (6.3). The closed-loop system can be represented in the form of (6.5)–(6.6), where x, e and matrices are replaced by the ones with bars as follows:

$$\bar{x} = [x^T \; x_c^T]^T, \; \bar{A} = \begin{bmatrix} A & BC_c \\ 0_{n_c \times n} & A_c \end{bmatrix}, \; \bar{B}_i = \begin{bmatrix} BD_c \\ B_c \end{bmatrix},$$
$$\bar{A}_1 = \begin{bmatrix} BD_cC \; 0_{n \times n_c} \\ B_cC \quad 0_{n_c \times n_c} \end{bmatrix}, \; \bar{C} = \begin{bmatrix} \bar{C}_1^T \cdots \bar{C}_N^T \end{bmatrix}^T,$$
$$\bar{C}_1 = \begin{bmatrix} C_1^T & 0 \\ 0 & 0 \end{bmatrix}^T, \; \bar{C}_2 = \begin{bmatrix} 0_{n \times n_1} & C_2^T \; 0 \\ 0_{n_c \times n_1} & 0 \;\; 0 \end{bmatrix}^T, \cdots, \bar{C}_N = \begin{bmatrix} 0 & C_N^T \\ 0 & 0 \end{bmatrix}^T,$$

$$\bar{e}(t) = [\bar{e}_1^T(t) \; \cdots \; \bar{e}_N^T(t)]^T, \; \bar{e}_1(t) = [e_1^T(t) \; 0]^T, \; \bar{e}_2(t) = [0_{1 \times n_1} \; e_2^T(t) \; 0]^T, \cdots,$$
$$\bar{e}_N(t) = [0 \; e_N^T(t)]^T, \quad \bar{C}_i \in \mathbb{R}^{n_y \times (n+n_c)}, \quad \bar{e}_i(t) \in \mathbb{R}^{n_y}, \; i = 1, \ldots, N.$$

6.1.3 Stochastic Protocols

In the sequel, we will consider two classes of stochastic protocols, which are defined by iid and Markovian process, respectively.

6.1.3.1 Iid Protocol

The choice of σ_k is assumed to be iid with the probabilities given by

$$\text{Prob}\{\sigma_k = i\} = \beta_i, \; i \in \mathscr{I}, \tag{6.7}$$

where $\beta_i, i = 0, 1, \ldots, N$, are nonnegative scalars and $\sum_{i=0}^{N} \beta_i = 1$. Here, $\beta_j, j = 1, \ldots, N$, are the probabilities of the measurement $y_j(s_k)$ to be transmitted at s_k, whereas β_0 is the probability of collision.

6.1.3.2 Markovian Protocol

The protocol determines σ_k through a Markov chain. The conditional probability that the node $j \in \mathscr{I}$ gets access to the network at time s_k, given the values of $\sigma_{k-1} \in \mathscr{I}$, is defined by

$$\text{Prob}\{\sigma_k = j | \sigma_{k-1} = i\} = \pi_{ij}, \tag{6.8}$$

where $0 \le \pi_{ij} \le 1$ for all $i, j \in \mathscr{I}$, $\sum_{j=0}^{N} \pi_{ij} = 1$ for all $i \in \mathscr{I}$ and $\sigma_0 \in \mathscr{I}$ is assumed to be given. The transition probability matrix is denoted by $\Pi = \{\pi_{ij}\} \in \mathbb{R}^{(N+1)\times(N+1)}$.

Remark 6.3 The iid scheduling is a special case of the Markovian scheduling. For instance, assume that there are $N = 2$ sensor nodes and the collisions do not occur; the Markovian scheduling with $\Pi = \begin{bmatrix} p & 1-p \\ p & 1-p \end{bmatrix}$, $0 \le p \le 1$, is an iid scheduling with $\beta_1 = p$, $\beta_2 = 1 - p$.

Definition 6.1 The hybrid system (6.5)–(6.6) is said to be exponentially mean-square stable with respect to x if there exist constants $b > 0$, $\alpha > 0$ such that the following bound

$$\mathbb{E}\{|x(t)|^2\} \le be^{-2\alpha(t-t_0)}\mathbb{E}\{\|x_{t_0}\|_W^2 + |e(t_0)|^2\},\ t \ge t_0$$

holds for the solutions of the stochastic impulsive system (6.5)–(6.6) initialized with $e(t_0) \in \mathbb{R}^{n_y}$ and $x(t) = \phi(t),\ t \in [t_0 - \tau_M,\ t_0]$. The hybrid system (6.5)–(6.6) is exponentially mean-square stable if additionally the following bound

$$\mathbb{E}\{|e(t)|^2\} \le be^{-2\alpha(t-t_0)}\mathbb{E}\{\|x_{t_0}\|_W^2 + |e(t_0)|^2\},\ t \ge t_0$$

is valid.

6.2 NCSs Under Iid Stochastic Protocol

6.2.1 Stochastic Impulsive Time-Delay Model with Bernoulli-Distributed Parameters

Following [249], we introduce the indicator functions

$$\pi_{\{\sigma_k=i\}} = \begin{cases} 1,\ \sigma_k = i \\ 0,\ \sigma_k \ne i, \end{cases} \quad i \in \mathcal{I},\ k \in \mathbb{Z}^+.$$

Thus, from (6.7) it follows that

$$\begin{aligned} &\mathbb{E}\{\pi_{\{\sigma_k=i\}}\} = \mathbb{E}\{[\pi_{\{\sigma_k=i\}}]^2\} = \mathrm{Prob}\{\sigma_k = i\} = \beta_i, \\ &\mathbb{E}\{[\pi_{\{\sigma_k=i\}} - \beta_i][\pi_{\{\sigma_k=j\}} - \beta_j]\} = \begin{cases} -\beta_i\beta_j,\ i \ne j, \\ \beta_i(1-\beta_i),\ i = j. \end{cases} \end{aligned} \tag{6.9}$$

Therefore, the stochastic impulsive system model (6.5)–(6.7) can be rewritten as

$$\begin{cases} \dot{x}(t) = Ax(t) + A_1x(t_k - \eta_k) + \sum_{i=1}^{N}(1 - \pi_{\{\sigma_k=i\}})B_ie_i(t), \\ \dot{e}(t) = 0,\ t \in [t_k, t_{k+1}) \end{cases} \tag{6.10}$$

with the delayed reset system

$$\begin{cases} x(t_{k+1}) = x(t_{k+1}^-), \\ e_i(t_{k+1}) = (1 - \pi_{\{\sigma_k=i\}})e_i(t_{k+1}^-) + C_i[x(t_k - \eta_k) - x(t_{k+1} - \eta_{k+1})],\ i = 1, \dots, N. \end{cases} \tag{6.11}$$

Remark 6.4 Applying the Bernoulli-distributed stochastic variables $\pi_{\{\sigma_k=i\}}$, $i = 0, 1, \ldots, N$, the closed-loop system (6.10)–(6.11) is expressed as an impulsive time-delay system with stochastic parameters in the system matrices. Note that the Bernoulli distribution has been applied to NCSs with probabilistic measurements missing [95, 235], stochastic sampling intervals [67], time-delay system with stochastic interval delays [249], output tracking control under unreliable communication [126] and fuzzy control for nonlinear NCSs [119].

6.2.2 Exponential Mean-Square Stability of Stochastic Impulsive Delayed System

This section is devoted to derive LMI conditions for the exponential mean-square stability of the stochastic impulsive system (6.10)–(6.11). Consider the LKF (5.12). The piecewise-continuous in time term V_G is inserted to cope with the delays in the reset conditions. It is continuous on $[t_k, t_{k+1})$ and does not grow at the jumps $t = t_{k+1}$, since

$$\begin{aligned}\mathbb{E}\{V_{G|t=t_{k+1}} - V_{G|t=t_{k+1}^-}\} &\le -(\tau_M-\eta_m)e^{-2\alpha\tau_M}\sum_{i=1}^{N}\int_{t_k-\eta_k}^{t_{k+1}-\eta_{k+1}} \mathbb{E}\{|\sqrt{G_i}C_i\dot{x}(s)|^2\}ds\\ &\le -e^{-2\alpha\tau_M}\sum_{i=1}^{N}\mathbb{E}\Big\{\Big|\sqrt{G_i}C_i[x(t_k-\eta_k) - x(t_{k+1}-\eta_{k+1})]\Big|^2\Big\},\end{aligned} \tag{6.12}$$

where we applied Jensen's inequality. The infinitesimal operator $\mathscr{L}$ of $V_e(t)$ is defined as

$$\mathscr{L}V_e(t) = \lim_{\Delta\to 0^+}\frac{1}{\Delta}\Big\{\mathbb{E}\{V_e(t+\Delta)|t\} - V_e(t)\Big\}. \tag{6.13}$$

The following lemma gives the sufficient conditions for the exponential stability of (6.10)–(6.11) in the mean-square sense.

Lemma 6.1 *If there exist positive constant α, $0 < Q_i \in \mathbb{R}^{n_i\times n_i}$, $0 < U_i \in \mathbb{R}^{n_i\times n_i}$, $0 < G_i \in \mathbb{R}^{n_i\times n_i}$, $i = 1, \ldots, N$, and $V_e(t)$ of (5.12) such that along (6.10) for $t \in [t_k, t_{k+1})$*

$$\mathbb{E}\Big\{\mathscr{L}V_e(t) + 2\alpha V_e(t) - \frac{1}{\tau_M - \eta_m}\sum_{i=1}^{N} e_i^T(t)U_i e_i(t)\Big\} \le 0 \tag{6.14}$$

with

$$\Omega_i = \begin{bmatrix} -\beta_i Q_i + U_i & (1-\beta_i)Q_i \\ * & Q_i - G_i e^{-2\alpha\tau_M}\end{bmatrix} \le 0,\ i = 1, \ldots, N. \tag{6.15}$$

Then, $V_e(t)$ does not grow in the jumps along (6.10)–(6.11)

$$\Theta = \mathbb{E}\Big\{V_e(t_{k+1}) - V_e(t_{k+1}^-) + \sum_{i=1}^{N} e_i^T(t_k)U_i e_i(t_k)\Big\} \le 0. \tag{6.16}$$

Moreover, the following bounds hold for the solutions of (6.10)–(6.11) with the initial condition x_{t_0}, $e(t_0)$*:*

$$\begin{aligned} &\mathbb{E}\{V(t, x_t, \dot{x}_t)\} \le e^{-2\alpha(t-t_0)}\mathbb{E}\{V_e(t_0)\}, \ t \ge t_0, \\ &V_e(t_0) = V(t_0, x_{t_0}, \dot{x}_{t_0}) + \sum_{i=1}^{N} e_i^T(t_0)Q_i e_i(t_0), \end{aligned} \tag{6.17}$$

and

$$\sum_{i=1}^{N} \mathbb{E}\Big\{|\sqrt{Q_i}e_i(t)|^2\Big\} \le \tilde{c}e^{-2\alpha(t-t_0)}\mathbb{E}\{V_e(t_0)\}, \tag{6.18}$$

where $\tilde{c} = e^{2\alpha(\tau_M - \eta_m)}$*, implying exponential mean-square stability of (6.10)–(6.11).*

Proof Since $\int_{t_k}^{t} e^{-2\alpha(t-s)}ds \le \tau_M - \eta_m, \ t \in [t_k, t_{k+1})$ and

$$\mathscr{L}[e^{2\alpha t}V_e(t)] = e^{2\alpha t}[2\alpha V_e(t) + \mathscr{L}V_e(t)], \ \alpha > 0,$$

then (6.14) implies

$$\mathbb{E}\{V_e(t)\} \le e^{-2\alpha(t-t_k)}\mathbb{E}\{V_e(t_k)\} + \sum_{i=1}^{N} \mathbb{E}\{e_i^T(t_k)U_i e_i(t_k)\}, \ t \in [t_k, t_{k+1}). \tag{6.19}$$

Because (6.15) yields $U_i \le \beta_i Q_i < Q_i, \ i = 1, \dots, N$, we have

$$\mathbb{E}\{V(t, x_t, \dot{x}_t)\} \le e^{-2\alpha(t-t_k)}\mathbb{E}\{V_e(t_k)\}, \ t \in [t_k, t_{k+1}). \tag{6.20}$$

Note that

$$\mathbb{E}\{V_e(t_{k+1})\} = \mathbb{E}\Big\{\tilde{V}_{|t=t_{k+1}} + V_{G\,|t=t_{k+1}} + \sum_{i=1}^{N} e_i^T(t_{k+1})Q_i e_i(t_{k+1})\Big\}$$

and

$$\begin{aligned} &\mathbb{E}\{e_i^T(t_{k+1})Q_i e_i(t_{k+1})\} \\ &= \mathbb{E}\Big\{\Big|\sqrt{Q_i}[(1 - \pi_{\{\sigma_k = i\}})e_i(t_k) + C_i x(t_k - \eta_k) - C_i x(t_{k+1} - \eta_{k+1})]\Big|^2\Big\} \\ &= \mathbb{E}\Big\{(1 - \beta_i)e_i^T(t_k)Q_i e_i(t_k) + 2(1 - \beta_i)e_i^T(t_k)Q_i C_i[x(t_k - \eta_k) - x(t_{k+1} - \eta_{k+1})] \\ &\quad + \Big|\sqrt{Q_i}C_i[x(t_k - \eta_k) - x(t_{k+1} - \eta_{k+1})]\Big|^2\Big\}, \ i = 1, \dots, N. \end{aligned}$$

Taking (6.12) and (6.15) into account, we obtain

$$
\begin{aligned}
\Theta &= \mathbb{E}\Big\{\sum_{i=1}^{N}\Big[|\sqrt{Q_i}e_i(t_{k+1})|^2-|\sqrt{Q_i}e_i(t_k)|^2\Big]+V_{G|t=t_{k+1}}-V_{G|t=t_{k+1}^-}+\sum_{i=1}^{N}e_i^T(t_k)U_ie_i(t_k)\Big\}\\
&\le \mathbb{E}\Big\{\sum_{i=1}^{N}\Big[|\sqrt{Q_i}e_i(t_{k+1})|^2-|\sqrt{Q_i}e_i(t_k)|^2\Big]\Big\}\\
&\quad -e^{-2\alpha\tau_M}\sum_{i=1}^{N}\mathbb{E}\Big\{\Big|\sqrt{G_i}C_i[x(t_k-\eta_k)-x(t_{k+1}-\eta_{k+1})]\Big|^2\Big\}\\
&\quad +\mathbb{E}\Big\{\sum_{i=1}^{N}e_i^T(t_k)U_ie_i(t_k)\Big\}\\
&\le \sum_{i=1}^{N}\mathbb{E}\{\zeta_i(t_k)^T\Omega_i\zeta_i(t_k)\}\le 0,
\end{aligned}
$$

where $\zeta_i(t_k)=\mathrm{col}\{e_i(t_k), C_i[x(t_k-\eta_k)-x(t_{k+1}-\eta_{k+1})]\}$ and Ω_i is given by (6.15).

Therefore, the inequalities (6.16) and (6.19) with $t=t_{k+1}^-$ imply

$$
\begin{aligned}
\mathbb{E}\{V_e(t_{k+1})\} &\le e^{-2\alpha(t_{k+1}-t_k)}\mathbb{E}\{V_e(t_k)\}\\
&\le e^{-2\alpha(t_{k+1}-t_{k-1})}\mathbb{E}\{V_e(t_{k-1})\}\\
&\le e^{-2\alpha(t_{k+1}-t_0)}\mathbb{E}\{V_e(t_0)\}.
\end{aligned}
\tag{6.21}
$$

The latter inequality, with $k+1$ replaced by k, and (6.20) give (6.17). The inequality (6.17) implies the exponential mean-square stability of (6.10)–(6.11) with respect to x because

$$
\lambda_{\min}(P)\mathbb{E}\{|x(t)|^2\}\le\mathbb{E}\{V(t,x_t,\dot{x}_t)\},\ \ \mathbb{E}\{V(t_0,x_{t_0},\dot{x}_{t_0})\}\le v\mathbb{E}\{\|x_{t_0}\|_W^2\}
$$

for some scalar $v>0$. Moreover, the inequality (6.21) with $k+1$ replaced by k implies (6.18) since for $t\in[t_k,t_{k+1})$,

$$
e^{-2\alpha(t_k-t_0)}=e^{-2\alpha(t-t_0)}e^{-2\alpha(t_k-t)}\le\tilde{c}e^{-2\alpha(t-t_0)}.
$$

By using Lemma 6.1 and the standard arguments for delay-dependent analysis, we derive LMI conditions for the exponential mean-square stability of (6.10)–(6.11).

Theorem 6.1 *Given* $0\le\eta_m<\tau_M$, $\alpha>0$, $\beta_0\ge 0$, $\beta_i\ge 0$, $\sum_{i=0}^{N}\beta_i=1$ *and* K_i, $i=1,\dots,N$. *Suppose there exist* $n\times n$ *matrices* $P>0$, $S_j>0$, $R_j>0$, $j=0,1$, S_{12} *and* $n_i\times n_i$ *matrices* $Q_i>0$, $U_i>0$, $G_i>0$, $i=1,\dots,N$, *such that (6.15) and*

$$
\Phi=\begin{bmatrix}R_1 & S_{12}\\ * & R_1\end{bmatrix}\ge 0,
\tag{6.22}
$$

$$\Sigma + \Xi^T H \Xi + \sum_{i=1}^{N} \beta_i \Xi_i^T H \Xi_i < 0 \tag{6.23}$$

are feasible, where

$$\begin{aligned}
H &= \eta_m^2 R_0 + (\tau_M - \eta_m)^2 R_1 + (\tau_M - \eta_m) \sum_{l=1}^{N} C_l^T G_l C_l, \\
\Sigma &= \mathrm{He}(F_1^T P \Xi) + \Upsilon - F_2^T R_0 F_2 e^{-2\alpha\eta_m} - F^T \Phi F e^{-2\alpha\tau_M}, \\
F_1 &= [I_n \; 0_{n\times(3n+n_y)}], \; F_2 = [I_n \; -I_n \; 0_{n\times(2n+n_y)}], \\
F &= \begin{bmatrix} 0_{n\times n} & I_n & -I_n & 0_{n\times n} & 0_{n\times n_y} \\ 0_{n\times n} & 0_{n\times n} & I_n & -I_n & 0_{n\times n_y} \end{bmatrix}, \\
\Xi &= [A \; 0_{n\times n} \; A_1 \; 0_{n\times n} \; (1-\beta_1)B_1 \; \cdots \; (1-\beta_N)B_N], \\
\Xi_1 &= [0_{n\times 4n} \; -B_1 \; 0], \; \Xi_2 = [0_{n\times(4n+n_1)} \; -B_2 \; 0], \cdots, \\
\Xi_N &= [0 \; -B_N], \; \Xi_j \in \mathbb{R}^{n\times(4n+n_y)}, \\
\Upsilon &= \mathrm{diag}\{S_0 + 2\alpha P, -(S_0 - S_1)e^{-2\alpha\eta_m}, 0, -S_1 e^{-2\alpha\tau_M}, \psi_1, \cdots, \psi_N\}, \\
\psi_j &= -\frac{1}{\tau_M - \eta_m} U_j + 2\alpha Q_j, \; j = 1, \ldots, N.
\end{aligned} \tag{6.24}$$

Then, the solutions of (6.10)–(6.11) satisfy the bounds (6.17) and (6.18). Hence, the closed-loop system (6.10)–(6.11) with initial condition x_{t_0}, $e(t_0)$ is exponentially mean-square stable. If the aforementioned matrix inequalities are feasible with $\alpha = 0$, then the bounds (6.17) and (6.18) hold also for a sufficiently small $\alpha_0 > 0$.

6.3 NCSs Under Markovian Stochastic Protocol

In this section, we will derive LMI conditions for the exponential mean-square stability of the stochastic Markovian jump impulsive system (6.5), (6.6), (6.8) with respect to x. Note that the differential equation for x given by (6.5) depends on $e_j(t) = e_j(t_k)$, $t \in [t_k, t_{k+1})$ with $j \neq \sigma_k$, $j \in \mathcal{I}\setminus\{0\}$ only. Consider the following Lyapunov functional:

$$\begin{aligned}
V_e(t) &= V(t, x_t, \dot{x}_t) + \sum_{j=1, j\neq\sigma_k}^{N} e_j^T(t) Q_j e_j(t), \; \sigma_k \in \mathcal{I}, \\
V(t, x_t, \dot{x}_t) &= \tilde{V}(t, x_t, \dot{x}_t) + V_Q, \\
V_Q &= (\tau_M - \eta_m) \int_{s_k}^{t} e^{2\alpha(s-t)} |\sqrt{Q}\dot{x}(s)|^2 ds,
\end{aligned} \tag{6.25}$$

where $Q > 0$, $Q_j > 0$, $j = 1, \ldots, N$, $t \in [t_k, t_{k+1})$, $k \in \mathbb{Z}^+$ and $\tilde{V}(t, x_t, \dot{x}_t)$ is given by (5.12). The following statement holds.

Lemma 6.2 *If there exist positive constant α, matrices $0 < Q \in \mathbb{R}^{n\times n}$, $0 < Q_j \in \mathbb{R}^{n_i\times n_i}$, $0 < U_j \in \mathbb{R}^{n_i\times n_i}$, $j = 1, \dots, N$, and $V_e(t)$ of (6.25) such that for any $i \in \mathscr{I}$ along (6.5)*

$$\mathbb{E}\Big\{\mathscr{L}V_e(t) + 2\alpha V_e(t) - \frac{1}{\tau_M - \eta_m}\sum_{j=1, j\neq i}^{N} e_j^T(t)(Q_j - U_j)e_j(t)\Big\} \le 0,\ t \in [t_k, t_{k+1}), \tag{6.26}$$

with

$$\tilde{\Omega}_i = \begin{bmatrix} \Phi_{11}^i & \Phi_{12}^i \\ * & \Phi_{22}^i \end{bmatrix} \le 0, \tag{6.27}$$

holds, where

$$\begin{aligned}
\Phi_{11}^i &= \sum_{l=1}^{N}\sum_{j=0, j\neq l}^{N} \pi_{ij} C_l^T Q_l C_l - e^{-2\alpha\tau_M} Q,\\
\Phi_{12}^i &= \Big[\sum_{l=2}^{N}(\pi_{i0}+\pi_{il})C_1^T Q_1 \ \cdots \sum_{l=0, l\neq j}^{N} \pi_{il}(C_j^T Q_j)_{|j\neq i} \ \cdots \sum_{l=0}^{N-1}\pi_{il}C_N^T Q_N\Big],\\
\Phi_{22}^i &= \mathrm{diag}\Big\{\sum_{l=2}^{N}(\pi_{i0}+\pi_{il})Q_1 - U_1, \cdots, \sum_{l=0, l\neq j}^{N}\pi_{il}Q_{j_{|j\neq i}} - U_{j_{|j\neq i}}, \cdots, \sum_{l=0}^{N-1}\pi_{il}Q_N - U_N\Big\}.
\end{aligned}$$

Then, $V_e(t)$ satisfies

$$\mathbb{E}\Big\{V_e(t_{k+1}) - V_e(t_{k+1}^-) + \sum_{j=1, j\neq i}^{N} e_j^T(t_k)(Q_j - U_j)e_j(t_k)\Big\} \le 0,\ i \in \mathscr{I}. \tag{6.28}$$

The bound (6.17) is valid for the solutions of (6.5), (6.6), (6.8) with the initial condition x_{t_0}, $e(t_0)$, implying exponential mean-square stability of (6.5), (6.6), (6.8) with respect to x.

Proof Consider $t \in [t_k, t_{k+1})$ and assume that $\sigma_k = i \in \mathscr{I}$. Following the proof of Lemma 6.1, we obtain from (6.26)

$$\mathbb{E}\{V_e(t)\} \le e^{-2\alpha(t-t_k)}\mathbb{E}\{V_e(t_k)\} + \sum_{j=1, j\neq i}^{N} \mathbb{E}\{e_j^T(t_k)(Q_j - U_j)e_j(t_k)\},\ t \in [t_k, t_{k+1}). \tag{6.29}$$

Therefore, we have

$$\mathbb{E}\{V(t, x_t, \dot{x}_t)\} \le e^{-2\alpha(t-t_k)}\mathbb{E}\{V_e(t_k)\},\ t \in [t_k, t_{k+1}).$$

Note that

$$\mathbb{E}\{V_e(t_{k+1})\} = \mathbb{E}\Big\{\tilde{V}_{|t=t_{k+1}} + V_{Q\,|t=t_{k+1}} + \sum_{j=1, j\neq\sigma_{k+1}}^{N} e_j^T(t_{k+1})Q_j e_j(t_{k+1})\Big\}$$

and

$$\sum_{j=1, j\neq\sigma_{k+1}}^{N} \mathbb{E}\Big\{e_j^T(t_{k+1})Q_j e_j(t_{k+1})|\sigma_k = i\Big\} = \sum_{l=1}^{N}\sum_{j=0, j\neq l}^{N} \pi_{ij}\mathbb{E}\{e_l^T(t_{k+1})Q_l e_l(t_{k+1})\}.$$

Taking (6.27) into account, we obtain

$$\begin{aligned}
&\mathbb{E}\Big\{V_e(t_{k+1}|\sigma_k = i) - V_e(t_{k+1}^-)_{|\sigma_k=i} + \sum_{j=1, j\neq i}^{N} e_j^T(t_k)(Q_j - U_j)e_j(t_k)\Big\}\\
&\leq \mathbb{E}\Big\{\sum_{l=1}^{N}\sum_{j=0, j\neq l}^{N} \pi_{ij}e_l^T(t_{k+1})Q_l e_l(t_{k+1}) - \sum_{j=1, j\neq i}^{N} e_j^T(t_k)U_j e_j(t_k)\\
&\quad -e^{-2\alpha\tau_M}\Big|\sqrt{Q}[x(t_k - \eta_k) - x(t_{k+1} - \eta_{k+1})]\Big|^2\Big\}\\
&= \mathbb{E}\Big\{\tilde{\zeta}_i^T(t_k)\tilde{\Omega}_i\tilde{\zeta}_i(t_k)\Big\} \leq 0,
\end{aligned}$$

where $\tilde{\zeta}_i(t_k) = \mathrm{col}\{x(t_k - \eta_k) - x(t_{k+1} - \eta_{k+1}), e_1(t_k), \cdots, e_{j_{|j\neq i}}(t_k), \cdots, e_N(t_k)\}$ and $\tilde{\Omega}_i$ is given by (6.27). Therefore, the inequalities (6.28) and (6.29) with $t = t_{k+1}^-$ imply $\mathbb{E}\{V_e(t_{k+1})\} \leq e^{-2\alpha(t_{k+1}-t_k)}\mathbb{E}\{V_e(t_k)\} \leq e^{-2\alpha(t_{k+1}-t_0)}\mathbb{E}\{V_e(t_0)\}$. The latter inequality, with $k+1$ replaced by k, and (6.29) give (6.17), which implies the exponential mean-square stability of (6.5), (6.6), (6.8) with respect to x.

Remark 6.5 Different from Lemma 6.1, in Lemma 6.2 the inequality $\mathbb{E}\{V_e(t_{k+1})\} \leq e^{-2\alpha(t_{k+1}-t_0)}\mathbb{E}\{V_e(t_0)\}$ does not give a bound on $e_{\sigma_k}(t_k)$ since $V_e(t)$ of (6.25) for $t \in [t_k, t_{k+1})$ does not depend on $e_{\sigma_k}(t_k)$. That is why Lemma 6.2 only guarantees the mean-square stability with respect to x.

By using the above lemma and the arguments of Theorem 6.1, we arrive at the following result.

Theorem 6.2 *Given* $0 \leq \eta_m < \tau_M$, $\alpha > 0$, $0 \leq \pi_{ij} \leq 1$, $\sum_{j=0}^{N}\pi_{ij} = 1$, $i, j \in \mathscr{I}$ *and* K_l, $l = 1, \ldots, N$. *Suppose there exist* $n \times n$ *matrices* $P > 0$, $Q > 0$, $S_j > 0$, $R_j > 0$, $j = 0, 1$, S_{12} *and* $n_l \times n_l$ *matrices* $Q_l > 0$, $U_l > 0$, $l = 1, \ldots, N$, *such that for any* $i \in \mathscr{I}$, *the matrix inequalities (6.22), (6.27) and* $\tilde{\Sigma}_i + \tilde{\Xi}_i^T \tilde{H}\tilde{\Xi}_i < 0$ *are feasible, where the notation* Φ *is given by (6.22) and where*

$$
\begin{aligned}
\tilde{H} &= \eta_m^2 R_0 + (\tau_M - \eta_m)^2 R_1 + (\tau_M - \eta_m) Q, \\
\tilde{\Sigma}_i &= \mathrm{He}((\tilde{F}_1^i)^T P \Xi_i) + \tilde{\Upsilon}_i - (\tilde{F}_2^i)^T R_0 \tilde{F}_2^i e^{-2\alpha\eta_m} - (\tilde{F}^i)^T \Phi \tilde{F}^i e^{-2\alpha\tau_M}, \\
\Xi_i &= [A \ 0_{n\times n} \ A_1 \ 0_{n\times n} \ B_1 \ \cdots \ B_{j|j\neq i} \ \cdots \ B_N], \\
\tilde{F}_1^i &= [I_n \ 0_{n\times(3n+n_y-n_i)}], \ \tilde{F}_2^i = [I_n \ -I_n \ 0_{n\times(2n+n_y-n_i)}], \\
\tilde{F}^i &= \begin{bmatrix} 0_{n\times n} & I_n & -I_n & 0_{n\times n} & 0_{n\times(n_y-n_i)} \\ 0_{n\times n} & 0_{n\times n} & I_n & -I_n & 0_{n\times(n_y-n_i)} \end{bmatrix}, \\
\tilde{\Upsilon}_i &= \mathrm{diag}\{S_0 + 2\alpha P, -(S_0 - S_1)e^{-2\alpha\eta_m}, 0, -S_1 e^{-2\alpha\tau_M}, \tilde{\psi}_1, \cdots, \tilde{\psi}_{j|j\neq i}, \cdots, \tilde{\psi}_N\}, \\
\tilde{\psi}_j &= -\frac{1}{\tau_M - \eta_m}(Q_j - U_j) + 2\alpha Q_j, \ j = 1, \ldots, N.
\end{aligned}
$$

Then, the solutions of (6.5), (6.6), (6.8) satisfy the bound (6.17), implying exponential mean-square stability with respect to x. If the aforementioned matrix inequalities are feasible with $\alpha = 0$, then the solution bound holds also for a sufficiently small $\alpha_0 > 0$.

Remark 6.6 Note that Theorem 6.1 under iid scheduling protocol guarantees the exponential mean-square stability with respect to the full state $\mathrm{col}\{x, e\}$, while Theorem 6.2 under Markovian scheduling protocol only guarantees the exponential mean-square stability with respect to x. The LMI conditions in Theorems 6.1 and 6.2 are different. In the special case when the Markovian scheduling protocol is iid, the conditions in Theorem 6.2 give more conservative results (MATI and MAD) than those in Theorem 6.1.

Remark 6.7 Assume that the collisions do not occur. Let $N = 2$ and compare the number of scalar decision variables and the resulting LMIs (application of Schur complements) under different protocols. See Table 6.1 for the complexity of the LMI conditions for different protocols. Note that Theorem 6.2 achieves less conservative results than Theorem 6.1 at the price of more LMIs (see example in the next section).

Remark 6.8 The application of Schur complements leads the matrix inequalities of Theorems 6.1 and 6.2 to the LMIs that are affine in the system matrices. Therefore, for the case of system matrices from the uncertain time-varying polytope

Table 6.1 Complexity of stability conditions under different protocols (for $y_1, y_2 \in \mathbb{R}^{n/2}$)

Method	Decision variables	Number and order of LMIs
Theorem 4.1 (Round-Robin)	$8.5n^2 + 2.5n$	Two of $6n \times 6n$, two of $3n \times 3n$
Theorem 5.2 (TOD/Round-Robin)	$3.75n^2 + 3n$	Two of $5.5n \times 5.5n$, one of $2n \times 2n$
Theorem 6.1 (iid)	$4.25n^2 + 4n$	One of $8n \times 8n$, two of $2n \times 2n$
Theorem 6.2 (Markovian)	$4.5n^2 + 4n$	Two of $5.5n \times 5.5n$, one of $2n \times 2n$, two of $1.5n \times 1.5n$

$$\tilde{\Theta} = \sum_{j=1}^{M} g_j(t)\tilde{\Theta}_j, \ 0 \le g_j(t) \le 1,$$
$$\sum_{j=1}^{M} g_j(t) = 1, \ \tilde{\Theta}_j = \left[A^{(j)} \ B^{(j)} \right],$$

the LMIs need to be solved simultaneously for all M vertices $\tilde{\Theta}_j$, using the same decision matrices.

6.4 Simulation Results

Example 6.1 (*Uncertain inverted pendulum*) We illustrate the efficiency of the given conditions on the benchmark example of uncertain cart–pendulum given in Example 5.1. Assume that $\beta_0 = 0$, $\pi_{0i} = \pi_{i0} = 0$, $i = 0, 1, 2$, which means that the collisions do not occur. Choose $\beta_1 = 0.4$ and the transition matrix of Markov chain $\sigma_k \in \{1, 2\}$ as $\Pi_0 = \begin{bmatrix} 0.1 & 0.9 \\ 0.6 & 0.4 \end{bmatrix}$. For different values of η_m given in Table 6.2, by applying Theorems 6.1 and 6.2 ($\alpha = 0$) with Remark 6.8, we obtain the maximum values

Table 6.2 Example 6.1: estimated maximum values of τ_M = MATI + η_M for different η_m

$\tau_M \backslash \eta_m$	0	0.005	0.01	0.02	0.04
Theorem 4.1 (Round-Robin)	0.023	0.026	0.029	0.035	0.046
Theorem 5.2 (TOD/Round-Robin)	0.025	0.028	0.031	0.036	0.047
Theorem 6.1 ($\beta_1 = 0.4$)	0.021	0.024	0.027	0.033	0.045
Theorem 6.2 (Π_0)	0.027	0.029	0.032	0.037	0.047

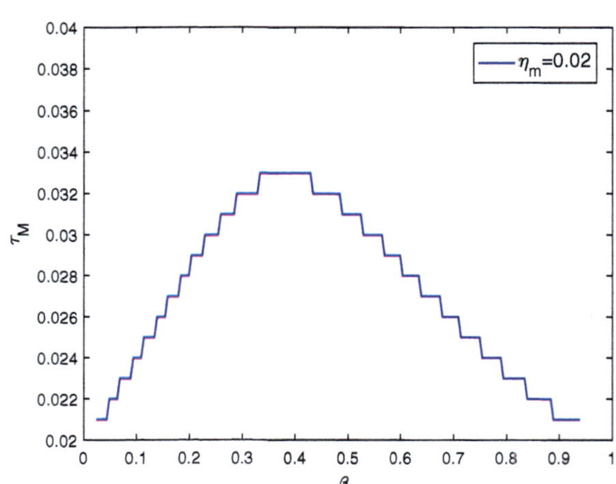

Fig. 6.2 Example 6.1: estimated maximum values of $\tau_M(\beta_1)$ by Theorem 6.1 with $\alpha = 0$ and Remark 6.8

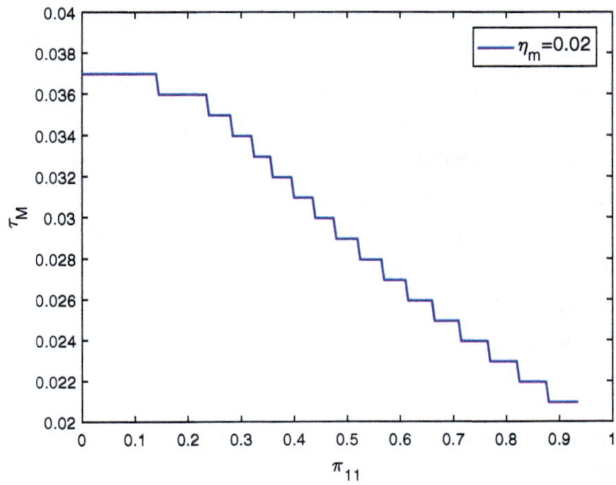

Fig. 6.3 Example 6.1: estimated maximum values of $\tau_M(\pi_{11})$ by Theorem 6.2 with $\alpha = 0$ and Remark 6.8

of $\tau_M = \text{MATI} + \eta_M$ that preserve mean-square stability of the impulsive system (6.5)–(6.6). From Table 6.2, it is seen that in this example the Markovian stochastic scheduling protocol stabilizes the system for larger τ_M than Round-Robin and TOD protocols, which have been proposed in Theorem 4.1 and Theorem 5.2, respectively. Note that when $\eta_m > \tau_M/2 (\eta_m = 0.02, 0.04)$, i.e., the network-induced delays are larger than the sampling intervals, our method is still applicable.

Choosing $\eta_m = 0.02$, by Theorem 6.1 with $\alpha = 0$ and Remark 6.8, we obtain the corresponding maximum values of τ_M shown in Fig. 6.2 for different β_1. Choosing $\eta_m = 0.02$ and $\pi_{21} = 0.6$, by Theorem 6.2 with $\alpha = 0$ and Remark 6.8, we obtain the corresponding maximum values of τ_M shown in Fig. 6.3 for different π_{11}.

Example 6.2 (*batch reactor*) Consider the benchmark example of a batch reactor under the dynamic output feedback given in Example 5.2. Assume that $\beta_0 = 0$, $\pi_{0i} = \pi_{i0} = 0, i = 0, 1, 2$, which means that the collisions do not occur. Let $\beta_1 = 0.6$ and the transition matrix of Markov chain $\sigma_k \in \{1, 2\}$ as $\Pi_1 = \begin{bmatrix} 0.4 & 0.6 \\ 0.9 & 0.1 \end{bmatrix}$. For the values of η_m given in Table 6.3, by applying Theorems 6.1 and 6.2 with $\alpha = 0$, we obtain the maximum values of $\tau_M = \text{MATI} + \text{MAD}$ that preserve mean-square stability of the impulsive system (6.5)–(6.6) (see Table 6.3). From Table 6.3, it is seen that for small transmission delays, our method essentially improves the results of [84], but is more conservative than the results obtained via the discrete-time approach. However, the latter approach becomes complicated for uncertain systems. Polytopic uncertainties in the system model can be easily included in our analysis. When $\eta_m > \tau_M/2$ ($\eta_m = 0.03, 0.04$), note that our method is still applicable.

Choosing $\eta_m = 0.02$, by Theorem 6.1 with $\alpha = 0$, we obtain the corresponding maximum values of τ_M shown in Fig. 6.4 for different β_1. Choosing $\eta_m = 0.02$ and $\pi_{21} = 0.9$, by Theorem 6.2 with $\alpha = 0$, we obtain the corresponding maximum values of τ_M shown in Fig. 6.5 for different π_{11}.

Table 6.3 Example 6.2: estimated maximum values of τ_M = MATI + MAD for different η_m

$\tau_M \backslash \eta_m$	0	0.004	0.02	0.03	0.04
Reference [84](MAD = 0.004, TOD)	0.0108	0.0133	–	–	–
Reference [84] (MAD = 0.004, Round-Robin)	0.0088	0.0088	–	–	–
Reference [29](MAD = 0.03, TOD)	0.069	0.069	0.069	0.069	–
Reference [29] (MAD = 0.03, Round-Robin)	0.068	0.068	0.068	0.068	–
Theorem 5.2 (TOD/Round-Robin)	0.035	0.037	0.047	0.053	0.059
Theorem 4.1 (Round-Robin)	0.042	0.044	0.053	0.058	0.063
Theorem 6.1 ($\beta_1 = 0.6$)	0.022	0.025	0.039	0.048	0.056
Theorem 6.2 (Π_1)	0.035	0.038	0.049	0.055	0.061

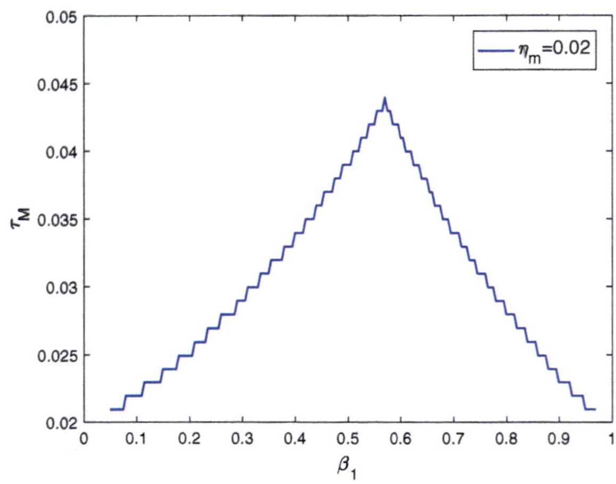

Fig. 6.4 Example 6.2: estimated maximum values of $\tau_M(\beta_1)$ by Theorem 6.1 with $\alpha = 0$

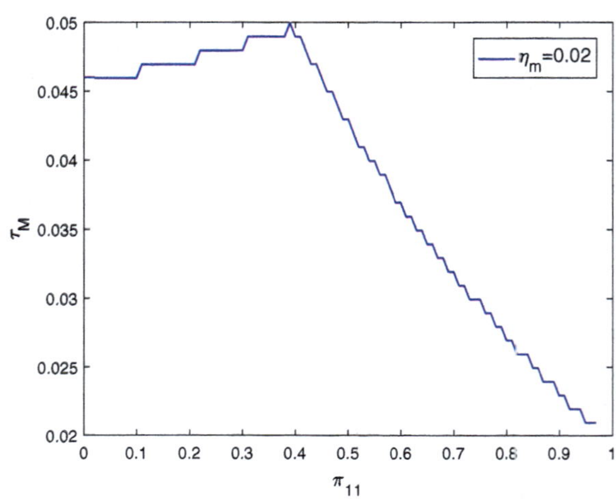

Fig. 6.5 Example 6.2: estimated maximum values of $\tau_M(\pi_{11})$ by Theorem 6.2 with $\alpha = 0$

6.5 Notes

In this chapter, the stabilization of NCSs under stochastic protocol was studied by a time-delay approach. Two types of stochastic protocols, which are defined by the iid and Markovian processes, were proposed. By developing appropriate Lyapunov methods, the exponential mean-square stability conditions for the delayed stochastic impulsive system were derived in terms of LMIs. Simulation results verify the effectiveness of the proposed method.

Chapter 7
Decentralized NCSs with Local Networks Under TOD/Round-Robin Protocol

In Chaps. 4–6, we assume that there is one controller and one global communication network. However, it is commonly accepted in the industry that the total plant to be controlled consists of a large number of interacting subsystems [152]. Usually, the control of the plant is designed in a decentralized manner with local control stations allocated to individual subsystems. In the control of large-scale systems, it is more efficient to use local controllers and local networks instead of the global ones. This leads to large-scale NCSs with independent and asynchronous local networks. Another application of NCSs with asynchronous local networks is platoons of vehicles that communicate wirelessly without timing coordination between members of the whole string [81]. Decentralized networked control of large-scale interconnected systems with local independent networks was studied in the framework of hybrid systems [14, 81], where variable sampling or/and small communication delays were taken into account. Distributed estimation in the presence of synchronous sampling of local networks and Round-Robin protocol was analyzed in [226] in the framework of time-delay approach.

This chapter is devoted to extend the time-delay approach to decentralized NCSs with multiple local communication networks connecting sensors, controllers and actuators. The local networks operate asynchronously and independently of each other in the presence of variable sampling intervals, transmission delays and scheduling protocols (from sensors to controllers). The communication delays are allowed to be greater than the sampling intervals. Note that the direct extension of the switched system modeling under Round-Robin protocol of Chap. 4 to large-scale system would lead to numerous LMIs. The Lyapunov–Krasovskii method of Chap. 5 developed for hybrid time-delay model of the closed-loop systems under TOD and Round-Robin protocols involves complicated conditions on the derivative and on the jumps of Lyapunov functionals that cannot be directly extended to large-scale systems.

In this chapter, a novel Lyapunov–Krasovskii method is proposed for the exponential stability analysis of the closed-loop large-scale system. In the case of networked control of a single plant, the results lead to simplified conditions in terms of reduced-order LMIs comparatively to the results in Chaps. 4 and 5. Numerical examples from

K. Liu et al., *Networked Control Under Communication Constraints*, Advances in Delays and Dynamics 11, https://doi.org/10.1007/978-981-15-4230-5_7

the literature illustrate the efficiency of the results. Throughout this chapter, denote by δ_{nm} the Kronecker delta meaning $\delta_{nm} = 0,\ n \neq m$ and $\delta_{nn} = 1\ (n, m \in \mathbb{N})$. The subscript or superscript j stands for a subsystem index, while subscript i denotes the sensor index.

7.1 Problem Formulation

Consider the large-scale system in Fig. 7.1, consisting of M physically coupled linear continuous-time plant j, controlled by M local controllers j $(j = 1, \ldots, M)$. The dynamics of the plant j is given by subsystems:

$$\dot{x}_j(t) = A_j x_j(t) + \sum_{l \neq j} F_{lj} x_l(t) + B_j u_j(t),\ t \geq 0, \tag{7.1}$$

where $j = 1, \ldots M$, is the subsystem index, $x_j(t) \in \mathbb{R}^{n^j}$ is the state, $u_j(t) \in \mathbb{R}^{m^j}$ is the control input, A_j, B_j and F_{lj} are matrices of appropriate dimensions. The subsystem j has several nodes (N_j distributed sensors, a controller node and an actuator node) connected via a local communication network. The initial condition is given by $x_j(0) = x_{0j}$. The measurements are given by

$$y_{ij}(t) = C_{ij} x_j(t) \in \mathbb{R}^{n_i^j},\ i = 1, \ldots, N_j,\ \sum_{i=1}^{N_j} n_i^j = n_y^j.$$

The jth subsystem is assumed to have an independent sequence of sampling instants

$$0 = s_0^j < s_1^j < \cdots < s_k^j < \cdots,\ \lim_{k \to \infty} s_k^j = \infty$$

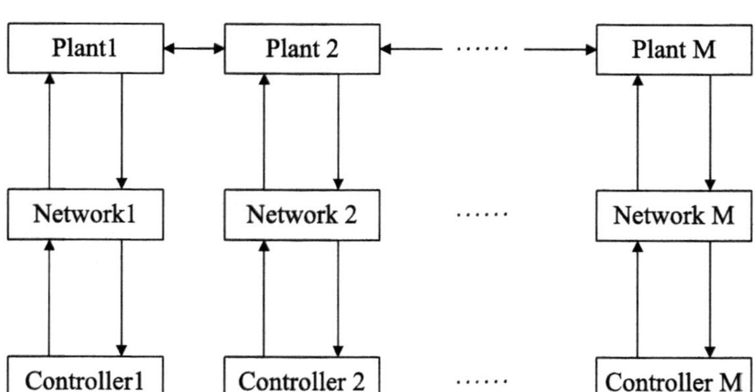

Fig. 7.1 Decentralized control of systems with local networks

with bounded sampling intervals $s_{k+1}^j - s_k^j \leq \text{MATI}_j$. At each s_k^j, one of the outputs $y_{ij}(s_k^j) \in \mathbb{R}^{n_i^j}$ is transmitted via the sensor network to controller j.

Suppose that the data loss is not possible and that the transmission of the information over the networks from sensors to actuators is subject to a variable delay η_k^j. Then, $t_k^j = s_k^j + \eta_k^j$ is the updating time instant of the ZOH. The communication delay is assumed to be bounded $\eta_k^j \in [\eta_m^j, \eta_M^j]$, where $\eta_M^j \triangleq \text{MAD}_j$.

We also allow the delay to be non-small provided that the old sample cannot get to the same destination (same controller or same actuator) after the most recent one. We suppose that the controllers and the actuators are event-driven.

It is assumed that there exist M gain matrices $K_j = [K_{1j} \cdots K_{N_j j}]$, $K_{ij} \in \mathbb{R}^{m^j \times n_i^j}$ such that the matrices $A_j + B_j K_j C_j$ are Hurwitz, where $C_j = [C_{1j}^T \cdots C_{N_j j}^T]^T$. This assumption means that the "nominal system" $\dot{x}_j = A_j + B_j u_j$ is stabilizable by a static output-feedback $u_j = K_j C_j x_j$. Note that in the case of only one network (from sensors to controller) in each subsystem, the presented results can be easily adapted to decentralized observer-based control of large-scale systems as shown for the case of one plant in Chaps. 4–6.

We will consider TOD and Round-Robin protocols that orchestrate the sensor data transmission to the controller. Denote $J = \{1, \ldots, M\}$, $J_{RR} = \{j \in J|\ j\text{th subsystem is under Round-Robin}\}$ and $J_{TOD} = \{j \in J|\ j\text{th subsystem is under TOD}\}$. Note that if for some j, there is no scheduling from the sensors ($N_j = 1$) to the controller we will refer to it as $j \in J_{RR}$, where $N_j = 1$. Thus, $J = J_{RR} \bigcup J_{TOD}$. Denote by

$$\hat{y}_j(s_k^j) = \left[\hat{y}_{1j}^T(s_k^j) \cdots \hat{y}_{N_j j}^T(s_k^j) \right]^T \in \mathbb{R}^{n_y^j} \tag{7.2}$$

the most recent output information submitted to the scheduling protocol of the jth subsystem (i.e., the most recent information at the jth controller side) at the sampling instant s_k^j. Then, the resulting static output feedbacks are given by

$$u_j(t) = \sum_{i=1}^{N_j} K_{ij} \hat{y}_{ij}(s_k^j),\ t \in [t_k^j, t_{k+1}^j),\ k \in \mathbb{Z}^+,\ j = 1, \ldots, M. \tag{7.3}$$

Denote

$$\begin{aligned} &T \triangleq \max\{\{t_{N_j-1}^j\}_{|j \in J_{RR}}, \{t_0^j\}_{|j \in J_{TOD}}\}, \\ &x(t) = \text{col}\{x_1(t), \ldots, x_M(t)\}. \end{aligned} \tag{7.4}$$

We will define next a notion of solution to the closed-loop system (7.1), (7.3) and justify its existence. Monotonically increasing for each $j = 1, \ldots, M$ sequences of updating times $t_0^j < t_1^j < \ldots$ can be reordered in one monotonically increasing sequence $\tilde{t}_0 < \tilde{t}_1 < \ldots$, where $\tilde{t}_{k^*} = T$ for some $k^* \in \mathbb{Z}^+$. For any initial condition $x_T \triangleq x(T + \cdot) \in W[-T, 0]$, by applying the step method for $t \in [\tilde{t}_k, \tilde{t}_{k+1}]$ ($k \geq k^*$) one can show that there exists a unique absolutely continuous function $x : [T, \infty) \to$

$\mathbb{R}^{\sum_{j=1}^{M} n_j}$ satisfying (7.1), (7.3) almost for all $t \geq T$. This function is called a solution of (7.1), (7.3) initialized by x_T.

Definition 7.1 The closed-loop large-scale system (7.1), (7.3) is called exponentially stable with a decay rate $\alpha_0 > 0$ if for any initial condition $x_T \in W[-T, 0]$ there exists $c > 0$ such that the solutions of the system initiated by x_T satisfy the following inequality

$$|x(t)| \leq ce^{-\alpha_0(t-T)}\|x_T\|_W, \ \forall t \geq T. \tag{7.5}$$

This chapter is devoted to derive the sufficient conditions for the exponential stability of the closed-loop system (7.1), (7.3).

7.2 NCSs Under Scheduling Protocols

As mentioned in the previous section, at each sampling instant s_k^j, one of the system nodes $i \in \{1, \dots, N_j\}$ is active, that is only one of $\hat{y}_{ij}(s_k^j)$ values is updated with the recent output $y_{ij}(s_k^j)$. Let $i_k^{*j} \in \{1, \dots, N_j\}$ denote the active output node at the sampling instant s_k^j, which will be chosen due to Round-Robin or TOD scheduling protocol. Then, we have

$$\hat{y}_{ij}(s_k^j) = \begin{cases} y_{ij}(s_k^j), & i = i_k^{*j}, \\ \hat{y}_{ij}(s_{k-1}^j), & i \neq i_k^{*j}. \end{cases} \tag{7.6}$$

For simplicity, we will omit j in i_k^{*j}.

7.2.1 *Round-Robin Protocol and the Closed-Loop Model*

The periodic choice of i_k^* corresponds to Round-Robin protocol. The measurements are sent in a periodic manner one after another. Then, the components of the most recent output on the controller side $\hat{y}_j(s_k^j)$ given by (7.2) can be rewritten as

$$\hat{y}_{ij}(s_k^j) = y_{ij}(s_{k-\Delta_k^i}^j), \ i = 1, \dots, N_j$$

with some $\Delta_k^i \in \{0, \dots, N_j - 1\}$. Following the time-delay approach to NCSs and denoting

$$\tau_{ij}(t) = t - s_{k-\Delta_k^i}^j, \ t \in [t_k^j, t_{k+1}^j),$$

we have

$$\begin{aligned}\eta_m^j \le \tau_{ij}(t) \le t_{k+1}^j - s_{k-\Delta_k^i}^j &= s_{k+1}^j - s_{k-\Delta_k^i}^j + \eta_{k+1}^j\\ &\le (\Delta_k^i + 1)\cdot \mathrm{MATI}_j + \mathrm{MAD}_j\\ &\le N_j \cdot \mathrm{MATI}_j + \mathrm{MAD}_j \triangleq \tau_M^j.\end{aligned}$$

Therefore, for $t \ge t_{N_j-1}^j$ (when all the measurements are transmitted at least once) the static output feedback (7.3) under Round-Robin protocol can be expressed as

$$u(t) = \sum_{i=1}^{N_j} K_{ij} y_{ij}(t - \tau_{ij}(t)),\ t \ge t_{N_j-1}^j. \tag{7.7}$$

The resulting closed-loop model is a system with multiple delays

$$\dot{x}_j(t) = A_j x_j(t) + \sum_{i=1}^{N_j} A_{ij} C_{ij} x_j(t - \tau_{ij}(t)) + \sum_{l \ne j} F_{lj} x_l(t),\ t \ge t_{N_j-1}^j, \tag{7.8}$$

where $A_{ij} = B_j K_{ij}$ and $\tau_{ij}(t) \in [\eta_m^j, \tau_M^j]$.

Note that since there exist K_{ij} such that (7.8) with $\tau_{ij} = 0$ and $F_{lj} = 0$ is exponentially stable, then for small enough τ_{ij} the system (7.8) with the same K_{ij} is ISS (where $x_{l|l\ne j}$ are the inputs).

Remark 7.1 A more accurate model of the closed-loop system under Round-Robin protocol has been presented in Chap. 4 in the form of switched N_j subsystems with ordered multiple delays. Then, the simplified model (7.8) (one system instead of N_j, but with independent delays from the maximum delay interval $[\eta_m^j, \tau_M^j]$) leads to reduced-order LMI conditions.

7.2.2 TOD Protocol and the Closed-Loop Model

In TOD protocol, the choice of i_k^* at the sampling instant s_k^j depends on the transmission error

$$\mathscr{E}_{ij}(s_k^j) = \hat{y}_{ij}(s_{k-1}^j) - y_{ij}(s_k^j),\ i \in \{1, \ldots, N_j\}.$$

The output node i with the greatest weighted error $\mathscr{E}_{ij}(s_k^j)$ will be granted the access to the network.

Definition 7.2 (*Weighted TOD protocol*) Let $Q_{i,j} > 0,\ i = 1, \ldots, N_j$, be some weighting matrices. At the sampling instant s_k^j, the weighted TOD protocol is a protocol for which the active output node with the index i_k^* is defined as any index that satisfies

$$\left|\sqrt{Q_{i_k^*,j}}\mathscr{E}_{i_k^*}(s_k^j)\right|^2 \geq \left|\sqrt{Q_{i,j}}\mathscr{E}_{ij}(s_k^j)\right|^2, \ k \in \mathbb{Z}^+, \ i = 1, \ldots, N_j. \tag{7.9}$$

Here, the weighting matrices $Q_{1,j}, \ldots, Q_{N_j,j}$ are variables to be designed. Then, the feedback can be represented as

$$u_j(t) = K_{i_k^*} y_{i_k^* j}(s_k^j) + \sum_{i=1, i \neq i_k^*}^{N_j} K_{ij}\hat{y}_{ij}(s_{k-1}^j), \ t \in [t_k^j, t_{k+1}^j), \ k \in \mathbb{Z}^+ \tag{7.10}$$

with $u_j(t) = 0, \ 0 \leq t < t_0^j$.

Note that for K_{ij} and small enough MATI_j and MAD_j, the closed-loop system (7.1), (7.10) is ISS, where $x_{l|l\neq j}$ are the inputs (cf., Remark 7.5). Denote

$$\tau_j(t) = t - s_k^j, \ t \in [t_k^j, t_{k+1}^j), \ k \in \mathbb{Z}^+.$$

Then we have

$$\eta_m^j \leq \tau_j(t) \leq \text{MATI}_j + \text{MAD}_j \overset{\Delta}{=} \tau_M^j.$$

In order to obtain the impulsive closed-loop model we define as in (5.5) the piecewise-continuous error

$$e_{ij}(t) = \mathscr{E}_{ij}(s_k^j), \ t \in [t_k^j, t_{k+1}^j), \ i = 1, \ldots, N_j,$$

where we assume $\hat{y}_{ij}(s_{-1}^j) = 0$, implying $e_{ij}(t_0^j) = -y_{ij}(s_0^j)$. Then, the closed-loop model has the following continuous dynamics:

$$\begin{cases} \dot{x}_j(t) = A_j x(t) + A_{1j} C_j x_j(t - \tau_j(t)) + \sum_{i=1, i\neq i_k^*}^{N_j} B_{ij} e_{ij}(t) + \sum_{l \neq j} F_{lj} x_l(t), \\ \dot{e}_{ij}(t) = 0, i = 1, \ldots, N_j, \ t \geq t_0^j, \end{cases} \tag{7.11}$$

where $A_{1j} = B_j K_j$, $B_{ij} = B_j K_{ij}$.

Following Chap. 5, we obtain for $i = i_k^*$

$$\begin{aligned} e_{ij}(t_{k+1}^j) &= \hat{y}_{ij}(s_k^j) - y_{ij}(s_{k+1}^j) \\ &= C_{ij}[x_j(s_k^j) - x_j(s_{k+1}^j)], \end{aligned}$$

and for $i \neq i_k^*$

$$\begin{aligned} e_{ij}(t_{k+1}^j) &= \hat{y}_{ij}(s_{k-1}^j) - y_{ij}(s_{k+1}^j) \\ &= e_{ij}(t_k^j) + C_{ij}[x_j(s_k^j) - x_j(s_{k+1}^j)]. \end{aligned}$$

Thus, the delayed reset system is given by

$$\begin{cases} x_j(t^j_{k+1}) = x_j(t^{j-}_{k+1}), \\ e_{ij}(t^j_{k+1}) = [1-\delta(i,i^*_k)]e_{ij}(t^j_k) + C_{ij}[x_j(t^j_k - \eta^j_k) - x_j(t^j_{k+1} - \eta^j_{k+1})], \end{cases} \tag{7.12}$$

where $i = 1, \dots, N_j$, $k \in \mathbb{Z}^+$, and δ is Kronecker delta.

Note that in our model, the first updating time t^j_0 corresponds to the time instant when the first data is received by the actuator. We define $x_j(t) = x_j(0)$ for $t < 0$. Thus, the initial conditions for (7.11)–(7.12) are given by

$$x_j(t^j_0 + \cdot) \in W[-\tau^j_M, 0],\ e_j(t^j_0) = -C_j x_j(t^j_0 - \eta^j_0). \tag{7.13}$$

7.2.3 Lyapunov-Based Analysis Under Round-Robin Protocol

Assume that the jth subsystem (7.1) is under Round-Robin protocol, i.e., $j \in J_{RR}$. Consider the closed-loop model (7.8) and the following Lyapunov functional:

$$\begin{aligned} V_j(t) &= x^T_j(t)P_j x_j(t) + V_{0j}(t) + V_{1j}(t), \\ V_{0j}(t) &= \sum_{i=1}^{N_j}\Big[\int_{t-\eta^j_m}^{t} e^{2\alpha(s-t)} x^T_j(s) C^T_{ij} S_{0i,j} C_{ij} x_j(s) ds \\ &\quad + \eta^j_m \int_{-\eta^j_m}^{0}\int_{t+\theta}^{t} e^{2\alpha(s-t)} \dot{x}^T_j(s) C^T_{ij} R_{0i,j} C_{ij} \dot{x}_j(s) ds d\theta\Big], \\ V_{1j}(t) &= \sum_{i=1}^{N_j}\Big[\int_{t-\tau^j_M}^{t-\eta^j_m} e^{2\alpha(s-t)} x^T_j(s) C^T_{ij} S_{1i,j} C_{ij} x_j(s) ds \\ &\quad + h_j \int_{-\tau^j_M}^{-\eta^j_m}\int_{t+\theta}^{t} e^{2\alpha(s-t)} \dot{x}^T_j(s) C^T_{ij} R_{1i,j} C_{ij} \dot{x}_j(s) ds d\theta\Big], \end{aligned} \tag{7.14}$$

where $h_j \triangleq (\tau^j_M - \eta^j_m)$, $\alpha > 0$, $P_j > 0$, $S_{mi,j} > 0$, $R_{mi,j} > 0$, $m = 0, 1$, and where we define (for simplicity) $x_j(t) = x_{0j}$, for $t < 0$. Note that different from conventional Lyapunov functionals for the stability of systems with interval delays (see, e.g., Chap. 4 and [173]), the one given by (7.14) contains C_{ij} in integral terms with the reduced-order matrices $S_{mi,j}$ and $R_{mi,j}$. The latter matrices will be decision variables of the resulting LMIs.

Proposition 7.1 *Consider the jth subsystem given by* (7.8). *Given tuning parameters $\alpha > \varepsilon > 0$ and $(M-1)$ $n_l \times n_l$ $(l \neq j)$ matrices $P_l > 0$, let there exist a $n_j \times n_j$ matrix $P_j > 0$, $n^j_i \times n^j_i$ matrices $R_{0i,j} > 0$, $R_{1i,j} > 0$, $S_{0i,j} > 0$, $S_{1i,j} > 0$ and $W_{i,j}$, $i = 1 \dots N_j$, that satisfy*

$$\Gamma_{i,j} = \begin{bmatrix} R_{1i,j} & W_{i,j} \\ * & R_{1i,j} \end{bmatrix} \geq 0,\ i = 1, \dots, N_j, \tag{7.15}$$

and

$$\hat{\Sigma}_j = \begin{bmatrix} \Sigma_j & \Xi_j \\ * & \Pi_j \end{bmatrix} < 0, \tag{7.16}$$

where $\boldsymbol{\Sigma}_j = \begin{bmatrix} \Phi & D_2^T C_j^T H_j \\ * & -H_j \end{bmatrix}$ *and*

$$\begin{aligned}
\Phi &= D_1^T(2\alpha P_j + C_j^T \hat{S}_{0j} C_j) D_1 + \mathrm{He}(D_1^T P_j D_2) - e^{-2\alpha\eta_m^j} D_3^T(\hat{S}_{0j} - \hat{S}_{1j}) D_3 \\
&\quad - e^{-2\alpha\eta_m^j} D_4^T \hat{R}_{0j} D_4 - e^{-2\alpha\tau_M^j} D_5^T \hat{S}_{1j} D_5 - e^{-2\alpha\tau_M^j} D_6^T \hat{\Gamma}_j D_6, \\
H_{i,j} &= \eta_m^2 R_{0i,j} + (\tau_M - \eta_m)^2 R_{1i,j}, \\
H_j &= \mathrm{diag}\{H_{1,j}, \ldots, H_{N_j,j}\}, \\
\hat{S}_{pj} &= \mathrm{diag}\{S_{p1,j}, \ldots, S_{pN_j,j}\}, \ p = 0, 1, \\
\hat{R}_{0j} &= \mathrm{diag}\{R_{01,j}, \ldots, R_{0N_j,j}\}, \\
\hat{\Gamma}_j &= \mathrm{diag}\{\Gamma_{1,j}, \ldots, \Gamma_{N_j,j}\}, \quad D_1 = [I_{n^j} \ 0_{n^j \times 3n_y^j}], \\
D_2 &= \left[A_j, [0\ 1\ 0] \otimes A_{1j}, \ldots, [0\ 1\ 0] \otimes A_{N_j j}\right], \\
D_3 &= \left[0_{n_y^j \times n^j} \quad \mathrm{diag}\left\{[1\ 0\ 0] \otimes I_{n_1^j}, \ldots, [1\ 0\ 0] \otimes I_{n_{N_j}^j}\right\}\right], \\
D_4 &= \left[C_j \quad \mathrm{diag}\left\{[-1\ 0\ 0] \otimes I_{n_1^j}, \ldots, [-1\ 0\ 0] \otimes I_{n_{N_j}^j}\right\}\right], \\
D_5 &= \left[0_{n_y^j \times n^j} \quad \mathrm{diag}\left\{[0\ 0\ 1] \otimes I_{n_1^j}, \ldots, [0\ 0\ 1] \otimes I_{n_{N_j}^j}\right\}\right], \\
D_6 &= \left[0_{2n_y^j \times n^j} \quad \mathrm{diag}\left\{\left[\begin{smallmatrix} 1 & -1 & 0 \\ 0 & 1 & -1 \end{smallmatrix}\right] \otimes I_{n_1^j}, \ldots, \left[\begin{smallmatrix} 1 & -1 & 0 \\ 0 & 1 & -1 \end{smallmatrix}\right] \otimes I_{n_{N_j}^j}\right\}\right], \\
\Xi_j^T &= \mathscr{F}_j^T \left[P_j D_1 \ C_j^T H_j\right], \quad \mathscr{F}_j = row_{\, l=1,\ldots,M}\{F_{lj}, \ l \neq j\} \\
\Pi_j &= -\frac{2\varepsilon}{M-1} \mathrm{diag}_{\, l=1,\ldots,M}\{P_l, \ l \neq j\}.
\end{aligned}$$

Then, the Lyapunov functional $V_j(t)$ *given by* (7.14) *satisfies the following inequality*

$$\dot{V}_j(t) + 2\alpha V_j(t) \leq \frac{2\varepsilon}{M-1} \sum_{l \neq j} x_l^T(t) P_l x_l(t), \ t \geq t_{N_j-1}^j \tag{7.17}$$

along the solutions of (7.8). *Moreover, in the case where the* j*th subsystem* (7.8) *is independent of other subsystems (i.e.,* $F_{lj}|_{l \neq j} = 0$, $l = 1, \ldots, M$*), if* $\Sigma_j < 0$*, then* (7.8) *is exponentially stable with the decay rate* α.

Proof We follow the standard arguments for exponential stability analysis via Krasovskii method (see, e.g., [47, 173]). Differentiating $V_j(t)$, we have

$$
\begin{aligned}
\dot{V}_j(t) + 2\alpha V_j(t) \leq\ & 2\alpha x_j^T(t) P_j x_j(t) + 2\dot{x}_j^T(t) P_j x_j(t) \\
& + \sum_{i=1}^{N_j} |\sqrt{S_{0i,j}} C_{ij} x_j(t)|^2 - |\sqrt{S_{0i,j}} e^{-\alpha \eta_m^j} C_{ij} x_j(t-\eta_m^j)|^2 \\
& + \rho_m |\sqrt{S_{1i,j}} C_{ij} x_j(t-\eta_m^j)|^2 - \rho |\sqrt{S_{1i,j}} C_{ij} x_j(t-\tau_M^j)|^2 \\
& - \eta_m^j \int_{t-\eta_m^j}^{t} |\sqrt{R_{0i,j}} e^{\alpha(s-t)} C_{ij} \dot{x}_j(s)|^2 ds \\
& - h_j \int_{t-\tau_M^j}^{t-\eta_m^j} |\sqrt{R_{1i,j}} e^{\alpha(s-t)} C_{ij} \dot{x}_j(s)|^2 ds + |\sqrt{H_{i,j}} C_{ij} \dot{x}_j(t)|^2.
\end{aligned}
$$

By Jensen's inequality, we have

$$
\eta_m^j \int_{t-\eta_m^j}^{t} |\sqrt{R_{0i,j}} e^{\alpha(s-t)} C_{ij} \dot{x}_j(s)|^2 ds \geq \rho_m |\sqrt{R_{0i,j}} C_{ij} (x_j(t) - x_j(t-\eta_m^j)|^2,
$$

whereas by the arguments of [173] and (7.15), we obtain

$$
\begin{aligned}
& e^{2\alpha \tau_M^j} h_j \int_{t-\tau_M^j}^{t-\eta_m^j} e^{-2\alpha \tau_M^j} |\sqrt{R_{1i,j}} e^{\alpha(s-t)} C_{ij} \dot{x}_j(s)|^2 ds \\
& \geq \begin{bmatrix} C_{ij}(x_j(t-\eta_m^j) - x_j(t-\tau_{ij}(t))) \\ C_{ij}(x_j(t-\tau_{ij}(t)) - x_j(t-\tau_M^j)) \end{bmatrix}^T \Gamma_{i,j} \begin{bmatrix} C_{ij}(x_j(t-\eta_m^j) - x_j(t-\tau_{ij}(t))) \\ C_{ij}(x_j(t-\tau_{ij}(t)) - x_j(t-\tau_M^j)) \end{bmatrix}.
\end{aligned}
$$

Denote

$$
\begin{aligned}
\zeta_j(t) &= \Big[x_j^T(t), [x_j(t-\eta_m^j)\ x_j(t-\tau_{1j}(t))\ x_j(t-\tau_M^j)]^T C_{1j}^T, \\
& \qquad \ldots, [x_j(t-\eta_m^j)\ x_j(t-\tau_{N_j j}(t))\ x_j(t-\tau_M^j)]^T C_{N_j j}^T \Big]^T, \\
X_j(t) &= \mathrm{col}_{\,l=1,\ldots,M}\{x_l(t), l \neq j\}.
\end{aligned}
$$

Substituting $\dot{x}_j(t) = D_2 \zeta_j(t) + \mathscr{F}_j X_j(t)$, we arrive at

$$
\begin{aligned}
& \dot{V}_j(t) + 2\alpha V_j(t) - \frac{2\varepsilon}{M-1} \sum_{l \neq j} x_l^T(t) P_l x_l(t) \\
& \leq \left[\zeta_j^T(t)\ X_j^T(t) \right] \begin{bmatrix} \Phi & D_1^T P_j \mathscr{F}_j \\ * & \Pi_j \end{bmatrix} \begin{bmatrix} \zeta_j(t) \\ X_j(t) \end{bmatrix} \\
& \quad + [D_2 \zeta_j(t) + \mathscr{F}_j X_j(t)]^T C_j^T H_j C_j [D_2 \zeta_j(t) + \mathscr{F}_j X_j(t)].
\end{aligned}
$$

Then by Schur complements, (7.16) implies (7.17).

For the case of the single jth subsystem, $\Sigma_j < 0$ implies $\dot{V}_j(t) + 2\alpha V_j(t) \leq 0$, i.e., by comparison principle

$$
x_j^T(t) P_j x_j(t) \leq V_j(t) \leq e^{-2\alpha(t - t_{N_j-1}^j)} V_j(t_{N_j-1}^j).
$$

The latter guarantees the exponential stability since $V_j(t^j_{N_j-1}) \leq \gamma_j \|x_{t^j_{N_j-1}}\|_W$ for some $\gamma_j > 0$.

Remark 7.2 Since there exists $P_j > 0$ such that

$$P_j\Big(A_j + \sum_{i=1}^{N_j} A_{ij}C_{ij}\Big) + \Big(A_j + \sum_{i=1}^{N_j} A_{ij}C_{ij}\Big)^T P_j < 0,$$

then by the standard arguments for delay-dependent conditions [47], for small enough τ_{ij} and $\alpha > 0$ there exist $R_{0i,j} > 0$, $R_{1i,j} > 0$, $S_{0i,j} > 0$, $S_{1i,j} > 0$ and $W_{i,j}$ $(i = 1 \ldots N_j)$ that satisfy (7.15) and $\Xi_j < 0$ with the same P_j. Therefore, by Schur complements, (7.16) is feasible for given $\varepsilon > 0$ and small enough F_{lj}.

Remark 7.3 The LMIs of Proposition 7.1 and of Theorem 7.1 (see Sect. 7.3 below) are affine in the system matrices. Therefore, in the case of system matrices from an uncertain time-varying polytope, one has to solve these LMIs simultaneously for all the vertices of the polytope applying the same decision matrices.

Example 7.1 Consider an inverted pendulum mounted on a cart given in Example 5.1. The linearized model can be written as (7.1) with one subsystem ($F_{lj} = 0$, $M = 1$), where $A_1 = E^{-1}A_f$, $B_1 = E^{-1}B_0$ with E, A_f, B_0 given by (5.44). Here $f_c(t) \in [0.15, 0.25]$ and $f_b(t) \in [0.15, 0.25]$ are uncertain parameters. Thus, A_1 belongs to uncertain polytope, defined by four vertices A_1^p $(p = 1, \ldots, 4)$ corresponding to $f_c/f_b = 0.15$ and $f_c/f_b = 0.25$. The pendulum can be stabilized by a state feedback $u_1(t) = K_1x_1(t)$ with

$$K_1 = [11.2062 \quad -128.8597 \quad 10.7823 \quad -22.2629].$$

Suppose that the state variables are not accessible simultaneously. Consider the case of $N_1 = 2$ measurements, where

$$C_{11} = \begin{bmatrix} 1 & 0 & 0 & 0 \\ 0 & 1 & 0 & 0 \end{bmatrix}, \quad C_{21} = \begin{bmatrix} 0 & 0 & 1 & 0 \\ 0 & 0 & 0 & 1 \end{bmatrix}.$$

For the values of η_m given in Table 7.1, we apply Remark 7.3, where the LMI (7.15) and four LMIs (7.16) corresponding to four vertices A_1^p (with A_1 substituted by A_1^p) are solved with the same decision variables and with $\alpha = 0.015$. Table 7.2 presents the MATI for a given MAD $= 0.024$ (the case of large communication delay). It is observed that under Round-Robin protocol the LMI conditions of Proposition 7.1 possess essentially less decision variables (the reduction is more than twice) and are given in terms of smaller LMIs than [133, 134] but in some cases guarantee the exponential stability for larger MATI.

Table 7.1 Example 7.1: maximum value of MATI for a given $\eta_M \triangleq$ MAD

MATI$\backslash\eta_m$	0	0.01	0.02
Reference [133]	$0.023 - \eta_M$	$0.029 - \eta_M$	$0.035 - \eta_M$
Reference [134]	$0.025 - \eta_M$	$0.031 - \eta_M$	$0.036 - \eta_M$
Proposition 7.1	$0.022 - \frac{\eta_M}{2}$	$0.0221 - \frac{\eta_M}{2}$	$0.0223 - \frac{\eta_M}{2}$

Table 7.2 Example 7.1: maximum value of MATI for MAD = 0.024

MATI$\backslash\eta_m$	0	0.01	0.02	Decision variables	LMI rows
Reference [133]	–	0.005	0.011	146	236
Reference [134]	0.001	0.007	0.012	72	208
Proposition 7.1	0.01	0.0101	0.0103	42	108

7.2.4 Lyapunov-Based Analysis Under TOD Protocol

In this section, we assume that the jth subsystem (7.1) is under TOD scheduling protocol, i.e., $j \in J_{TOD}$. Consider the closed-loop model (7.11)–(7.12) and the following Lyapunov functional:

$$V_j^e(t) = V_j(t) + \sum_{i=1}^{N_j} e_{ij}^T(t) Q_{i,j} e_{ij}(t) + \mathscr{W}_j^e(t), \tag{7.18}$$

where

$$\begin{aligned}
\mathscr{W}_j^e(t) &= 2\alpha(t_k^j - t) e_{i_k^* j}^T(t) Q_{i_k^*,j} e_{i_k^* j}(t) + \sum_{i=1, i \neq i_k^*}^{N_j} \frac{t_k^j - t}{t_{k+1}^j - t_k^j} e_{ij}^T(t) U_{i,j} e_{ij}(t), \\
V_j(t) &= \tilde{V}_j(t) + V_j^G(t), \\
V_j^G &= \sum_{i=1}^{N_j} h_j \int_{s_k^j}^{t} e^{2\alpha(s-t)} |\sqrt{G_{i,j}} C_{ij} \dot{x}_j(s)|^2 ds, \\
\tilde{V}_j(t) &= x_j^T(t) P_j x_j(t) + \int_{t-\eta_m^j}^{t} e^{2\alpha(s-t)} x_j^T(s) C_j^T S_{0j} C_j x_j(s) ds \\
&\quad + \int_{t-\tau_M^j}^{t-\eta_m^j} e^{2\alpha(s-t)} x_j^T(s) C_j^T S_{1j} C_j x_j(s) ds \\
&\quad + \eta_m^j \int_{-\eta_m^j}^{0} \int_{t+\theta}^{t} e^{2\alpha(s-t)} \dot{x}_j^T(s) C_j^T R_{0j} C_j \dot{x}_j(s) ds d\theta \\
&\quad + h_j \int_{-\tau_M^j}^{-\eta_m^j} \int_{t+\theta}^{t} e^{2\alpha(s-t)} \dot{x}_j^T(s) C_j^T R_{1j} C_j \dot{x}_j(s) ds d\theta,
\end{aligned}$$

where $P_j > 0$, $S_{0j} > 0$, $S_{1j} > 0$, $R_{0j} > 0$, $R_{1j} > 0$, $G_{i,j} > 0$, $Q_{i,j} > 0$, $U_{i,j} > 0$, $\alpha > 0$, $h_j = \tau_M^j - \eta_m^j$, $i = 1, \dots, N_j$, $t \in [t_k^j, t_{k+1}^j)$, $k \in \mathbb{Z}^+$.

Remark 7.4 Different from Lyapunov functional (5.12), the Lyapunov functional (7.18) contains novel negative terms $\mathscr{W}_j^e(t)$ that essentially simplify the exponential stability analysis of the hybrid system. Indeed, denoting $\hat{V}_j^e(t) = V_j^e(t)_{|\mathscr{W}_j^e=0}$, we have

$$\begin{aligned}\dot{V}_j^e(t) + 2\alpha V_j^e(t) &\le \dot{\hat{V}}_j^e(t) + 2\alpha \hat{V}_j^e(t) - \frac{1}{\tau_M^j - \eta_m^j} \sum_{i=1, i\neq i_k^*}^{N_j} |\sqrt{U_{i,j}} e_{ij}(t)|^2 \\ &\quad -2\alpha \left|\sqrt{Q_{i_k^*}} e_{i_k^*}(t)\right|^2, \; t \in [t_k^j, t_{k+1}^j), \\ V_j^e(t_{k+1}^j) - V_j^e(t_{k+1}^{j\,-}) &\le \hat{V}_j^e(t_{k+1}^j) - \hat{V}_j^e(t_{k+1}^{j\,-}) + \sum_{i=1, i\neq i_k^*}^{N} |\sqrt{U_{i,j}} e_{ij}(t_k^j)|^2 \\ &\quad +2\alpha h_j \left|\sqrt{Q_{i_k^*}} e_{i_k^*}(t_k^j)\right|^2. \end{aligned} \tag{7.19}$$

In Chap. 5, the stability of (7.11)–(7.12) with $F_{lj} = 0$ is guaranteed if the right-hand sides of (7.19) are non-positive along the system for some $\alpha > 0$. By using the novel functional (7.18), under the same LMIs as in Chap. 5 up to the order reduction due to C_j in $\tilde{V}_j(t)$ (see LMIs of Proposition 7.2 below), we will guarantee that V_e^j is positive, does not grow at t_k^j and satisfies

$$\dot{V}_j^e(t) + 2\alpha V_j^e(t) \le 0, \; t \in [t_k^j, t_{k+1}^j) \tag{7.20}$$

along the jth hybrid system with $F_{lj} = 0$. The inequality (7.20) immediately implies the exponential stability of the jth hybrid subsystem, that essentially simplifies the proof of the stability (which is crucial for the extension of the results to large-scale hybrid systems).

The terms

$$e_{ij}^T(t) Q_{i,j} e_{ij}(t) \equiv e_{ij}^T(t_k) Q_{i,j} e_{ij}(t_k^j), \; t \in [t_k^j, t_{k+1}^j)$$

are piecewise-constant, the term $\tilde{V}_j(t)$ presents a Lyapunov functional (with reduced-order decision matrices) for systems with interval delays $\tau_j(t) \in [\eta_m^j, \tau_M^j]$. The piecewise-continuous in time term V_j^G has been introduced in (5.12) (the term V_G) to cope with the delays in the reset conditions:

$$\begin{aligned} V_j^G(t_{k+1}^j) - V_j^G(t_{k+1}^{j-}) &= \sum_{i=1}^{N_j} h_j \int_{s_{k+1}^j}^{t_{k+1}^j} e^{2\alpha(s-t_{k+1}^j)} |\sqrt{G_{i,j}} C_{ij} \dot{x}_j(s)|^2 ds \\ &\quad - \sum_{i=1}^{N_j} h_j \int_{s_k^j}^{t_{k+1}^{j-}} e^{2\alpha(s-t_{k+1}^j)} |\sqrt{G_{ij}} C_{ij} \dot{x}_j(s)|^2 ds \\ &\leq - \sum_{i=1}^{N_j} h_j e^{-2\alpha\tau_M^j} \int_{s_k^j}^{s_{k+1}^j} |\sqrt{G_{i,j}} C_{ij} \dot{x}_j(s)|^2 ds \\ &\leq - \sum_{i=1}^{N_j} e^{-2\alpha\tau_M^j} |\sqrt{G_{i,j}} C_{ij} [x(s_k^j) - x(s_{k+1}^j)]|^2, \end{aligned} \tag{7.21}$$

where we applied Jensen's inequality. The function $V_j^e(t)$ is thus continuous and differentiable over $[t_k^j, t_{k+1}^j)$. The following lemma gives the sufficient conditions for the positivity of $V_j^e(t)$ and for the fact that it does not grow in the jumps t_k^j.

Lemma 7.1 *Given a tuning parameter* $\alpha > 0$, *let there exist matrices* $0 < Q_{i,j} \in \mathbb{R}^{n_i^j \times n_i^j}$, $0 < U_{i,j} \in \mathbb{R}^{n_i^j \times n_i^j}$ *and* $0 < G_{i,j} \in \mathbb{R}^{n_i^j \times n_i^j}$, $i = 1, \ldots, N_j$, *that satisfy the LMIs*

$$\Omega_{ij} \triangleq \begin{bmatrix} -\frac{1 - 2\alpha(\tau_M^j - \eta_m^j)}{N_j - 1} Q_{i,j} + U_{i,j} & Q_{i,j} \\ * & Q_{i,j} - G_{i,j} e^{-2\alpha\tau_M^j} \end{bmatrix} < 0, \; i = 1, \ldots, N_j. \tag{7.22}$$

Then, $V_j^e(t)$ *of (7.18) is positive in the sense that*

$$V_j^e(t) \geq \beta\Big(|x_j(t)|^2 + |e_j(t)|^2\Big), \; t \geq t_0^j, \; e_j(t) \triangleq \text{col}\{e_{1,j}(t), \ldots, e_{N_j j}(t)\} \tag{7.23}$$

for some $\beta > 0$. *Moreover,* $V_j^e(t)$ *does not grow in the jumps along* (7.11)–(7.12)*:*

$$\Theta \triangleq V_j^e(t_{k+1}^j) - V_j^e(t_{k+1}^{j-}) \leq 0. \tag{7.24}$$

Proof It can be seen that (7.22) implies

$$2\alpha(\tau_M^j - \eta_m^j) < 1 \text{ and } U_{i,j} < Q_{i,j}$$

yielding the positivity of $V_j^e(t)$.

We show next that $V_j^e(t)$ does not grow in the jumps. Since $\tilde{V}_j(t_{k+1}^j) = \tilde{V}_j(t_{k+1}^{j-})$ and $e_j(t_{k+1}^{j-}) = e_j(t_k^j)$, we obtain by taking into account (7.21)

$$
\begin{aligned}
\Theta &= \sum_{i=1}^{N_j}\left[|\sqrt{Q_{i,j}}e_{ij}(t_{k+1}^j)|^2 - |\sqrt{Q_{i,j}}e_{ij}(t_k^j)|^2\right] + 2\alpha(t_{k+1}^j - t_k^j)\left|\sqrt{Q_{i_k^*,j}}e_{i_k^*j}(t_k^j)\right|^2 \\
&\quad + \sum_{i\neq i_k^*}|\sqrt{U_{i,j}}e_{ij}(t_k^j)|^2 + V_j^G(t_{k+1}^j) - V_j^G(t_{k+1}^{j-}) \\
&\leq \left|\sqrt{Q_{i_k^*,j}}e_{i_k^*j}(t_{k+1}^j)\right|^2 + \sum_{i=1,i\neq i_k^*}^{N_j}\left[|\sqrt{Q_{i,j}}e_{ij}(t_{k+1}^j)|^2 - |\sqrt{Q_{i,j}-U_{i,j}}e_{ij}(t_k^j)|^2\right] \\
&\quad -(1-2\alpha h_j)\left|\sqrt{Q_{i_k^*,j}}e_{i_k^*j}(t_k^j)\right|^2 - \sum_{i=1}^{N_j}e^{-2\alpha\tau_M^j}\left|\sqrt{G_{i,j}}C_{ij}[x(s_k^j) - x(s_{k+1}^j)]\right|^2.
\end{aligned}
$$

Under TOD protocol, we have

$$
-\left|\sqrt{Q_{i_k^*,j}}e_{i_k^*j}(t_k^j)\right|^2 \leq -\frac{1}{N_j-1}\sum_{i\neq i_k^*}\left|\sqrt{Q_{i,j}}e_{ij}(t_k^j)\right|^2.
$$

Denote $\zeta_i = \mathrm{col}\{e_{ij}(t_k^j), C_{ij}[x_j(s_k^j) - x_j(s_{k+1}^j)]\}$. Then employing (7.12), we arrive at

$$
\begin{aligned}
\Theta &\leq -\left|\sqrt{G_{i_k^*,j}e^{-2\alpha\tau_M^j} - Q_{i_k^*,j}}C_{i_k^*j}[x_j(s_k^j) - x_j(s_{k+1}^j)]\right|^2 \\
&\quad + \sum_{i\neq i_k^*}\Bigg[\left|\sqrt{Q_{i,j}}\left[C_{ij}[x_j(s_k^j) - x_j(s_{k+1}^j)] + e_{ij}(t_k^j)\right]\right|^2 \\
&\quad -\left|\sqrt{Q_{i,j} + \frac{1-2\alpha h_j}{N_j-1}Q_{i,j} - U_{i,j}}e_{ij}(t_k^j)\right|^2 \\
&\quad -e^{-2\alpha\tau_M^j}|\sqrt{G_{i,j}}C_{ij}[x(s_k^j) - x(s_{k+1}^j)]|^2\Bigg] \\
&= -\left|\sqrt{G_{i_k^*,j}e^{-2\alpha\tau_M^j} - Q_{i_k^*,j}}C_{i_k^*j}[x_j(s_k^j) - x_j(s_{k+1}^j)]\right|^2 + \sum_{i\neq i_k^*}\zeta_i^T\Omega_{i_j}\zeta_i.
\end{aligned}
$$

Therefore, under (7.22) $\Theta \leq 0$.

By applying Lemma 7.1 and by modifying the derivations in Chap. 5 for $F_{lj} \neq 0$, we arrive at

Proposition 7.2 *Consider the* j*th hybrid subsystem* (7.11)–(7.12). *Given tuning parameters* $\alpha > \varepsilon > 0$ *and* $(M-1)$ $n_l \times n_l$ $(l \neq j)$ *matrices* $P_l > 0$, *let there exist a* $n_j \times n_j$ *matrix* $P_j > 0$, $n_y^j \times n_y^j$ *matrices* $R_{0j} > 0$, $R_{1j} > 0$, $S_{0j} > 0$, $S_{1j} > 0$, W_j, *and* $n_i^j \times n_i^j$ *matrices* $Q_{i,j} > 0$, $U_{i,j} > 0$, $G_{i,j} > 0$, $i = 1, \ldots, N_j$, *that satisfy the LMIs* (7.22) *and*

$$
\Gamma_j = \begin{bmatrix} R_{1j} & W_j \\ W_j^T & R_{1j} \end{bmatrix} \geq 0, \quad \Sigma_j^i < 0, \; i = 1, \ldots, N_j, \tag{7.25}
$$

where

$$
\begin{aligned}
\Sigma_j^i &= \begin{bmatrix} \phi_{ij} & D_2^T C_j^T H_j \\ * & -H_j \end{bmatrix}, \\
\phi_{ij} &= D_1^T(2\alpha P_j + C_j^T S_{0j} C_j)D_1 + \mathrm{He}(D_2^T P_j D_1) \\
&\quad - e^{-2\alpha\eta_m^j} D_3^T(S_{0j} - S_{1j})D_3 - e^{-2\alpha\eta_m^j} D_4^T R_{0j} D_4 - e^{-2\alpha\tau_M^j} D_5^T S_{1j} D_5 \\
&\quad - e^{-2\alpha\tau_M^j} D_6^T \Gamma_j D_6 + D_7^T \Psi_i^j D_7 + D_8^T \Pi_j D_8, \\
H_j &= \eta_m^2 R_{0,j} + (\tau_M - \eta_m)^2 R_{1,j} + h_j \cdot \mathrm{diag}\{G_{1,j}, \ldots, G_{N_j,j}\}, \\
\Psi_j^i &= \mathrm{diag}\{\psi_1^j, \ldots, \psi_{r\neq i}^j, \ldots\}, \\
\psi_r^j &\triangleq 2\alpha Q_{r,j} - \frac{1}{h_j} U_{r,j}, \quad \Pi_j = \frac{-2\varepsilon}{M-1}\mathrm{diag}\{P_l, l \neq j\}, \\
D_1 &= [I_{n^j} \;\; 0_{n^j\times(4n_y^j - n_i^j + n - n^j)}], \quad D_2 = [A_j \;\; [0\;1\;0]\otimes A_{1j} \;\; B_j K_j^i \;\; \mathscr{F}_j], \\
K_j^i &= row_{\, r=1,\ldots,N_j}\{K_{rj},\; r \neq i\}, \quad \mathscr{F}_j = row_{\, l=1,\ldots,M}\{F_{1j}, l \neq j\}, \\
D_3 &= \begin{bmatrix} 0_{n_y^j \times n^j} & [1\;0\;0]\otimes I_{n_y^j} & Z_{ij} \end{bmatrix}, \\
D_4 &= \begin{bmatrix} C_j & [-1\;0\;0]\otimes I_{n_y^j} & Z_{ij} \end{bmatrix}, \\
D_5 &= \begin{bmatrix} 0_{n_y^j \times n^j} & [0\;0\;1]\otimes I_{n_y^j} & Z_{ij} \end{bmatrix}, \\
D_6 &= \begin{bmatrix} 0_{2n_y^j \times n^j} & \left[\begin{smallmatrix} 1 & -1 & 0 \\ 0 & 1 & -1 \end{smallmatrix}\right] \otimes I_{n_y^j} & \left[\begin{smallmatrix} 1 \\ 1 \end{smallmatrix}\right] \otimes Z_{ij} \end{bmatrix}, \\
D_7 &= \begin{bmatrix} 0\;0\;0\;0\; I_{n_y^j - n_i^j} & 0 \end{bmatrix}, \\
D_8 &= \begin{bmatrix} 0 & I_{n-n^j} \end{bmatrix}, \\
Z_{ij} &\triangleq 0_{n_y^j \times (n_y^j - n_i^j + n - n^j)}.
\end{aligned}
$$

Then, the Lyapunov functional $V_j^e(t)$ *given by* (7.18) *is positive (i.e.,* (7.23) *holds); it does not grow in the jumps (i.e.,* (7.24) *holds) and satisfies the following inequality*

$$
\dot{V}_j^e(t) + 2\alpha V_j^e(t) \leq \frac{2\varepsilon}{M-1}\sum_{l\neq j} x_l^T(t) P_l x_l(t), \; t \in [t_k^j, t_{k+1}^j), \; k \in \mathbb{Z}^+ \tag{7.26}
$$

along (7.11).

Proof Denote

$$
\begin{aligned}
\zeta_j(t) &= \left[x_j^T(t), \; [x_j(t-\eta_m^j)\; x_j(t-\tau_{1j}(t))\; x_j(t-\tau_M^j)]^T C_j^T\right]^T, \\
\xi_{ij}(t) &= \mathrm{col}_{\, i=1,\ldots,N_j}\{e_{rj}(t), r \neq i\}, \\
X_j(t) &= \mathrm{col}_{\, l=1,\ldots,M}\{x_l(t), l \neq j\}.
\end{aligned}
$$

By using the arguments of Proposition 7.1, where

$$\dot{x}_j^T(t) = \begin{bmatrix} \zeta_j^T(t) & \xi_{i_k^* j}^T(t) & X_j^T(t) \end{bmatrix} D_2^T,$$

we arrive at for $t \in [t_k^j, t_{k+1}^j)$, $k \in \mathbb{Z}^+$

$$\dot{V}_j^e(t) + 2\alpha V_j^e(t) - \frac{2\varepsilon}{M-1} \sum_{l \neq j} x_l^T(t) P_l x_l(t) \leq \begin{bmatrix} \zeta_j^T(t) \xi_{i_k^* j}^T(t) X_j^T(t) \end{bmatrix} \Phi_{i_k^* j} \begin{bmatrix} \zeta_j(t) \\ \xi_{i_k^* j}(t) \\ X_j(t) \end{bmatrix} + \dot{x}_j^T(t) C_j^T H_j C_j \dot{x}_j(t).$$

Then by Schur complements, (7.25) implies (7.26).

Remark 7.5 For $F_{lj} = 0$, the LMIs (7.22) and (7.25) are feasible (see Remark 5.5). Then for given $\varepsilon > 0$ and small enough F_{lj}, the above LMIs are feasible.

7.3 Decentralized Networked Control

Consider the decentralized NCS given by (7.1) where every plant is controlled over a communication network and is either under Round-Robin or under TOD scheduling protocol. The controllers u_j are given by (7.7) and (7.10) respectively. We are in a position to formulate the main result.

Theorem 7.1 *Given tuning parameters $\alpha > \varepsilon > 0$, let there exist $n_j \times n_j$ matrices $P_j > 0 (j \in J)$, $n_i^j \times n_i^j$ matrices $R_{0i,j} > 0$, $R_{1i,j} > 0$, $S_{0i,j} > 0$, $S_{1i,j} > 0$ and $W_{i,j}$ $(i = 1 \ldots N_j,\ j \in J_{RR})$ that satisfy the LMIs (7.15) and (7.16) for all $j \in J_{RR}$, and $n_y^j \times n_y^j$ matrices $R_{0j} > 0$, $R_{1j} > 0$, $S_{0j} > 0$, $S_{1j} > 0$, W_j, and $n_i^j \times n_i^j$ matrices $Q_{i,j} > 0$, $U_{i,j} > 0$, $G_{i,j} > 0$ $(i = 1 \ldots N_j,\ j \in J_{TOD})$ that satisfy the LMIs (7.22) and (7.25) for all $j \in J_{TOD}$. Then, the closed-loop large-scale system (7.1), (7.3) is exponentially stable with a decay rate $\alpha_0 = \alpha - \varepsilon$.*

Proof Let the LMIs of the theorem be feasible. We choose the following Lyapunov functional for the large-scale system (7.1), (7.3):

$$V(t) = \sum_{j \in J_{RR}} V_j(t) + \sum_{j \in J_{TOD}} V_j^e(t), \ t \geq 0,$$

where $\{V_j(t)\}_{j \in J_{RR}}$ and $\{V_j^e(t)\}_{j \in J_{TOD}}$ are given by (7.14) and (7.18), respectively. Define $x(t) = x(0)$ for $t < 0$ and denote

$$\Delta_{TOD} = \{t \geq 0 | \ t = t_k^j, \ j \in J_{TOD}, \ k \in \mathbb{Z}^+\}.$$

Let T be given by (7.4). We apply further Propositions 7.1 and 7.2. Then for some constants $0 < \beta_m < \beta_M$, the functional V satisfies the following bounds:

$$\beta_m\Big[|x(t)|^2 + \sum_{j\in J_{TOD}} |e_j(t)|^2\Big] \le V(t) \le \beta_M\Big[\|x_t\|^2_{W[-\tau_M,0]} + \sum_{j\in J_{TOD}} |e_j(t)|^2\Big], \tag{7.27}$$

where $\tau_M = \max_{j\in J} \tau_M^j$. Moreover, by summing the inequalities (7.17) and (7.26) in $j = 1, \dots, M$, we obtain

$$\dot V(t) + 2\alpha V(t) \le 2\varepsilon \sum_{l=1}^{M} x_l^T(t) P_l x_l(t)$$

for all $t \ge T$ and $t \notin \Delta_{TOD}$, implying

$$\dot V(t) + 2(\alpha - \varepsilon)V(t) \le 0,\ \forall t \ge T,\ t \notin \Delta_{TOD}. \tag{7.28}$$

Additionally, we have

$$V(t) - V(t^-) \le 0,\ \forall t \ge T,\ t \in \Delta_{TOD} \tag{7.29}$$

along (7.1), (7.3). The inequalities (7.28) and (7.29) yield

$$V(t) \le e^{-2\alpha_0(t-T)} V(T),\ t \ge T. \tag{7.30}$$

Then from (7.27), (7.30) for some $\gamma > 0$, we have

$$|x(t)|^2 + \sum_{j\in J_{TOD}} |e_j(t)|^2 \le \gamma e^{-2\alpha_0(t-T)}\Big[\|x_T\|^2_{W[-T,0]} + \sum_{j\in J_{TOD}} |e_j(T)|^2\Big],\ t \ge T. \tag{7.31}$$

Next, we will show that for some $\gamma_0 > 0$

$$\sum_{j\in J_{TOD}} |e_j(T)|^2 \le \gamma_0 \|x_T\|^2_{W[-T,0]}. \tag{7.32}$$

Indeed, from (7.24) and (7.26) we obtain that for some $\gamma_1 > 0$, the following

$$\begin{aligned} V_e^j(t) &\le e^{-2\alpha(t-t_k^j)} V_e^j(t_k^j) + \gamma_1 \int_{t_k^j}^{t} e^{-2\alpha(t-s)} |x(s)|^2 ds \\ &\le \dots \le e^{-2\alpha(t-t_0^j)} V_e^j(t_0^j) + \gamma_1 \int_{t_0^j}^{t} e^{-2\alpha(t-s)} |x(s)|^2 ds \end{aligned} \tag{7.33}$$

holds for $t \in [t_k^j, t_{k+1}^j)$. Taking into account the initial conditions (7.13) and $x(t) = x(0)$ for $t < 0$, we arrive at

$$V_e^j(t_0^j) \le \beta_{jM} \|x(t_0^j + \cdot)\|_{W[-t_0^j,0]}$$

with some $\beta_{jM} > 0$. Moreover, $\beta_{jm}|e^j(t)|^2 \leq V_e^j(t)$ for some $\beta_{jm} > 0$ that together with (7.33) yields (7.32). The inequalities (7.31) and (7.32) imply (7.5) with $c = \sqrt{\gamma + \gamma_0}$.

Remark 7.6 The inequalities (7.31) and (7.32) imply the exponentially converging bound on the errors $e_j(t)$, $j \in J_{TOD}$ meaning the exponential stability of the large-scale hybrid system given by (7.11)–(7.12) for $j \in J_{TOD}$ and by (7.8) for $j \in J_{RR}$.

Example 7.2 [14] Consider two coupled inverted pendulums under the scenario of decentralized networked control, where $M = 2$, $N_j = 2$ or $N_j = 4$ $(j = 1, 2)$. The system matrices are given by

$$A_1 = A_2 = \begin{bmatrix} 0 & 1 & 0 & 0 \\ 2.9156 & 0 & -0.0005 & 0 \\ 0 & 0 & 0 & 1 \\ -1.6663 & 0 & 0.0002 & 0 \end{bmatrix},$$

$$B_1 = B_2 = \begin{bmatrix} 0 \\ -0.0042 \\ 0 \\ 0.0167 \end{bmatrix},$$

$$F_{12} = F_{21} = \begin{bmatrix} 0 & 0 & 0 & 0 \\ 0.0011 & 0 & 0.0005 & 0 \\ 0 & 0 & 0 & 0 \\ -0.0003 & 0 & -0.0002 & 0 \end{bmatrix},$$

$$K_1 = [k_{11}\ k_{21}\ k_{31}\ k_{41}] = [11396\ \ 7196.2\ \ 573.96\ \ 1199.0],$$

$$K_2 = [k_{12}\ k_{22}\ k_{32}\ k_{42}] = [29241\ \ 18135\ \ 2875.3\ \ 3693.9].$$

In the case of $N_j = 2$, we consider

$$C_{1j} = \begin{bmatrix} 1\,0\,0\,0 \\ 0\,1\,0\,0 \end{bmatrix}, \quad C_{2j} = \begin{bmatrix} 0\,0\,1\,0 \\ 0\,0\,0\,1 \end{bmatrix},$$

$$K_{1j} = [k_{1j}\ k_{2j}], \quad K_{2j} = [k_{3j}\ k_{4j}], \quad j = 1, 2.$$

In the case of $N_j = 4$, $C_{1j}, \ldots, C_{4j}$ are the rows of I_4 and $K_{1j}, \ldots, K_{4j}$ are the entries of K_j.

We analyze the exponential stability for $\eta_m^j = 0$ by applying LMI conditions of Theorem 7.1 with $\alpha = 0.015$ and $\varepsilon = 0.002$ for the case where both pendulums are either under Round-Robin or under TOD protocol (the resulting decay rate α_0 is 0.013). The maximum values of τ_M^j that preserve the stability are given in Table 7.3. Then for $\text{MAD}_j = 0$ and $\text{MAD}_j = 0.005$ ($\text{MAD}_j = 0.005$ is larger than the maximum value of MATI_j achieved in [14]), the resulting maximum values of MATI_j that preserve the stability are given in Tables 7.4 and 7.5, respectively.

It is seen that the proposed method leads to essentially larger values of maximum MATI_j comparatively to [14] and allows larger values of MAD_j. More-

Table 7.3 Example 7.2: maximum τ_M^j for $\eta_m^j = 0$ ($\tau_M^j(Round - Robin) = N_j \cdot \mathrm{MATI}_j + \mathrm{MAD}_j$, $\tau_M^j(TOD) = \mathrm{MATI}_j + \mathrm{MAD}_j$)

N_j	2		4	
	τ_M^1	τ_M^2	τ_M^1	τ_M^2
Theorem 7.1 (Round-Robin)	0.0209	0.0074	0.0202	0.0073
Theorem 7.1 (TOD)	0.01	0.0039	0.0029	0.001

Table 7.4 Example 7.2: maximum MATI_j for $\mathrm{MAD}_j = \eta_m^j = 0$ $(j = 1, 2)$

N_j	2		4	
	MATI_1	MATI_2	MATI_1	MATI_2
Reference [14] (TOD)	both $< 2 \times 10^{-6}$		both $< 1 \times 10^{-6}$	
Theorem 7.1 (Round-Robin)	0.0104	0.0037	0.005	0.0018
Theorem 7.1 (TOD)	0.01	0.0039	0.0029	0.001

Table 7.5 Example 7.2: maximum MATI_j for $\mathrm{MAD}_j = 0.005$, $\eta_m^j = 0$

N_j	2		4	
	MATI_1	MATI_2	MATI_1	MATI_2
Theorem 7.1 (Round-Robin)	0.0079	0.0012	0.0038	0.00057
Theorem 7.1 (TOD)	0.095	–	–	–

over, our method is applicable in this example with a much stronger coupling. Thus, for $F_{12} = F_{21} = 40 \cdot A_{12}$ by Theorem 7.1 the stability is preserved, e.g., for $\mathrm{MATI}_j = \mathrm{MAD}_j = 0.001$ $(j = 1, 2)$ (either under Round-Robin or under TOD protocol).

7.4 Notes

In this chapter, a time-delay approach was developed for the decentralized exponential stabilization of large-scale NCSs with local networks, where asynchronous variable sampling intervals, large bounded variable communication delays and Round-Robin/TOD scheduling protocols are taken into account. The proposed novel Lyapunov–Krasovskii method leads to LMI conditions for the exponential stability of the closed-loop large-scale system. Being applied to the example of two coupled pendulums with local networks, our results are favorably compared to the existing ones. The presented new technique may be useful for decentralized control of microgrids with islanded generators.

Chapter 8
Dynamic Quantization of Uncertain Linear NCSs

In this chapter, the time-delay approach is applied to uncertain linear NCSs under dynamic quantization, variable sampling intervals and large communication delays. We follow Liberzon's framework [123] and model the closed-loop quantized system as a system with bounded disturbances. The sensor quantization is the focus of our study, cf., [65]. The communication delays lead to additional challenges: (1) The initial and level sets are defined in infinite-dimensional spaces, though the saturation condition is given in terms of the delayed output vector. (2) The closed-loop system and the resulting solution bounds are formulated in terms of updating time instants at the actuators, while the zooming algorithm should be given in terms of sampling instants at the sensors. (3) The solution bounds include additional bounds on the first time interval of the delay length [130]. In this chapter, we first suggest a time-triggered zooming algorithm for uncertain linear NCSs, which is implemented at the sensors although the solution bounds of the closed-loop system are given in terms of the updating time instants at the actuators. The zooming algorithm is formulated in terms of LMIs. Then, we propose a direct Lyapunov approach to the initialization of the zoom variable. More precisely, given a bound on the initial state conditions and the values of the quantizer range and error, we derive conditions for finding the initial value of the zoom variable to guarantee the exponential stability of the closed-loop system. The proposed framework can easily incorporate polytopic-type uncertainties in the system model.

K. Liu et al., *Networked Control Under Communication Constraints*, Advances in Delays and Dynamics 11, https://doi.org/10.1007/978-981-15-4230-5_8

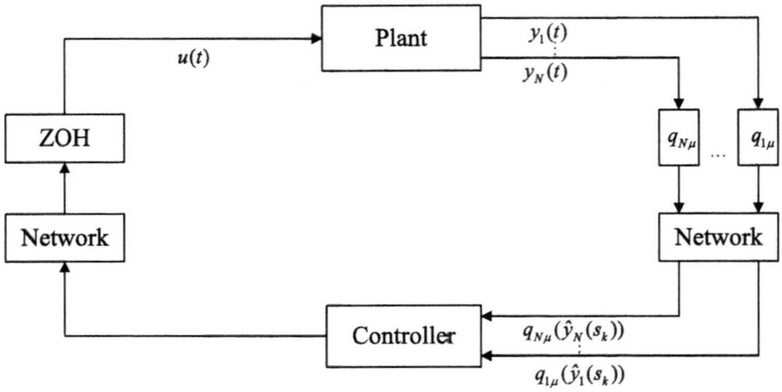

Fig. 8.1 Architecture of NCSs with quantizers

8.1 System Model and Preliminaries

8.1.1 *Quantized NCSs*

Consider the system architecture in Fig. 8.1 with plant

$$\dot{x}(t) = Ax(t) + Bu(t), \tag{8.1}$$

where $x(t) \in \mathbb{R}^n$ is the state vector, $u(t) \in \mathbb{R}^m$ the control input, and A and B system matrices with appropriate dimensions. These matrices can be uncertain with polytopic-type uncertainties. The NCS has N distributed sensors and quantizers, a controller node and an actuator node, which are all connected via two wireless networks. The measurements are given by $y_i(t) = C_i x(t) \in \mathbb{R}^{n_i}$, $i = 1, \ldots, N$, $\sum_{i=1}^{N} n_i = n_y$. Denote $C = [C_1^T \ \cdots \ C_N^T]^T$ and $y(t) = [y_1^T(t) \ \cdots \ y_N^T(t)]^T \in \mathbb{R}^{n_y}$. Following [65], in this chapter, the quantization is performed at the sensor side.

Let $z_i \in \mathbb{R}^{n_i}$, $i = 1, \ldots, N$, be the vectors being quantized. The quantizers are piecewise-constant functions q_i: $\mathbb{R}^{n_i} \to \mathbb{D}_i$, where $\mathbb{D}_i$ is a finite subset of $\mathbb{R}^{n_i}$, $i = 1, \ldots, N$. Following [123], we assume that there exist real numbers $M_i > \Delta_i > 0$, $i = 1, \ldots, N$, such that the following two conditions hold:

(a) If $|z_i| \le M_i$, then $|q_i(z_i) - z_i| \le \Delta_i$,
(b) If $|z_i| > M_i$, then $|q_i(z_i)| > M_i - \Delta_i$,

where Δ_i and M_i are the quantization error bounds and ranges, respectively.

We consider quantized measurements of the form

$$q_{i\mu}(z_i) := \mu q_i\left(\frac{z_i}{\mu}\right), \ i = 1, \ldots, N, \tag{8.2}$$

where $\mu > 0$ is the zoom variable. The range of the quantizer $q_{i\mu}$, $i = 1, \ldots, N$, is μM_i, and the quantization error is $\mu \Delta_i$. The quantized measurements $q_{i\mu}(y_i)$ of the output y_i, $i = 1, \ldots, N$, are available at the controller. The zoom variable μ will change dynamically at some discrete-time sampling instants in order to achieve exponential stability.

Let s_k denote the unbounded and monotonously increasing sequence of sampling instants satisfying (2.2). At each sampling instant s_k, all the outputs $y_i(t) \in \mathbb{R}^{n_i}$, $i = 1, \ldots, N$, are sampled, quantized and transmitted over the networks. Assume that the data

$$q_\mu(y(s_k)) = \left[q_{1\mu}^T(y_1(s_k)) \cdots q_{N\mu}^T(y_N(s_k)) \right]^T, \ k \in \mathbb{Z}^+,$$

is transmitted in packets. We suppose that there is no data loss but the transmission over the two networks is subject to a variable delay η_k. Then, $t_k = s_k + \eta_k$ is the updating time instant of the ZOH device. We allow the delay to be large provided that the order of transmission of $q_\mu(y(s_k))$ is maintained at the reception. Assume that the network-induced delay η_k and the time span between the updating instant t_{k+1} and the current sampling instant s_k are bounded:

$$t_{k+1} - t_k + \eta_k \leq \tau_M, \ 0 \leq \eta_m \leq \eta_k \leq \eta_M, \ k \in \mathbb{Z}^+, \tag{8.3}$$

where η_m and η_M (i.e., MAD) are known bounds and $\tau_M = \text{MATI} + \eta_M$.

We suppose that the controller and the actuator are event-driven. The first updating time t_0 corresponds to the time instant when the first data packet is received by the actuator, which means that $u(t) = 0$, $t \in [0, t_0)$. Then for $t \in [0, t_0)$, (8.1) is given by

$$\dot{x}(t) = Ax(t), \ x(0) = x_0, \ t \in [0, t_0). \tag{8.4}$$

We assume that x_0 may be unknown, but satisfies the bound $|x_0| < X_0$, where $X_0 > 0$ is known. Note that this assumption is common, e.g., for interval observer design [178].

8.1.2 The Closed-Loop Model and Solution Bounds

Assume that there exists a matrix $K = [K_1 \ \cdots \ K_N]$, $K_i \in \mathbb{R}^{m \times n_i}$ such that $A + BKC$ is Hurwitz. Consider the static output-feedback controller of the form

$$u(t) = \sum_{i=1}^{N} K_i q_{i\mu}(y_i(t_k - \eta_k)), \ t \in [t_k, t_{k+1}),$$

where η_k is the communication delay. We thus obtain the closed-loop model as follows:

$$\dot{x}(t) = Ax(t) + A_1 x(t_k - \eta_k) + \sum_{i=1}^{N} B_i \omega_i(t), \ t \in [t_k, t_{k+1}), \tag{8.5}$$

where $A_1 = BKC$, $B_i = BK_i$, $i = 1, \ldots, N$, and $\omega_i(t) = q_{i\mu}(y_i(s_k)) - y_i(s_k)$, $i = 1, \ldots, N$, represent the quantization errors. If $|y_i(s_k)| \leq \mu M_i$, then $|\omega_i(t)| \leq \mu \Delta_i$, $i = 1, \ldots, N$, for $t \in [t_k, t_{k+1})$.

We apply the time-delay approach to NCSs and denote $\tau(t) = t - t_k + \eta_k$, $t \in [t_k, t_{k+1})$. Then, $\tau(t) \in [\eta_m, \tau_M]$ (cf., (8.3)) and $x(t_k - \eta_k) = x(t - \tau(t))$, $t \in [t_k, t_{k+1})$. The initial conditions for (8.5) are given by (8.4).

Consider first static quantizers with a constant zoom variable μ. We apply the following LKF for delay-dependent analysis [50, 173]:

$$\begin{aligned} V(t, x_t, \dot{x}_t) &= x^T(t)Px(t) + \int_{t-\eta_m}^{t} e^{2\alpha(s-t)} x^T(s) S_0 x(s) ds \\ &+ \int_{t-\tau_M}^{t-\eta_m} e^{2\alpha(s-t)} x^T(s) S_1 x(s) ds, \\ &+ \eta_m \int_{-\eta_m}^{0} \int_{t+\theta}^{t} e^{2\alpha(s-t)} \dot{x}^T(s) R_0 \dot{x}(s) ds d\theta \\ &+ (\tau_M - \eta_m) \int_{-\tau_M}^{-\eta_m} \int_{t+\theta}^{t} e^{2\alpha(s-t)} \dot{x}^T(s) R_1 \dot{x}(s) ds d\theta, \end{aligned} \tag{8.6}$$

where $P > 0$, $S_j > 0$, $R_j > 0$, $\alpha > 0$, $j = 0, 1$, $t \in [t_k, t_{k+1})$, $k \in \mathbb{Z}^+$ and where (following [130]) we define $x(t) = x_0$, $t < 0$. Following [50] and using the convex analysis of [173], we derive the following result (see appendix for the proof).

Lemma 8.1 *Given* $0 \leq \eta_m < \tau_M$, $\alpha > 0$, *assume that there exist positive scalars* $b_i, i = 1, \ldots, N$, $n \times n$ *matrices* $P > 0$, $S_0 > 0$, $R_0 > 0$, $S_1 > 0$, $R_1 > 0$, S_{12}, *such that the following LMIs*

$$\Phi = \begin{bmatrix} R_1 & S_{12} \\ * & R_1 \end{bmatrix} \geq 0, \tag{8.7}$$

$$\Psi = \begin{bmatrix} \Sigma - F^T \Phi F e^{-2\alpha\tau_M} & \Xi^T H \\ * & -H \end{bmatrix} < 0, \tag{8.8}$$

are feasible, where

$$\begin{aligned} \Sigma &= \mathrm{He}(F_1^T P \Xi) + \Upsilon - F_2^T R_0 F_2 e^{-2\alpha\eta_m}, \\ F_1 &= [I_n \ 0_{n\times(3n+n_y)}], \ F_2 = [I_n \ -I_n \ 0_{n\times(2n+n_y)}], \\ F &= \begin{bmatrix} 0_{n\times n} & I_n & -I_n & 0_{n\times n} & 0_{n\times n_y} \\ 0_{n\times n} & 0_{n\times n} & I_n & -I_n & 0_{n\times n_y} \end{bmatrix}, \\ H &= \eta_m^2 R_0 + (\tau_M - \eta_m)^2 R_1, \\ \Xi &= [A \ 0_{n\times n} \ A_1 \ 0_{n\times n} \ B_1 \cdots B_N], \end{aligned} \tag{8.9}$$

and $\Upsilon = \mathrm{diag}\{S_0+2\alpha P, -(S_0 - S_1)e^{-2\alpha\eta_m}, 0_{n\times n}, -S_1 e^{-2\alpha\tau_M}, -b_1 I_{n_1}, \dots, -b_N I_{n_N}\}$. *Let* $\mu > 0$ *be constant and* $|\omega_i(t)| \le \mu\Delta_i$, $i = 1, \dots, N$. *Then, the solutions of system (8.5) with the initial conditions* $x_{t_0} \in W[-\tau_M, 0]$ *satisfy the following inequality for* $t \ge t_0$*:*

$$V(t, x_t, \dot{x}_t) \le e^{-2\alpha(t-t_0)} V(t_0, x_{t_0}, \dot{x}_{t_0}) + \frac{\mu^2}{2\alpha} \sum_{i=1}^{N} b_i \Delta_i^2. \tag{8.10}$$

Lemma 8.1 gives the sufficient conditions for the ISS. It will play a key role in developing the "zooming-in" algorithm for dynamic quantization. In what follows, based on Lemma 8.1 we will present the main results on dynamic quantization of NCSs. By defining the initial and level sets in Sect. 8.2.1, in Sect. 8.2.2 we will find an LMI-based time-triggered zooming algorithm (i.e., the choice of μ) for the stabilization of the closed-loop system (8.5). In Sect. 8.2.3, we will develop a novel Lyapunov-based method for the initialization of the zoom parameter.

8.2 Dynamic Quantization of NCSs

8.2.1 *Initial and Level Sets*

We first define the initial and level sets. Given a positive scalar σ, define the region of initial conditions (initial set)

$$\begin{aligned} \mathscr{W}_\sigma = \{x_{t_0} \in W[-\tau_M, 0] : V(t_0, x_{t_0}, \dot{x}_{t_0}) < \sigma, \\ x^T(t)Px(t) < \sigma, \ t \in [t_0 - \eta_M, t_0]\}. \end{aligned} \tag{8.11}$$

Define the level set

$$\mathscr{X}_{t^*,\rho} = \{x_t \in W[-\tau_M, 0] : V(t, x_t, \dot{x}_t) < \rho, \ t \ge t^*\}.$$

Given positive numbers μ, M_0, $\beta < 1$ and $\nu < 1$, we derive conditions to guarantee the following: All the solutions of (8.5) with $x_{t_0} \in \mathscr{W}_{\mu^2 M_0^2}$ will stay inside the region $\mathscr{X}_{t_0,(1+\beta\nu^2)\mu^2 M_0^2}$ for all $t \ge t_0$ and will enter a smaller region $\mathscr{X}_{t_0+\mathrm{T},\nu^2\mu^2 M_0^2}$ in a finite time T (see appendix for the proof).

Lemma 8.2 *Given* $M_j > 0$, $j = 0, 1, \dots, N$, $\Delta_i > 0, i = 1, \dots, N, 0 \le \eta_m < \tau_M$ *and tuning parameters* $\alpha > 0$, $0 < \nu < 1$, *assume that there exist scalars* $0 < \beta < 1$, b_i, $i = 1, \dots, N$, $n \times n$ *matrices* $P > 0$, $S_0 > 0$, $R_0 > 0$, $S_1 > 0$, $R_1 > 0$, S_{12}, *such that the LMIs (8.7)–(8.8) and*

$$(1 + \beta\nu^2)M_0^2 C_i^T C_i < P M_i^2, \ i = 1, \dots, N, \tag{8.12}$$

$$\frac{1}{2\alpha}\sum_{i=1}^{N} b_i \Delta_i^2 < \beta \nu^2 M_0^2 \tag{8.13}$$

hold. Let $\mu > 0$ *be constant. Then, the solutions of (8.5) that start in the region* $\mathscr{W}_{\mu^2 M_0^2}$

(i) *Satisfy* $|C_i x(t_k - \eta_k)| = |y_i(t_k - \eta_k)| < \mu M_i,\ k \in \mathbb{Z}^+$, *(implying* $|\omega_i(t)| \leq \mu \Delta_i$ *for all* $t \geq t_0,\ i = 1, \ldots, N$*),*
(ii) *Remain in the set* $\mathscr{X}_{t_0,(1+\beta\nu^2)\mu^2 M_0^2}$,
(iii) *Enter a smaller set* $\mathscr{X}_{t_0+\mathrm{T},\nu^2\mu^2 M_0^2}$ *in a finite time* T, *where* T *is the solution of*

$$e^{-2\alpha \mathrm{T}} = (1-\beta)\nu^2. \tag{8.14}$$

Note that the second inequality in (8.11) with $\sigma = \mu^2 M_0^2$ allows us to guarantee the bounds on $y(s_k),\ s_k < t_0$ by verifying (8.12).

The LMIs of Lemma 8.2 are feasible for small enough delay bound τ_M, large enough quantization ranges $M_1, \ldots, M_N$ and small enough quantization errors $\Delta_1, \ldots, \Delta_N$. Indeed, the LMIs (8.7) and (8.8) are feasible for $\tau_M = 0$ (i.e., in the absence of delay) since $A + BKC$ is Hurwitz. Hence, (8.7) and (8.8) are feasible for small enough τ_M. The LMIs (8.12) and (8.13) are feasible for large enough quantization ranges and small enough quantization errors.

8.2.2 Dynamic Quantization and Zooming Algorithm

In this section, we consider dynamic quantizers with the zoom variable μ. The zooming is performed at the sensor side. Therefore, in the closed-loop system $\mu = \mu(s_k)$ is constant on $[t_k, t_{k+1})$.

Given $\mu_0 > 0$, let $\mu = \mu_0,\ x_{t_0} \in \mathscr{W}_{\mu^2 M_0^2} = \mathscr{W}_{\mu_0^2 M_0^2}$. We will show how to choose μ_0 in Theorem 8.1. Assume that the LMIs of Lemma 8.2 are feasible. We suggest a "zooming-in" algorithm shown in Fig. 8.2, where μ is decreased and, thus, the resulting quantization error is reduced in such a way as to drive the state of (8.5) to the origin exponentially.

Definition 8.1 The system (8.5) with $|\omega_i(t)| \leq \mu\Delta_i,\ i = 1, \ldots, N$, is said to be exponentially stable for some choice of the zoom variable $\mu > 0$ if there exist constants $b > 0,\ \kappa > 0$ such that

$$|x(t)|^2 \leq b e^{-2\kappa(t-t_0)} \mu_0^2 M_0^2,\ \forall t \geq t_0$$

for the solutions of the system (8.5) initialized with $x_{t_0} \in \mathscr{W}_{\mu_0^2 M_0^2}$.

Proposition 8.1 *Assume that the LMIs of Lemma 8.2 are feasible. Given* $\mu_0 > 0$, *let* $\mu = \mu_0,\ x_{t_0} \in \mathscr{W}_{\mu_0^2 M_0^2}$. *Then under the algorithm in Fig. 8.2 with the "zooming-in"*

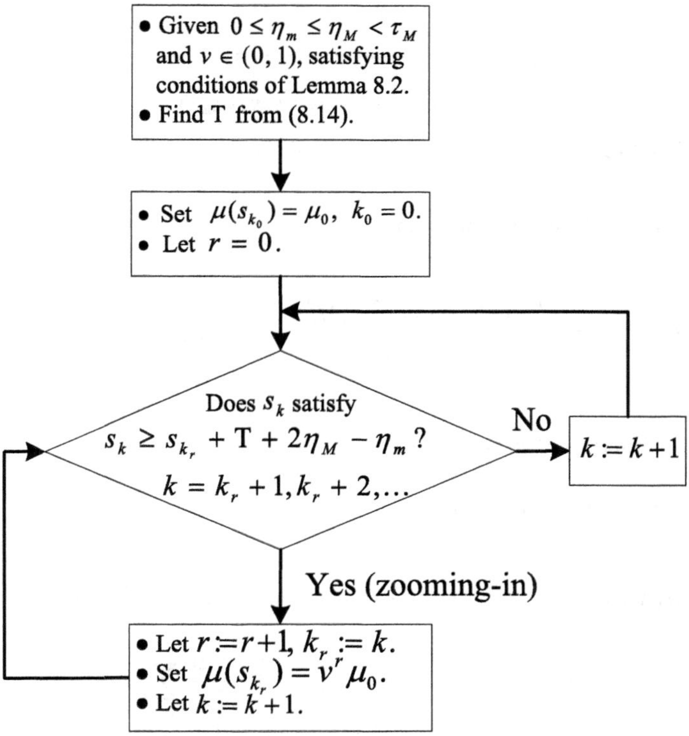

Fig. 8.2 "Zooming-in" algorithm for dynamic quantization

instants shown in Fig. 8.3, the system (8.5) is exponentially stable with a decay rate $\kappa = -\frac{\ln \nu}{\mathrm{T}+\tau_M+2\eta_M-2\eta_m}$.

Proof Set $r = 0$. Since

$$t_{k_1} - \eta_M = s_{k_1} + \eta_{k_1} - \eta_M \ge t_0 + \mathrm{T} + \eta_{k_1} - \eta_m \ge t_0 + \mathrm{T},$$

the application of Lemma 8.2 with $\mu = \mu_0$ leads to

$$x^T(t)Px(t) \le V(t, x_t, \dot{x}_t) < \nu^2 \mu_0^2 M_0^2, \ \forall t \ge t_{k_1} - \eta_M.$$

Set $r = 1$. After zooming-in at s_{k_1}, the resulting closed-loop system has the initial condition

$$x_{t_{k_1}} \in W[-\tau_M, 0] : V(t_{k_1}, x_{t_{k_1}}, \dot{x}_{t_{k_1}}) < \nu^2 \mu_0^2 M_0^2. \tag{8.15}$$

Then, Lemma 8.2 is applied with $\mu = \mu_0 \nu$, where t_0 and η_0 are changed by t_{k_1} and η_{k_1}, respectively. Thus, the solutions of (8.5) initiated by (8.15) remain in a region $\mathscr{X}_{t_{k_1},(1+\beta\nu^2)\nu^2\mu_0^2 M_0^2}$ for all $t \ge t_{k_1}$. Since $s_{k_1} = t_{k_1} - \eta_{k_1} \ge t_{k_1} - \eta_M$, from (8.12) it follows that

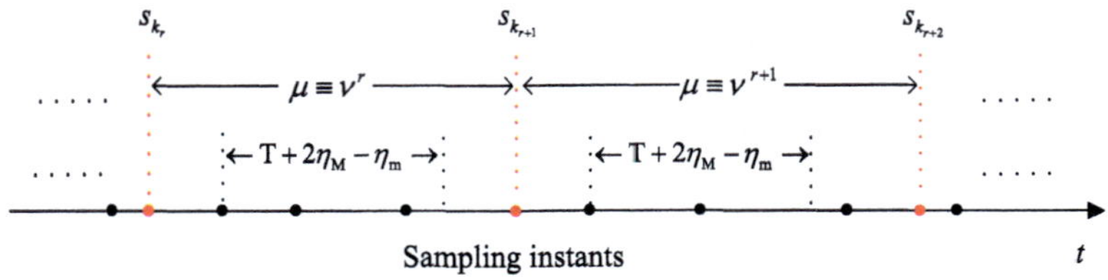

Fig. 8.3 Illustration of the "zooming-in" instants

$$x^T(s_k)C_i^T C_i x(s_k) < \frac{x^T(s_k)Px(s_k)\cdot \nu^2\mu_0^2 M_i^2}{(1+\beta\nu^2)\nu^2\mu_0^2 M_0^2} < \nu^2\mu_0^2 M_i^2,\ i{=}1,\ldots,N,\ \forall k \geq k_1,$$

and thus,

$$|\omega_i(t)| \leq \nu\mu_0\Delta_i,\ i=1,\ldots,N,\ t\geq t_{k_1}=s_{k_1}+\eta_{k_1}.$$

Therefore, for $t \geq t_{k_2}-\eta_M \geq t_{k_1}+\mathrm{T}$,

$$\begin{aligned}
V(t,x_t,\dot{x}_t) &\leq e^{-2\alpha(t-s_{k_1}-\eta_{k_1})}V(t,x_t,\dot{x}_t)_{|t=s_{k_1}+\eta_{k_1}} + \frac{\nu^2\mu_0^2}{2\alpha}\sum_{i=1}^{N} b_i\Delta_i^2 \\
&\leq e^{-2\alpha\mathrm{T}}V(t,x_t,\dot{x}_t)_{|t=s_{k_1}+\eta_{k_1}} + \frac{\nu^2\mu_0^2}{2\alpha}\sum_{i=1}^{N} b_i\Delta_i^2 \\
&< (1-\beta)\nu^2\cdot\nu^2\mu_0^2M_0^2 + \beta\nu^2\mu_0^2M_0^2\cdot\nu^2 \\
&= \nu^4\mu_0^2M_0^2.
\end{aligned}$$

Similarly, for $r=2,3,\ldots$ we have $V(t,x_t,\dot{x}_t) < \nu^{2r}\mu_0^2M_0^2$ for all $t\in[t_{k_r}-\eta_M, t_{k_{r+1}}-\eta_M)$. Noting that

$$\begin{aligned}
&r\mathrm{T}+(r-1)(2\eta_M-\eta_m)+t_0 \leq t_{k_r}-\eta_M \leq t \leq \\
&t_{k_{r+1}}-\eta_M < (r+1)(\mathrm{T}+\tau_M+2\eta_M-2\eta_m)+t_0,
\end{aligned}$$

we obtain

$$\begin{aligned}
V(t,x_t,\dot{x}_t) &< \nu^{2r}\mu_0^2M_0^2 < \nu^{2[\frac{t-t_0}{\mathrm{T}+\tau_M+2\eta_M-2\eta_m}-1]}\mu_0^2M_0^2 \\
&= \nu^{-2}e^{\frac{2\ln\nu}{\mathrm{T}+\tau_M+2\eta_M-2\eta_m}\cdot(t-t_0)}\mu_0^2M_0^2,\ t\in[t_{k_r}-\eta_M, t_{k_{r+1}}-\eta_M),\ r\in\mathbb{N}.
\end{aligned}$$

Then, the following inequality holds for $t\geq t_0$

$$|x(t)|^2 \leq \nu^{-2}[\lambda_{\min}(P)]^{-1}e^{\frac{2\ln\nu}{\mathrm{T}+\tau_M+2\eta_M-2\eta_m}\cdot(t-t_0)}\mu_0^2M_0^2,$$

which completes the proof.

For the implementation of the "zooming-in" algorithm, we set a counter $\theta(t)$ at the sensor in terms of sampling instants s_k. The counter triggers the zooming-in whenever $\theta(s_k) \geq \mathrm{T} + 2\eta_M - \eta_m$. At the triggering times ($t = s_{k_r}$), it is reset to zero.

Remark 8.1 In the above reasoning, we assumed that the packet loss does not occur. However, if the number of successive packet dropouts is upper bounded by $\bar{d}$, we could accommodate the packet dropouts by modeling them as prolongations of the transmission interval and replace T by $\mathrm{T} + 2\bar{d} \cdot \mathrm{MATI}$ in the algorithm.

8.2.3 *Initialization of the Zoom Variable*

The algorithm of the previous section is given in terms of the initial set $\mathscr{W}_{\mu_0^2 M_0^2}$ that involves the bound on $V(t_0, x_{t_0}, \dot{x}_{t_0})$. In this section, we find the ball of initial conditions $x(0) = x_0$, starting from which the solutions of (8.4)–(8.5) remain in the initial set $\mathscr{W}_{\mu_0^2 M_0^2}$. Following [130], we derive a bound on $V(t_0, x_{t_0}, \dot{x}_{t_0})$ in terms of x_0 in the next lemma.

Lemma 8.3 ([130]) *Consider LKF $\bar{V}(t) = V(t, x_t, \dot{x}_t)$ given by (8.6) and denote $V_0(t) = x^T(t)Px(t)$. Under the constant initial condition $x(t) = x_0, t < 0$, let there exist $\alpha > 0$ and $\delta > 0$ such that the following inequalities*

$$\dot{V}_0(t) - 2\delta V_0(t) \leq 0, \tag{8.16a}$$

$$\dot{\bar{V}}(t) + 2\alpha \bar{V}(t) - 2\delta V_0(t) \leq 0, \tag{8.16b}$$

hold for $0 \leq t < t_0$ along (8.4). Then, we have

$$\begin{aligned} V_0(t) &\leq \lambda_{\max}(e^{2\delta\eta_M} P)|x_0|^2, \ 0 \leq t < t_0, \\ \bar{V}(t_0) &\leq \lambda_{\max}(e^{2\delta\eta_M} P + \Omega)|x_0|^2, \end{aligned} \tag{8.17}$$

where

$$\Omega = \eta_m S_0 + e^{-2\alpha\eta_m}(\tau_M - \eta_m)S_1. \tag{8.18}$$

We are in a position to formulate our main result.

Theorem 8.1 *Given $M_j > 0,\ j = 0, 1, \ldots, N,\ \Delta_i > 0,\ i = 1, \ldots, N,\ 0 \leq \eta_m \leq \eta_M < \tau_M$ and tuning parameters $\alpha > 0,\ 0 < \nu < 1,\ \delta > 0$, assume that there exist scalars $0 < \beta < 1, b_i,\ i = 1, \ldots, N$, $n \times n$ matrices $P > 0,\ S_0 > 0,\ R_0 > 0,\ S_1 > 0,\ R_1 > 0,\ S_{12}$, such that LMIs (8.7)–(8.8), (8.12)–(8.13) and the following LMIs*

$$\tilde{\Psi}_1 = P(A - \delta I_n) + (A - \delta I_n)^T P < 0, \tag{8.19}$$

$$\tilde{\Psi}_2 = \begin{bmatrix} \tilde{\Sigma} - \tilde{F}^T \Phi \tilde{F} e^{-2\alpha\tau_M} & \tilde{\Xi}^T H \\ * & -H \end{bmatrix} < 0, \tag{8.20}$$

are feasible, where

$$
\begin{aligned}
\tilde{\Sigma} &= \mathrm{He}(\tilde{F}_1^T P \tilde{\Xi}) + \tilde{\Upsilon} - \tilde{F}_2^T R_0 \tilde{F}_2 e^{-2\alpha\eta_m}, \\
\tilde{F}_1 &= [I_n \ 0_{n\times 3n}], \ \tilde{F}_2 = [I_n \ -I_n \ 0_{n\times 2n}], \ \tilde{\Xi} = [A \ 0_{n\times 3n}], \\
\tilde{F} &= \begin{bmatrix} 0_{n\times n} & I_n & -I_n & 0_{n\times n} \\ 0_{n\times n} & 0_{n\times n} & I_n & -I_n \end{bmatrix}, \\
\tilde{\Upsilon} &= \mathrm{diag}\{S_0 + 2\alpha P - 2\delta P, -(S_0 - S_1)e^{-2\alpha\eta_m}, 0_{n\times n}, -S_1 e^{-2\alpha\tau_M}\},
\end{aligned}
\tag{8.21}
$$

and the notations Φ and H are given by (8.7) and (8.9), respectively. Assume that the initial condition satisfies the inequality $|x_0| < X_0$, where $X_0 > 0$ is known. Then, the "zooming-in" algorithm of Sect. 8.2.2 starting with $\mu(s_0) = \mu_0$ with μ_0 given by

$$
\mu_0^2 = \frac{\lambda_{\max}(e^{2\delta\eta_M} P + \Omega)}{M_0^2} X_0^2 \tag{8.22}
$$

exponentially stabilizes system (8.4)–(8.5), where Ω is given by (8.18).

Proof As follows from [130], the LMIs (8.7), (8.19) and (8.20) guarantee (8.16) along (8.4) for $0 \le t < t_0$. Therefore, if the initial condition satisfies the inequality $|x_0| < X_0$, then we have

$$
\begin{aligned}
\max\{V_0(t), \bar{V}(t_0)\} &\le \lambda_{\max}(e^{2\delta\eta_M} P + \Omega) X_0^2 \\
&= \mu_0^2 M_0^2, \ t \in [0, t_0],
\end{aligned}
$$

meaning that $x_{t_0} \in \mathscr{X}_{\mu_0^2 M_0^2}$. The result then follows from Proposition 8.1.

Remark 8.2 Given $\alpha > 0$, there always exists $\delta > 0$ such that $A - (\delta - \alpha)I_n$ is Hurwitz; i.e., $\dot{V}_0 + 2(\alpha + \varepsilon - \delta)V_0 \le 0$ holds for some $P > 0$ with small enough $\varepsilon > 0$. Then, LMI (8.19) is feasible, and for small enough $\tau_M > 0$, by the standard arguments for delay-dependent methods (see, e.g., [47]) the LMIs (8.7) and (8.20) are satisfied.

Remark 8.3 Note that given a bound $X_0 > 0$ on the initial state conditions and the values of the quantizer range $M_i > 0$ and the error $\Delta_i > 0$, $i = 1, \ldots, N$, the Eq. (8.22) defines the initial value of the zoom variable, starting from which the exponential stability is guaranteed by using "zooming-in" only.

Remark 8.4 If the initial value of the zoom variable is given by μ_0, then the "zooming-in" algorithm starting with $\mu(s_0) = \mu_0$ exponentially stabilizes all the solutions of (8.4)–(8.5) starting from the initial ball

$$
|x_0| < X_0, \quad X_0 = \frac{\mu_0 M_0}{\sqrt{\lambda_{\max}(e^{2\delta\eta_M} P + \Omega)}}. \tag{8.23}
$$

In order to maximize the initial ball (8.23), i.e., to minimize $\lambda_{\max}(e^{2\delta\eta_M} P + \Omega)$, the condition $e^{2\delta\eta_M} P + \Omega - \gamma I_n < 0$ can be added to the conditions of Theorem 8.1, where $\gamma > 0$ is to be minimized.

Remark 8.5 Consider the case where all the conditions of Theorem 8.1 are satisfied, but the initial ball $|x_0| < \bar{X}_0$ is larger: $\bar{X}_0 > X_0$, where X_0 is given by (8.23). Then, we change μ_0 in the algorithm by $\bar{\mu}_0 = M_0^{-1}\bar{X}_0\sqrt{\lambda_{\max}(e^{2\delta\eta_M}P + \Omega)}$ and zoom out by resetting $\bar{M}_i = \bar{\mu}_0 M_i$, $\bar{\Delta}_i = \bar{\mu}_0\Delta_i$, $i = 1, \ldots, N$. Therefore, we can start with the quantizer $q_{i\bar{\mu}_0}(z_i)$ (corresponding to $\mu(s_0) = \bar{\mu}_0$) whose range and quantization error are given by $\bar{M}_i$ and $\bar{\Delta}_i$, $i = 1, \ldots, N$, respectively. After this initial "zooming-out," "zooming-in" is used as suggested in the algorithm of Sect. 8.2.2. This "zooming-in"–"zooming-out" algorithm was originally proposed by Liberzon [123].

Since the LMIs (8.8), (8.19) and (8.20) are affine in the system matrices, the conditions of Theorem 8.1 can be applied to the case where these matrices are uncertain. Next, consider system (8.1) with the polytopic-type uncertainties. Denote $\Theta = [A \;\; B]$ and assume that

$$\Theta = \sum_{j=1}^{M} g_j(t)\Theta_j, \quad 0 \le g_j(t) \le 1, \quad \sum_{j=1}^{M} g_j(t) = 1, \tag{8.24}$$

where g_j are uncertain time-varying parameters and where the M vertices of the polytope are described by $\Theta_j = [A^{(j)} \;\; B^{(j)}]$, $j = 1, \ldots, M$.

Suppose that the following LMIs for $j = 1, \ldots, M$, are feasible with the same decision matrices:

$$\Psi^{(j)} = \begin{bmatrix} \Sigma^{(j)} - F^T\Phi F e^{-2\alpha\tau_M} & (\Xi^{(j)})^T H \\ * & -H \end{bmatrix} < 0, \tag{8.25}$$

$$\tilde{\Psi}_1^{(j)} = P(A^{(j)} - \delta I_n) + (A^{(j)} - \delta I_n)^T P < 0, \tag{8.26}$$

$$\tilde{\Psi}_2^{(j)} = \begin{bmatrix} \tilde{\Sigma}^{(j)} - \tilde{F}^T\Phi\tilde{F}e^{-2\alpha\tau_M} & (\tilde{\Xi}^{(j)})^T H \\ * & -H \end{bmatrix} < 0, \tag{8.27}$$

where

$$\begin{aligned} A_1^{(j)} &= B^{(j)}KC, \; B_i^{(j)} = B^{(j)}K_i, \; i = 1, \ldots, N, \\ \Sigma^{(j)} &= \mathrm{He}(F_1^T P \Xi^{(j)}) + \Upsilon - F_2^T R_0 F_2 e^{-2\alpha\eta_m}, \\ \Xi^{(j)} &= [A^{(j)} \, 0_{n\times n} \, A_1^{(j)} \, 0_{n\times n} \, B_1^{(j)} \cdots B_N^{(j)}], \\ \tilde{\Sigma}^{(j)} &= \mathrm{He}(\tilde{F}_1^T P \tilde{\Xi}^{(j)}) + \tilde{\Upsilon} - \tilde{F}_2^T R_0 \tilde{F}_2 e^{-2\alpha\eta_m}, \\ \tilde{\Xi}^{(j)} &= [A^{(j)} \, 0_{n\times 3n}], \end{aligned} \tag{8.28}$$

and where the notations are given by (8.7), (8.9), (8.21). Then, we obtain

$$\sum_{j=1}^{M} g_j(t)\Psi^{(j)} = \Psi < 0 \;\text{ and}$$
$$\sum_{j=1}^{M} g_j(t)\tilde{\Psi}_i^{(j)} = \tilde{\Psi}_i < 0, \; i = 1, 2,$$

which mean that (8.8), (8.19) and (8.20) are feasible. The following statement holds.

Theorem 8.2 *Given* $M_j > 0,\ j = 0, 1, \ldots, N,\ \Delta_i > 0,\ i = 1, \ldots, N,\ 0 \le \eta_m \le \eta_M < \tau_M$ *and tuning parameters* $\alpha > 0,\ 0 < \nu < 1,\ \delta > 0$, *assume that there exist scalars* $0 < \beta < 1,\ b_i,\ i = 1, \ldots, N$, $n \times n$ *matrices* $P > 0,\ S_0 > 0,\ R_0 > 0,\ S_1 > 0,\ R_1 > 0,\ S_{12}$, *such that LMIs (8.7), (8.12)–(8.13), (8.25)–(8.27) are feasible for* $j = 1, \ldots, M$, *where the notations are given by (8.7), (8.9), (8.21) and (8.28). Assume that the initial condition satisfies the inequality* $|x_0| < X_0$, *where* $X_0 > 0$ *is known. Then, the "zooming-in" algorithm of Sect. 8.2.2 starting with* $\mu(s_0) = \mu_0$ *with* μ_0 *given by (8.22) exponentially stabilizes the uncertain system (8.4)–(8.5) with (8.24).*

Remark 8.6 Theorems 8.1 and 8.2 focus on stability analysis of NCSs with dynamic quantization, variable communication delays and variable sampling intervals. For the static output-feedback stabilization problem, one possible solution is to apply the approach introduced in [128] together with the descriptor method.

8.3 Numerical Example: Uncertain Inverted Pendulum

In this section, an inverted pendulum mounted on a car given in Example 5.1 is to illustrate the effectiveness of the proposed techniques in the previous sections. The system matrices A and B are given by (5.43). Note that A can be described by a polytope with four vertices. The pendulum can be stabilized by a state feedback $u(t) = Kx(t) = YQ^{-1}x(t)$, where $Y \in \mathbb{R}^{1\times 4}$ and $0 < Q \in \mathbb{R}^{4\times 4}$ satisfy

$$AQ + QA^T + 2\alpha Q + BY + Y^T B^T < 0 \tag{8.29}$$

in the vertices of polytope for a tuning parameter $\alpha > 0$.

Consider $N = 2$ and

$$C_1 = \begin{bmatrix} 1\,0\,0\,0 \\ 0\,1\,0\,0 \end{bmatrix},\ C_2 = \begin{bmatrix} 0\,0\,1\,0 \\ 0\,0\,0\,1 \end{bmatrix}.$$

The quantizer is chosen as

$$q_\mu(y^i) = \begin{cases} 100\mu\ \mathrm{sgn}(y^i), & \text{if } |y^i| > 100\mu, \\ \mu\left\lfloor \frac{y^i}{\mu} + 0.1 \right\rfloor, & \text{if } |y^i| \le 100\mu, \end{cases}$$

where y^i is the ith component of y, $i = 1, \ldots, 4$. Therefore, we can take $M_1 = M_2 = 100$, $\Delta_1 = \Delta_2 = 0.1$. Choose $\mu_0 = 1$, $M_0 = 100$, $\nu = 0.8$, $\delta = 10$.

First, choosing $\alpha = 0.3$, from (8.29) we obtain the controller gain

$$K = [K_1\ \ K_2], \tag{8.30}$$

where

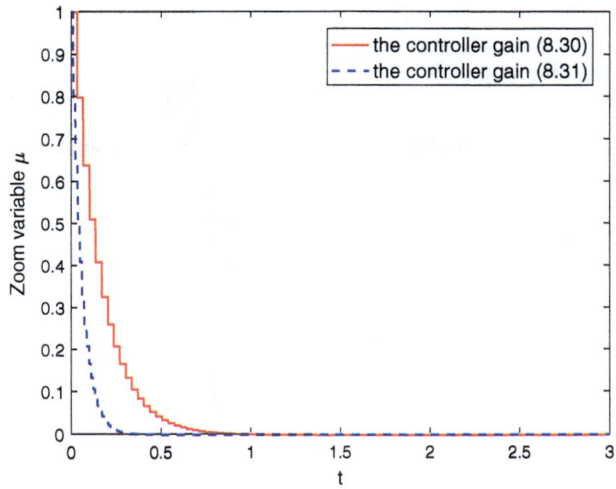

Fig. 8.4 Evolution of the zoom variable μ in the "zooming-in" algorithm

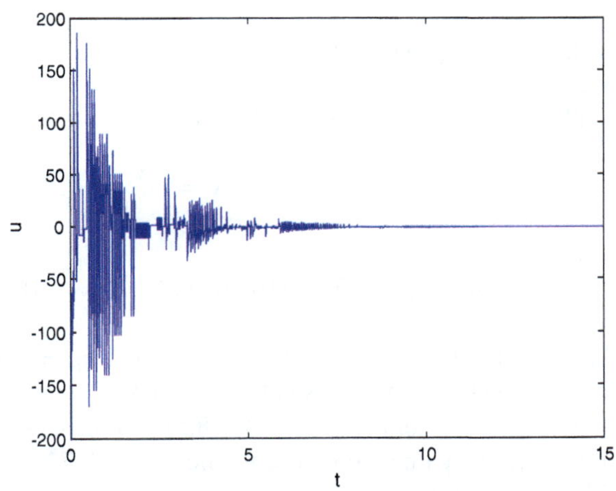

Fig. 8.5 Evolution of the control input with the gain (8.30)

$$K_1 = \begin{bmatrix} 25.1319 & -222.9722 \end{bmatrix}, \ K_2 = \begin{bmatrix} 28.7826 & -44.2075 \end{bmatrix}.$$

The application of Theorem 8.2 with $\tau_M = 0.02$, $\eta_m = 0.011$, $\eta_M = 0.015$ leads to $\mathrm{T} = -\frac{\ln(1-\beta)+2\ln\nu}{2\alpha} = 1.0002$ from (8.14). Then, the "zooming-in" algorithm of Sect. 8.2.2 with $\mathrm{T} = 1.0002$ and $\nu = 0.8$ exponentially stabilizes all the solutions of (8.4)–(8.5) with (8.24) starting from the initial ball $|x_0| < 5.2693$. The evolution of the zoom variable μ is shown in Fig. 8.4.

Moreover, we find that the system is exponentially stable with a decay rate $\kappa = -\frac{\ln\nu}{\mathrm{T}+\tau_M+2\eta_M-2\eta_m} = 0.2170$. Let the initial state $x_0 = [1\ 3\ 2\ -1]^T$. The evolution of the control input and the state is shown in Figs. 8.5 and 8.6, respectively.

If all the conditions of Theorem 8.2 are satisfied, but the initial ball is $|x_0| < 15$, which is out of $|x_0| < 5.2693$, we substitute $\mu_0 = 15/5.2693 = 2.85$ for 1 and zoom out by resetting $\bar{M}_i = \mu_0 M_i = 285$, $\bar{\Delta}_i = \mu_0 \Delta_i = 0.285$, $i = 1, 2$. After this

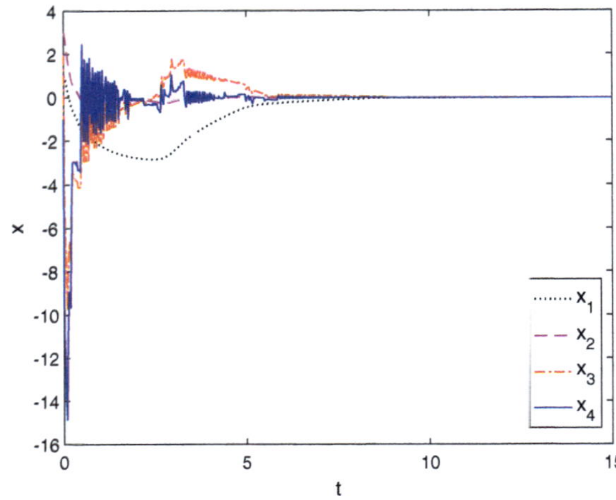

Fig. 8.6 Trajectory of the closed-loop system with the controller gain (8.30)

initial "zooming-out," "zooming-in" is used by Theorem 8.2 and the algorithm of Sect. 8.2.2.

Next, taking a larger $\alpha = 0.7$, from (8.29) we have another controller gain

$$\bar{K} = [\bar{K}_1 \;\; \bar{K}_2], \tag{8.31}$$

where

$$\bar{K}_1 = \begin{bmatrix} 15.6527 & -105.6658 \end{bmatrix}, \; \bar{K}_2 = \begin{bmatrix} 16.0894 & -22.0086 \end{bmatrix}.$$

We find that given $\tau_M = 0.02$, $\eta_m = 0.011$, $\eta_M = 0.015$, the "zooming-in" algorithm of Sect. 8.2.2 with a smaller $\mathrm{T} = 0.3457$ and the same $\nu = 0.8$ exponentially stabilizes all the solutions of (8.4), (8.5) and (8.24) starting from a larger initial ball $|x_0| < 6.2782$ with a larger decay rate $\kappa = 0.5972$. The evolution of the zoom variable μ, the control input and the state with the initial state x_0 is shown in Figs. 8.4, 8.7 and 8.8, respectively (confirming the theoretical results).

The simulation results listed above show that the larger value of α and the resulting controller gain give rise to a larger decay rate. The choice of α to be 1.58 and the corresponding controller gain $\tilde{K} = [\tilde{K}_1 \;\; \tilde{K}_2]$, where $\tilde{K}_1 = [39.8549 \quad -141.5345]$, $\tilde{K}_2 = [30.1820 \quad -30.7271]$, lead to the maximum value of the decay rate $\kappa = 1.0689$.

8.4 Notes

In this chapter, we proposed a time-delay approach to the stability analysis of uncertain linear NCSs with dynamic quantization, variable communication delays and variable sampling intervals. An LMI-based time-triggered zooming algorithm was given for the dynamic quantization that leads to the exponential stability of the closed-

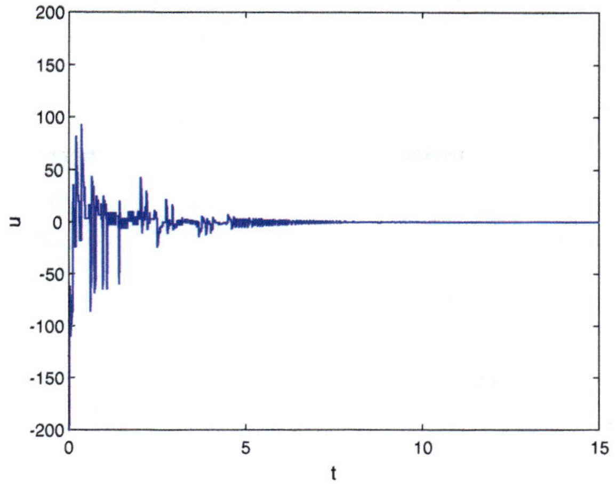

Fig. 8.7 Evolution of the control input with the gain (8.31)

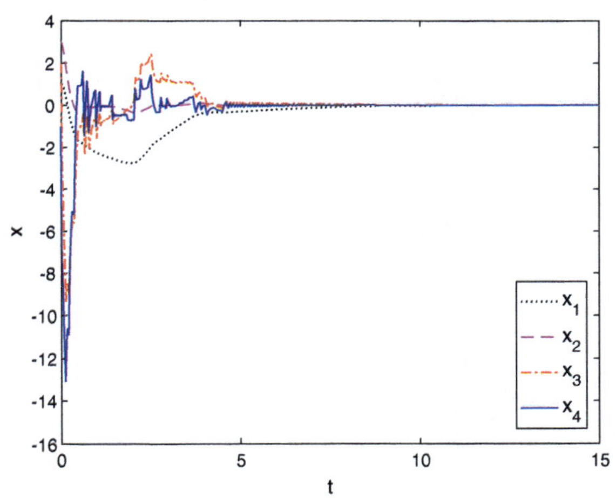

Fig. 8.8 Trajectory of the closed-loop system with the controller gain (8.31)

loop system. A novel Lyapunov-based method was provided for the initialization of the zoom parameter.

An improved "zooming-in" algorithm was proposed in [234] to find a shorter time T and, thus, to bring faster convergence of the closed-loop system.

Chapter 9
Discrete-Time Network-Based Control Under Scheduling and Actuator Constraints

In Chaps. 2–8 we have studied networked control for continuous-time plants. Chapters 9–12 will present our analysis on networked control for discrete-time plants. In this chapter, we consider discrete-time NCSs with communication constraints and actuator constraints. Systems with actuator constraints were extensively studied during the 1960s due to their intimate connection with optimal control. Concurrently, the design approaches, such as the describing function method, which dealt specifically with nonlinearities such as saturation were developed. Only very limited research into actuator saturation was carried out during the 1970s and 1980s with the emphasis being placed mostly on the development of the linear state space approach and its numerous offshoots. This situation changed during the late 1980s and early 1990s (see [11] for an extensive bibliography of the work carried out during this period) and has continued apace to the present time (see [182] for later developments). In terms of stabilization, the research can be classified as: global, semi-global (that guarantees that any given compact set of initial conditions, no matter how large, can be included in the domain of attraction of the closed-loop system) and local or regional (that estimates the domain of attraction). The main drawback with the global and the semi-global stabilizability approach lies in the requirements for the open-loop poles to be located in the closed left-half plane (see, e.g., [181]). To relax these assumptions, the regional stabilization has been investigated (see, e.g., [52, 100, 222]). The stabilization of sampled-data systems under variable samplings and actuator saturation was studied in [196], where scheduling protocols and delays were not included.

On the other hand, as shown in [130], when one deals with the solution bounds of time-delay systems via Lyapunov–Krasovskii method, the first time interval of the delay length needs a special analysis. The solution bounds are widely used for systems with input saturation caused by actuator saturation or by the quantizers with saturation. This first time interval does not influence on the stability and the

K. Liu et al., *Networked Control Under Communication Constraints*, Advances in Delays and Dynamics 11, https://doi.org/10.1007/978-981-15-4230-5_9

exponential decay rate analysis. The analysis of the first time interval of the delay length is important for nonlinear systems, e.g., for finding the domain of attraction.

In this chapter, we focus on the regional stability analysis of discrete-time NCSs in the presence of actuator saturation under Round-Robin or under TOD protocol. A linear (probably, uncertain) system with distributed sensors is considered. Following [130], we propose a direct Lyapunov approach to find the domain of attraction under both scheduling protocols. Throughout this chapter, for any matrix $A \in \mathbb{R}^{n\times n}$ and vector $x \in \mathbb{R}^n$, the notations A_j and x_j denote, respectively, the jth line of matrix A and the jth component of vector x. Given $\bar{u} = [\bar{u}_1, \ldots, \bar{u}_{n_u}]^T$, $0 < \bar{u}_i$, $i = 1, \ldots, n_u$, for any $u = [u_1, \ldots, u_{n_u}]^T$, we denote by $\text{sat}(u)$ the vector with coordinates $\text{sign}(u_i)\min(|u_i|, \ \bar{u}_i)$.

9.1 Discrete-Time NCSs with Actuator Saturation Under Round-Robin Protocol

9.1.1 *Problem Formulation and a Switched System Model*

Consider the system architecture in Fig. 9.1 with plant

$$x(t+1) = Ax(t) + Bu(t), \quad t \in \mathbb{Z}^+, \tag{9.1}$$

where $x(t) \in \mathbb{R}^n$ is the state vector, $u(t) \in \mathbb{R}^{n_u}$ is the control input, A and B are (probably, uncertain) system matrices with appropriate dimensions. The initial condition is given by $x(0) = x_0$. We suppose that the control input is subject to the following amplitude constraints

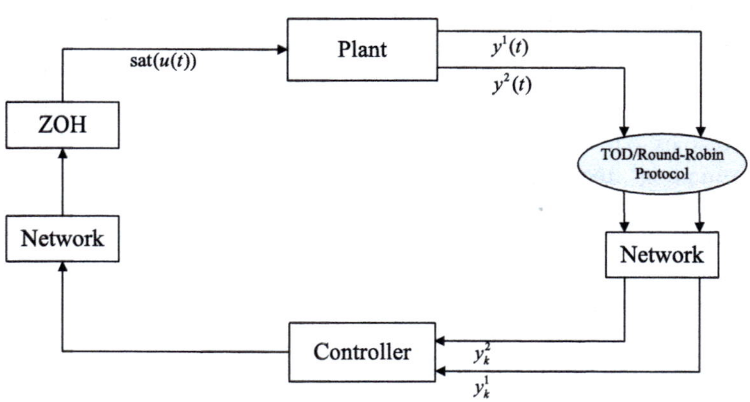

Fig. 9.1 NCSs with actuator saturation under Round-Robin protocol

$$|u_i(t)| \leq \bar{u}_i,\ 0 < \bar{u}_i,\ i = 1, \ldots, n_u,\ t \in \mathbb{Z}^+. \tag{9.2}$$

The NCS has several nodes (distributed sensors, a controller node and an actuator node) which are connected via networks. For the sake of simplicity, we consider two sensor nodes $y^i(t) = C^i x(t)$, $i = 1, 2$, and we denote

$$C = \begin{bmatrix} C^1 \\ C^2 \end{bmatrix},\ y(t) = \begin{bmatrix} y^1(t) \\ y^2(t) \end{bmatrix} \in \mathbb{R}^{n_y},\ t \in \mathbb{Z}^+.$$

Denote by s_k the unbounded and monotonously increasing sequence of sampling instants, i.e.,

$$0 = s_0 < s_1 < \cdots < s_k < \cdots,\ k \in \mathbb{Z}^+,\ \lim_{k\to\infty} s_k = \infty,\ s_{k+1} - s_k \leq \text{MATI}, \tag{9.3}$$

where $\{s_0, s_1, s_2, \ldots\}$ is a subsequence of $\mathbb{Z}^+$. At each sampling instant s_k, one of the outputs $y^i(t) \in \mathbb{R}^{n_i}$ $(n_1 + n_2 = n_y)$ is sampled and transmitted via the network. Firstly, we consider Round-Robin scheduling protocol for the choice of the active output node: The outputs are transmitted one after another, i.e., $y^i(t) = C^i x(t)$, $t \in \mathbb{Z}^+$, is transmitted only at the sampling instant $t = s_{2p+i-1}$, $p \in \mathbb{Z}^+$, $i = 1, 2$. After each transmission and reception, the values in $y^i(t)$ are updated with the newly received values, while the values of $y^j(t)$ for $j \neq i$ remain the same, as no additional information is received. This leads to the constrained data exchange expressed as

$$y_k^i = \begin{cases} y^i(s_k) = C^i x(s_k), & k = 2p + i - 1, \\ y_{k-1}^i, & k \neq 2p + i - 1, \end{cases} \quad p \in \mathbb{Z}^+.$$

It is assumed that no packet dropouts and no packet disorders will happen during the data transmission over the network. The transmission of the information over the two networks (between the sensor and the actuator) is subject to a variable delay $\eta_k = \eta_k^{sc} + \eta_k^{ca} \in \mathbb{Z}^+$, where η_k^{sc} and η_k^{ca} are the network-induced delays from the sensor to the controller and from the controller to the actuator, respectively. Then, $t_k = s_k + \eta_k$ is the updating time instant of the ZOH device.

As in the previous chapters, we allow the delays to be non-small provided that the old sample cannot get to the destination (to the controller or to the actuator) after the current one. Assume that the network-induced delay η_k and the time span between the updating and the current sampling instants are bounded:

$$t_{k+1} - 1 - t_k + \eta_k \leq \tau_M,\ 0 \leq \eta_m \leq \eta_k \leq \eta_M,\ k \in \mathbb{Z}^+, \tag{9.4}$$

where τ_M, η_m and η_M are known non-negative integers. Then, we have

$$\begin{aligned} &(t_{k+1} - 1) - s_k = s_{k+1} - s_k + \eta_{k+1} - 1 \leq \text{MATI} + \eta_M - 1 = \tau_M, \\ &(t_{k+1} - 1) - s_{k-1} = s_{k+1} - s_{k-1} + \eta_{k+1} - 1 \leq 2\,\text{MATI} + \eta_M - 1 = 2\tau_M - \eta_M + 1 \triangleq \bar{\tau}_M, \\ &t_{k+1} - t_k \leq \tau_M - \eta_m + 1. \end{aligned} \tag{9.5}$$

In this section, we consider the stability analysis of discrete-time NCSs with actuator saturation under Round-Robin protocol. Due to the control bounds defined in (9.2), the effective control signal to be applied to the system (9.1) is given by

$$u(t) = \mathrm{sat}(K^1 y_k^1 + K^2 y_k^2),\ t \in \left[t_k, t_{k+1} - 1\right],\ t \in \mathbb{N},\ k \in \mathbb{N},$$

where $K = [K^1\ K^2]$, $K^1 \in \mathbb{R}^{n_u \times n_1}$, $K^2 \in \mathbb{R}^{n_u \times n_2}$ such that $A + BKC$ is Schur.

We define the polyhedron

$$\mathscr{P}(K^j, \bar{u}) = \left\{x \in \mathbb{R}^n : |(K^j C^j)_i x| \le \frac{1}{2}\bar{u}_i,\ i = 1, \dots, n_u\right\},\ j = 1, 2.$$

If the control is such that $x(t) \in \mathscr{P}(K^1, \bar{u}) \cap \mathscr{P}(K^2, \bar{u})$, then $|(K^1 C^1)_i x + (K^2 C^2)_i x| \le \bar{u}_i$. We follow (4.7) and model the closed-loop system with Round-Robin protocol as the following switched system:

$$\begin{cases} x(t+1) = Ax(t) + A_1 x(t_{k-1} - \eta_{k-1}) + A_2 x(t_k - \eta_k),\ t \in [t_k, t_{k+1} - 1], \\ x(t+1) = Ax(t) + A_1 x(t_{k+1} - \eta_{k+1}) + A_2 x(t_k - \eta_k),\ t \in [t_{k+1}, t_{k+2} - 1], \end{cases} \tag{9.6}$$

where $k = 2p - 1$, $p \in \mathbb{N}$, $A_i = BK^i C^i$, $i = 1, 2$.

For $t \in [t_k, t_{k+1} - 1]$, we can represent $t_k - \eta_k = t - \tau_1(t)$, $t_{k-1} - \eta_{k-1} = t - \tau_2(t)$, where

$$\begin{aligned} &\tau_1(t) = t - t_k + \eta_k < \tau_2(t) = t - t_{k-1} + \eta_{k-1}, \\ &\tau_1(t) \in [\eta_m, \tau_M],\ \tau_2(t) \in [\eta_m,\ \bar{\tau}_M],\ t \in [t_k, t_{k+1} - 1]. \end{aligned}$$

Therefore, (9.6) for $t \in [t_k, t_{k+1} - 1]$ can be considered as a system with two time-varying interval delays, where $\tau_1(t) < \tau_2(t)$. Similarly, for $t \in [t_{k+1}, t_{k+2} - 1]$ (9.6) is a system with two time-varying delays, one of which is less than another.

9.1.2 Solution Bounds

We apply the following discrete-time LKF to system (9.6) with time-varying delay from the maximum delay interval $[\eta_m, \bar{\tau}_M]$:

$$\begin{aligned} V_{RR}(t) = {} & x^T(t) P x(t) + \sum_{s=t-\eta_m}^{t-1} \lambda^{t-s-1} x^T(s) S_0 x(s) \\ & + \eta_m \sum_{j=-\eta_m}^{-1} \sum_{s=t+j}^{t-1} \lambda^{t-s-1} \eta^T(s) R_0 \eta(s) + \sum_{s=t-\bar{\tau}_M}^{t-\eta_m-1} \lambda^{t-s-1} x^T(s) S_1 x(s) \\ & + (\bar{\tau}_M - \eta_m) \sum_{j=-\bar{\tau}_M}^{-\eta_m-1} \sum_{s=t+j}^{t-1} \lambda^{t-s-1} \eta^T(s) R_1 \eta(s),\ \eta(t) = x(t+1) - x(t), \end{aligned} \tag{9.7}$$

where $P > 0,\ S_i > 0,\ R_i > 0,\ i = 0, 1,\ 0 < \lambda < 1,\ t \geq 0$. Following [130], we define (for simplicity)

$$x(t) = x_0,\ t \leq 0. \tag{9.8}$$

We will find the conditions that guarantee

$$V_{RR}(t+1) - \lambda V_{RR}(t) \leq 0,\ t = t_1, t_1 + 1, \ldots, \tag{9.9}$$

which implies

$$V_{RR}(t) \leq \lambda^{t-t_1} V_{RR}(t_1),\ t = t_1, t_1 + 1, \ldots.$$

In order to derive a bound on $V_{RR}(t_1)$ in terms of x_0 in a simple way, on the controller side we need to wait for all (both) latest transmitted measurements $y_1(s_0)$, $y_2(s_1)$ and then send them together to the actuator side. Therefore, the first updating time $t_0 = t_1$ corresponds to the updating time instant when the first data is received by the actuator, which means that $u(t) = 0, t \in [0, t_1 - 1]$. Then for $t \in [0, t_1 - 1]$, (9.1) is given by

$$x(t+1) = Ax(t),\ t = 0, 1, \ldots, t_1 - 1,\ t \in \mathbb{Z}^+. \tag{9.10}$$

Let $x(t, x_0)$ denote the state trajectory of (9.6), (9.10) with the initial condition $x_0 \in \mathbb{R}^n$. The domain of attraction of the closed-loop system (9.6), (9.10) is the set

$$\mathscr{A} = \{x_0 \in \mathbb{R}^n :\ \lim_{t\to\infty} x(t, x_0) = 0\}.$$

Given K^1, K^2 and positive integers $0 \leq \eta_m \leq \eta_M < \tau_M$, the objective is to get an estimate $\mathscr{X}_\beta \subset \mathscr{A}$ (as large as we can get) on the domain of attraction, for which the exponential stability of the closed-loop system is ensured, where

$$\mathscr{X}_\beta = \{x_0 \in \mathbb{R}^n : x_0^T P x_0 \leq \beta^{-1}\}, \tag{9.11}$$

and where $\beta > 0$ is a scalar, $P > 0$ is an $n \times n$-matrix.

By extending the direct Lyapunov approach suggested in [130] for time-delay system to the switched system given by (9.6), (9.10), we obtain

Lemma 9.1 *Consider LKF $V_{RR}(t)$ given by (9.7) and denote $V_0(t) = x^T(t)Px(t)$. Under (9.8), let there exist $0 < \lambda < 1$ and $\mu > 1$ such that the following inequalities*

$$V_0(t+1) - \mu V_0(t) \leq 0,\ t = 0, 1, \ldots, t_1 - 1, \tag{9.12a}$$
$$V_{RR}(t+1) - \lambda V_{RR}(t) - (\mu - 1)V_0(t) \leq 0,\ t = 0, 1, \ldots, t_1 - 1 \tag{9.12b}$$

hold along (9.10). Then, the solutions of (9.6), (9.10) at time t_1 satisfy:

$$V_{RR}(t_1) \leq x_0^T[\lambda^{\eta_m} \Xi_{RR} + (\mu^{\tau_M+1} - 1)P]x_0, \tag{9.13}$$

where

$$\Xi_{RR} = P + \eta_m S_0 + \lambda^{\eta_m}(\bar{\tau}_M - \eta_m)S_1. \tag{9.14}$$

Proof From (9.12a), we have $V_0(t) \le \mu^t V_0(0)$ for $t = 0, 1, \ldots, t_1$. Under the constant initial condition (9.8) and $V_{RR}(t)$ of (9.7), we have for $t = 0$

$$V_{RR}(0) = x_0^T P x_0 + \sum_{s=-\eta_m}^{-1} \lambda^{-s-1} x^T(s) S_0 x(s) + \sum_{s=-\bar{\tau}_M}^{-\eta_m-1} \lambda^{-s-1} x_0^T S_1 x_0.$$

Hence, $V_{RR}(0) \le x_0^T \Xi_{RR} x_0$. Note that $\eta_m < t_1 \le \tau_M + 1$. It holds that (9.12b) yields (9.13) since

$$V_{RR}(t_1) \le \lambda^{t_1} V_{RR}(0) + (\mu^{t_1} - 1)x_0^T P x_0 \le x_0^T \Big[\lambda^{t_1}\Xi_{RR} + (\mu^{t_1} - 1)P\Big] x_0.$$

Lemma 9.1 implies the following result.

Theorem 9.1 *Given scalars* $0 < \lambda \le 1$, $\beta > 0$, $\mu > 1$, $\sigma > 0$, *positive integers* $0 \le \eta_m \le \eta_M < \tau_M$, *and* K^1, K^2, *let there exist scalars* $n \times n$ *matrices* $P > 0$, $S_\vartheta > 0$, $R_\vartheta > 0$, $\vartheta = 0, 1$, G_1^i, G_2^i, G_3^i, $i = 1, 2$, *such that* $\eta_m S_0 + \lambda^{\eta_m}(\bar{\tau}_M - \eta_m)S_1 \le \sigma P$ *and the following matrix inequalities*

$$\Omega_i = \begin{bmatrix} R_1 & G_1^i & G_2^i \\ * & R_1 & G_3^i \\ * & * & R_1 \end{bmatrix} \ge 0, \tag{9.15}$$

$$\begin{bmatrix} -\mu P & A^T P \\ * & -P \end{bmatrix} < 0, \tag{9.16}$$

$$\begin{bmatrix} P\rho_{RR}^{-1} & (K^i C^i)_j^T \\ * & \frac{1}{4}\beta \bar{u}_j^2 \end{bmatrix} \ge 0, \ j = 1, \ldots, n_u, \tag{9.17}$$

$$F_0^T P F_0 + \Sigma + F_{01}^T H F_{01} - \lambda^{\eta_m} F_{12}^T R_0 F_{12} - \lambda^{\bar{\tau}_M} F^T \Omega_i F < 0, \tag{9.18}$$

$$\bar{F}_0^T P \bar{F}_0 + \bar{\Sigma} + \bar{F}_{01}^T H \bar{F}_{01} - \lambda^{\eta_m} F_{12}^T R_0 F_{12} - \lambda^{\bar{\tau}_M} F^T \Omega_1 F < 0, \ i = 1, 2, \tag{9.19}$$

are feasible, where

$$
\begin{aligned}
F_0 &= [A\ 0\ A_{3-i}\ A_i\ 0],\ \ F_{01} = [A-I\ \ 0\ A_{3-i}\ A_i\ 0],\ i=1,2,\\
F_{12} &= [I\ \ -I\ \ 0\,0\,0],\ \ \bar{F}_0 = [A\ 0\,0\,0\,0],\ \ \bar{F}_{01} = [A-I\ \ 0\,0\,0\,0],\\
F &= \begin{bmatrix} 0 & I & -I & 0 & 0 \\ 0 & 0 & I & -I & 0 \\ 0 & 0 & 0 & I & -I \end{bmatrix},\\
\Sigma &= \operatorname{diag}\{S_0 - \lambda P,\ -\lambda^{\eta_m}(S_0 - S_1),\ 0,\ 0,\ -\lambda^{\bar{\tau}_M} S_1\},\\
\bar{\Sigma} &= \operatorname{diag}\{S_0 - \lambda P + (1-\mu)P,\ -\lambda^{\eta_m}(S_0 - S_1),\ 0,\ 0,\ -\lambda^{\bar{\tau}_M} S_1\},\\
\rho_{RR} &= \lambda^{\eta_m}(1+\sigma) + \mu^{\tau_M+1} - 1,\\
H &= \eta_m^2 R_0 + (\bar{\tau}_M - \eta_m)^2 R_1.
\end{aligned}
\tag{9.20}
$$

Then, for all initial conditions x_0 belonging to $\mathscr{X}_\beta$, the closed-loop system (9.6), (9.10) is exponentially stable.

Proof Suppose that $x(t) \in \mathscr{P}(K^1, \bar{u}) \cap \mathscr{P}(K^2, \bar{u})$. Following Chap. 4, we take advantage of the ordered delays and use the convex analysis of [173]. It is shown that (9.16) and (9.15), (9.19) guarantee, respectively, (9.12a) and (9.12b) for $t = 0, \ldots, t_1 - 1$; (9.15) and (9.18) guarantee (9.9) for $t = t_1, t_1 + 1, \ldots$.

Noting that (9.9), (9.13) and $\eta_m S_0 + \lambda^{\eta_m}(\bar{\tau}_M - \eta_m)S_1 \le \sigma P$, we arrive at

$$
\begin{aligned}
x^T(t)Px(t) &\le V_{RR}(t) \le \lambda^{t-t_1} V_{RR}(t_1)\\
&\le x_0^T[\lambda^{\eta_m} \Xi_{RR} + (\mu^{\tau_M+1} - 1)P]x_0\\
&\le x_0^T[\lambda^{\eta_m}(P + \sigma P) + (\mu^{\tau_M+1} - 1)P]x_0\\
&= \rho_{RR} x_0^T P x_0,\ \ t = t_1, t_1+1, \ldots.
\end{aligned}
$$

Therefore, for all $x(t)$, $x^T(t)Px(t) \le \rho_{RR}\beta^{-1}$ implies $x^T(t)(K^iC^i)_j^T(K^iC^i)_j x(t) \le \frac{1}{4}\bar{u}_j^2$, if

$$
x^T(t)(K^iC^i)_j^T(K^iC^i)_j x(t) \le \frac{1}{4}\beta\rho_{RR}^{-1} x^T(t)Px(t)\bar{u}_j^2,\ \ j = 1, \ldots, n_u,\ \ i = 1, 2.
$$

The latter inequality is guaranteed if $\frac{1}{4}\beta\rho_{RR}^{-1}P\bar{u}_j^2 - (K^iC^i)_j^T(K^iC^i)_j \ge 0$, and thus, by Schur complements if (9.17) is feasible. Hence, the trajectories of the system (9.6), (9.10) converge to the origin exponentially, provided that $x_0 \in \mathscr{X}_\beta$.

9.2 Discrete-Time NCSs with Actuator Saturation Under TOD Protocol

9.2.1 *Problem Formulation and a Hybrid Time-Delay Model*

In [132], a weighted TOD protocol was analyzed for the stabilization of continuous-time NCSs in the framework of time-delay approach. The actuator saturation was not taken into account. In this section, we consider discrete-time NCSs with actuator saturation under TOD protocol.

Consider (9.1) with two sensor nodes $y^i(t) = C^i x(t)$, $i = 1, 2$, under the saturated control input (9.2). Consider a sequence of sampling instants (9.3). At each sampling instant s_k, one of the outputs $y^i(t) \in \mathbb{R}^{n_i}$ $(n_1 + n_2 = n_y)$ is sampled and transmitted via the network. Let

$$\hat{y}(s_k) = \begin{bmatrix} \hat{y}^1(s_k) \\ \hat{y}^2(s_k) \end{bmatrix} \in \mathbb{R}^{n_y}$$

denote the output information submitted to the scheduling protocol. At each sampling instant s_k, one of $\hat{y}^i(s_k)$ values is updated with the recent output $y^i(s_k)$.

It is assumed that no packet dropouts and no packet disorders will happen during the data transmission over the network. The transmission of the information over the two networks (between the sensor and the actuator) is subject to a variable delay η_k. Then, $t_k = s_k + \eta_k$ is the updating time instant. We allow the delays to be non-small provided that the old sample cannot get to the destination (to the controller or to the actuator) after the current one. Assume that the network-induced delay η_k and the time span between the updating and the current sampling instants satisfy (9.4).

The choice of the active output node is ruled by a weighted TOD protocol. Following (5.5), consider the error between the system output $y(s_k)$ and the last available information $\hat{y}(s_{k-1})$:

$$\begin{aligned} &e(t) = \mathrm{col}\{e_1(t), e_2(t)\} \equiv \hat{y}(s_{k-1}) - y(s_k),\ t \in [t_k, t_{k+1} - 1], \\ &t \in \mathbb{Z}^+,\ k \in \mathbb{Z}^+,\ \hat{y}(s_{-1}) \triangleq 0,\ e(t) \in \mathbb{R}^{n_y}. \end{aligned}$$

Let $Q_i > 0$, $i = 1, 2$, be some weighting matrices (they will be found from matrix inequalities in Lemma 9.2 below). The node that has the largest error, $|\sqrt{Q}_i e_i(t)|^2$, $i = 1, 2$, is granted access to the network.

Let

$$i_k^* = \min\left\{\arg\max_{i\in\{1,2\}} |\sqrt{Q}_i\left(\hat{y}^i(s_{k-1}) - y^i(s_k)\right)|^2\right\} \tag{9.21}$$

be the index of the active output node at the sampling instant s_k. This can be rewritten as

$$\left|\sqrt{Q}_{i_k^*} e_{i_k^*}(t)\right|^2 \ge \left|\sqrt{Q}_i e_i(t)_{|i\ne i_k^*}\right|^2,\ t \in [t_k, t_{k+1} - 1],\ t \in \mathbb{Z}^+. \tag{9.22}$$

Due to the control bounds defined in (9.2), the effective control signal to be applied to the system (9.1) is given by

$$u(t) = \text{sat}\left[K^{i_k^*} y^{i_k^*}(t_k - \eta_k) + K^i \hat{y}^i(t_{k-1} - \eta_{k-1})_{|i \neq i_k^*}\right], \ t \in [t_k, t_{k+1} - 1], \ t \in \mathbb{Z}^+.$$

If the control is such that $x(t) \in \mathscr{P}(K^1, \bar{u}) \cap \mathscr{P}(K^2, \bar{u})$, then

$$|(K^1 C^1)_i x + (K^2 C^2)_i x(t)| \leq \bar{u}_i.$$

The closed-loop system can be written as

$$\begin{cases} x(t+1) = Ax(t) + A_1 x(t_k - \eta_k) + B_i e_i(t)_{|i \neq i_k^*}, \\ e(t+1) = e(t), \ t \in [t_k, t_{k+1} - 2], \ t \in \mathbb{Z}^+ \end{cases} \tag{9.23}$$

with the delayed reset system for $t = t_{k+1} - 1$

$$\begin{cases} x(t_{k+1}) = Ax(t_{k+1} - 1) + A_1 x(t_k - \eta_k) + B_i e_i(t_k)_{|i \neq i_k^*}, \\ e_i(t_{k+1}) = C^i[x(t_k - \eta_k) - x(t_{k+1} - \eta_{k+1})], \ i = i_k^*, \\ e_i(t_{k+1}) = e_i(t_k) + C^i[x(t_k - \eta_k) - x(t_{k+1} - \eta_{k+1})], \ i \neq i_k^*, \end{cases} \tag{9.24}$$

where $A_1 = BKC$, $B_i = BK^i$, $K = [K^1 \ K^2]$, $i = 1, 2$.

The initial condition for (9.22)–(9.24) has the form of $e(t_0) = -Cx(t_0 - \eta_0) = -Cx_0$ and

$$x(t+1) = Ax(t), \ t = 0, 1, \ldots, t_0 - 1, \ \ t \in \mathbb{Z}^+. \tag{9.25}$$

Definition 9.1 The hybrid system (9.22)–(9.24) is said to be partially exponentially stable with respect to x if there exist constants $b > 0$, $0 < \kappa < 1$ such that the following inequality

$$|x(t)|^2 \leq b\kappa^{t-t_0}\left[|x_0|^2 + |e(t_0)|^2\right], \ t \geq t_0$$

holds for the solutions of the hybrid system initialized with (9.25) and $e(t_0) \in \mathbb{R}^{n_y}$.

Given K^1, K^2 and positive integers $0 \leq \eta_m \leq \eta_M < \tau_M$, the objective is to obtain an estimate $\mathscr{X}_\beta \subset \mathscr{A}$ (as large as we can get) on the domain of attraction, for which the exponential stability of the closed-loop system (9.22)–(9.24) with respect to variable of interest x is ensured, where $\mathscr{X}_\beta$ is given by (9.11).

9.2.2 *Partial Exponential Stability of Hybrid Delayed System Without Actuator Saturation*

Consider the LKF of the form:

$$
\begin{aligned}
V_e(t) &= V_{TOD}(t) + \frac{t_{k+1}-t}{\tau_M-\eta_m+1} e_i^T(t_k) Q_i e_i(t_k)_{|i\neq i_k^*},\\
V_{TOD}(t) &= \tilde{V}(t) + V_Q(t),\\
V_Q(t) &= (\tau_M-\eta_m) \sum_{s=t_k-\eta_k}^{t-1} \lambda^{t-s-1} \eta^T(s) Q \eta(s),\\
\tilde{V}(t) &= x^T(t)Px(t) + \sum_{s=t-\eta_m}^{t-1} \lambda^{t-s-1} x^T(s) S_0 x(s) + \sum_{s=t-\tau_M}^{t-\eta_m-1} \lambda^{t-s-1} x^T(s) S_1 x(s)\\
&\quad + \eta_m \sum_{j=-\eta_m}^{-1} \sum_{s=t+j}^{t-1} \lambda^{t-s-1} \eta^T(s) R_0 \eta(s)\\
&\quad + (\tau_M-\eta_m) \sum_{j=-\tau_M}^{-\eta_m-1} \sum_{s=t+j}^{t-1} \lambda^{t-s-1} \eta^T(s) R_1 \eta(s),
\end{aligned}
\tag{9.26}
$$

where $P > 0,\ S_i > 0,\ R_i > 0,\ Q > 0,\ Q_j > 0,\ 0 < \lambda < 1,\ i = 0, 1,\ j = 1, 2,\ t \in [t_k, t_{k+1} - 1],\ t \in \mathbb{Z}^+,\ k \in \mathbb{Z}^+$, and where we define $x(t) = x_0,\ t \le 0$.

The objective is to guarantee that

$$
V_e(t+1) - \lambda V_e(t) \le 0,\ t \in [t_k, t_{k+1} - 1],\ t \in \mathbb{Z}^+, \tag{9.27}
$$

holds along (9.22)–(9.24). The inequality (9.27) implies the following bound

$$
\begin{aligned}
V_{TOD}(t) &\le V_e(t) \le \lambda^{t-t_0} V_e(t_0),\ t \ge t_0,\ t \in \mathbb{Z}^+,\\
V_e(t_0) &\le V_{TOD}(t_0) + \min_{i=1,2}\{|\sqrt{Q_i} e_i(t_0)|^2\}
\end{aligned}
\tag{9.28}
$$

for the solution of (9.22)–(9.24) with the initial condition (9.25) and $e(t_0) \in \mathbb{R}^{n_y}$. Here we took into account that for the case of two sensor nodes

$$
|\sqrt{Q_i} e_i(t_0)|^2_{|i\neq i_k^*} = \min_{i=1,2}\{|\sqrt{Q_i} e_i(t_0)|^2\}.
$$

From (9.28), it follows that the system (9.22)–(9.24) is exponentially stable with respect to x.

Following the construction of the term V_G in (5.12) to cope with the delays in the reset conditions in the continuous-time case, in LKF (9.26) we insert the term V_Q to cope with the delays in the reset conditions in the discrete-time case:

$$
\begin{aligned}
&V_Q(t_{k+1}) - \lambda V_Q(t_{k+1} - 1)\\
&= (\tau_M - \eta_m)\Big[\sum_{s=t_{k+1}-\eta_{k+1}}^{t_{k+1}-1} \lambda^{t_{k+1}-s-1} \eta^T(s) Q \eta(s) - \sum_{s=t_k-\eta_k}^{t_{k+1}-2} \lambda^{t_{k+1}-s-1} \eta^T(s) Q \eta(s) \Big]
\end{aligned}
$$

$$
\begin{aligned}
&\leq (\tau_M-\eta_m)\eta^T(t_{k+1}-1)Q\eta(t_{k+1}-1)-(\tau_M-\eta_m)\lambda^{\tau_M}\sum_{s=t_k-\eta_k}^{t_{k+1}-\eta_{k+1}-1}\eta^T(s)Q\eta(s)\\
&\leq (\tau_M-\eta_m)\eta^T(t_{k+1}-1)Q\eta(t_{k+1}-1)-\lambda^{\tau_M}|\sqrt{Q}[x(t_{k+1}-\eta_{k+1})-x(t_k-\eta_k)]|^2,
\end{aligned}
$$

where we applied Cauchy–Schwarz inequality [57]. The term

$$
\frac{t_{k+1}-t}{\tau_M-\eta_m+1}e_i^T(t_k)Q_ie_i(t_k)
$$

is inspired by the similar construction of LKF for the sampled-data systems [46].

We have

$$
\begin{aligned}
&V_e(t_{k+1})-\lambda V_e(t_{k+1}-1)\\
&=\tilde V(t_{k+1})-\lambda\tilde V(t_{k+1}-1)+\frac{t_{k+2}-t_{k+1}}{\tau_M-\eta_m+1}e_i^T(t_{k+1})Q_ie_i(t_{k+1})_{|i\neq i^*_{k+1}}\\
&\quad-\frac{\lambda}{\tau_M-\eta_m+1}e_i^T(t_k)Q_ie_i(t_k)_{|i\neq i^*_k}+(\tau_M-\eta_m)\eta^T(t_{k+1}-1)Q\eta(t_{k+1}-1)\\
&\quad-\lambda^{\tau_M}|\sqrt{Q}[x(t_{k+1}-\eta_{k+1})-x(t_k-\eta_k)]|^2.
\end{aligned}
\tag{9.29}
$$

Note that under TOD protocol for $i^*_{k+1}=i^*_k$

$$
e_i^T(t_{k+1})Q_ie_i(t_{k+1})_{|i\neq i^*_{k+1}}\leq e^T_{i^*_k}(t_{k+1})Q_{i^*_k}e_{i^*_k}(t_{k+1}),
$$

whereas for $i^*_{k+1}\neq i^*_k$, the latter relation holds with the equality. Hence,

$$
\begin{aligned}
\frac{t_{k+2}-t_{k+1}}{\tau_M-\eta_m+1}e_i^T(t_{k+1})Q_ie_i(t_{k+1})_{|i\neq i^*_{k+1}}&\leq e^T_{i^*_k}(t_{k+1})Q_ie_{i^*_k}(t_{k+1})\\
&=\left|\sqrt{Q_{i^*_k}}C^{i^*_k}[x(t_{k+1}-\eta_{k+1})-x(t_k-\eta_k)]\right|^2.
\end{aligned}
$$

Assume that

$$
\lambda^{\tau_M}Q>C^{iT}Q_iC^i,\ i=1,2. \tag{9.30}
$$

Then for $t=t_{k+1}-1$, we obtain

$$
\begin{aligned}
V_e(t+1)-\lambda V_e(t)\leq \tilde V(t+1)-\lambda\tilde V(t)&+(\tau_M-\eta_m)\eta^T(t)Q\eta(t)\\
&-\frac{\lambda}{\tau_M-\eta_m+1}e_i^T(t_k)Q_ie_i(t_k)_{|i\neq i^*_k}.
\end{aligned}
$$

Furthermore, due to

$$
-\frac{\lambda}{\tau_M-\eta_m+1}=-\frac{1}{\tau_M-\eta_m+1}+\frac{1-\lambda}{\tau_M-\eta_m+1}\leq-\frac{1}{\tau_M-\eta_m+1}+1-\lambda,
$$

for $t=t_{k+1}-1$ we arrive at

$$V_e(t+1) - \lambda V_e(t) \le \tilde{V}(t+1) - \lambda \tilde{V}(t) + (\tau_M - \eta_m)\eta^T(t)Q\eta(t) \\ -\Big[\frac{1}{\tau_M - \eta_m + 1} - (1-\lambda)\Big] e_i^T(t_k)Q_i e_i(t_k)_{|i \neq i_k^*} \triangleq \Psi(t). \tag{9.31}$$

For $t \in [t_k, t_{k+1} - 2]$, we have

$$V_e(t+1) - \lambda V_e(t) \le \tilde{V}(t+1) - \lambda \tilde{V}(t) + (\tau_M - \eta_m)\eta^T(t)Q\eta(t) \\ + \left[\frac{t_{k+1} - t - 1}{\tau_M - \eta_m + 1} - \lambda \frac{t_{k+1} - t}{\tau_M - \eta_m + 1}\right] e_i^T(t_k)Q_i e_i(t_k)_{|i \neq i_k^*}.$$

Since

$$\begin{aligned}\frac{t_{k+1} - t - 1}{\tau_M - \eta_m + 1} - \lambda \frac{t_{k+1} - t}{\tau_M - \eta_m + 1} &= -\frac{1}{\tau_M - \eta_m + 1} + (1-\lambda)\frac{t_{k+1} - t}{\tau_M - \eta_m + 1} \\ &\le -\frac{1}{\tau_M - \eta_m + 1} + 1 - \lambda,\end{aligned}$$

we conclude that (9.31) is valid also for $t \in [t_k, t_{k+1} - 2]$. Therefore, (9.27) holds if

$$\Psi(t) \le 0, \ t \in [t_k, t_{k+1} - 1]. \tag{9.32}$$

Note that $i \neq i_k^*$ for $i = 1, 2$, is the same as $i = 3 - i_k^*$. By using the standard arguments for the delay-dependent analysis [173], we derive the following conditions for (9.32) (and thus, for (9.28)).

Lemma 9.2 *Given scalar $0 < \lambda < 1$, positive integers $0 \le \eta_m \le \eta_M < \tau_M$, and K^1, K^2, if there exist $n \times n$ matrices $P > 0$, $Q > 0$, $S_j > 0$, $R_j > 0$, $j = 0, 1$, S_{12}, $n_i \times n_i$ matrices $Q_i > 0$, $i = 1, 2$, such that (9.30) and*

$$\hat{\Omega} = \begin{bmatrix} R_1 & S_{12} \\ * & R_1 \end{bmatrix} \ge 0, \tag{9.33}$$

$$\hat{F}_0^T P \hat{F}_0 + \hat{\Sigma} + \hat{F}_{01}^T W \hat{F}_{01} - \lambda^{\eta_m} F_{12}^T R_0 F_{12} - \lambda^{\tau_M} \hat{F}^T \hat{\Omega} \hat{F} < 0, \tag{9.34}$$

are feasible, where

$$\begin{aligned}\hat{F}_0 &= [A\ 0\ A_1\ 0\ B_{3-i}], \\ \hat{F}_{01} &= [A - I\ 0\ A_1\ 0\ B_{3-i}], \\ \hat{F} &= \begin{bmatrix} 0 & I & -I & 0 & 0 \\ 0 & 0 & I & -I & 0 \end{bmatrix}, \\ \hat{\Sigma} &= \mathrm{diag}\{S_0 - \lambda P, -\lambda^{\eta_m}(S_0 - S_1), 0, -\lambda^{\tau_M} S_1, \varphi\}, \\ \varphi &= -\Big[\frac{1}{\tau_M - \eta_m + 1} - (1-\lambda)\Big] Q_{3-i}, \\ W &= \eta_m^2 R_0 + (\tau_M - \eta_m)^2 R_1 + (\tau_M - \eta_m) Q, \quad i = 1, 2.\end{aligned} \tag{9.35}$$

Then, the solutions of the hybrid system (9.22)–(9.24) satisfy the bound (9.28) and are exponentially stable with respect to x. Moreover, if the above inequalities are feasible with $\lambda = 1$, then the bound (9.28) holds with $\lambda = 1 - \varepsilon$, where $\varepsilon > 0$ is small enough.

Remark 9.1 The inequality $V_e(t) \le \lambda^{t-t_0} V_e(t_0), t \ge t_0, t \in \mathbb{Z}^+$ in (9.28) guarantees that

$$\frac{t_{k+1} - t_k}{\tau_M - \eta_m + 1} e_i^T(t_k) Q_i e_i(t_k)_{|i \neq i_k^*}$$

is bounded, and it does not guarantee that $e(t_k)$ is bounded. That is why (9.28) only implies the partial stability with respect to x.

9.2.3 Partial Exponential Stability of Hybrid Delayed System with Actuator Saturation

This section is devoted to derive a bound on $V_{TOD}(t_0)$ in terms of x_0. By using the arguments similar to Lemma 9.1, we arrive at

Lemma 9.3 *Consider LKF $V_{TOD}(t)$ given by (9.26) and denote $V_0(t)=x^T(t)Px(t)$. Under (9.8) assume that there exist $0 < \lambda < 1$ and $\mu > 1$ such that*

$$V_0(t+1) - \mu V_0(t) \le 0, \; t = 0, 1, \ldots, t_0 - 1, \tag{9.36a}$$

$$V_{TOD}(t+1) - \lambda V_{TOD}(t) - (\mu - 1) V_0(t) \le 0, \; t = 0, 1, \ldots, t_0 - 1, \tag{9.36b}$$

hold along (9.25). Then, we have

$$V_{TOD}(t_0) \le x_0^T \left[\lambda^{\eta_m} \Xi_{TOD} + (\mu^{\eta_M} - 1) P \right] x_0, \tag{9.37}$$

where

$$\Xi_{TOD} = P + \eta_m S_0 + \lambda^{\eta_m} (\tau_M - \eta_m) S_1. \tag{9.38}$$

Proof From (9.36a), we have $V_0(t) \le \mu^t V_0(0)$ for $t = 0, 1, \ldots, t_0$. Under the constant initial condition (9.8) and $V_{TOD}(t)$ of (9.26), we have for $t = 0$

$$V_{TOD}(0) = x_0^T P x_0 + \sum_{s=-\eta_m}^{-1} \lambda^{-s-1} x_0^T S_0 x_0 + \sum_{s=-\tau_M}^{-\eta_m - 1} \lambda^{-s-1} x_0^T S_1 x_0.$$

Thus, $V_{TOD}(0) \le x_0^T \Xi_{TOD} x_0$. Note that $\eta_m \le t_0 = \eta_0 \le \eta_M$. It holds that the inequality (9.36b) implies

$$
\begin{aligned}
V_{TOD}(t_0) \le \lambda^{t_0} V_{TOD}(0) + (\mu^{t_0} - 1)x_0^T P x_0 &\le x_0^T\Big[\lambda^{t_0} \Xi_{TOD} + (\mu^{t_0} - 1)P\Big]x_0 \\
&\le x_0^T\Big[\lambda^{\eta_m} \Xi_{TOD} + (\mu^{\eta_M} - 1)P\Big]x_0.
\end{aligned} \tag{9.39}
$$

Lemmas 9.2 and 9.3 imply the following result.

Theorem 9.2 *Given scalars* $0 < \lambda < 1, \beta > 0, \mu > 1, \sigma > 0, \bar{\sigma} > 0$, *positive integers* $0 \le \eta_m \le \eta_M < \tau_M$, *and* K^1, K^2, *if there exist* $n \times n$ *matrices* $P > 0$, $Q > 0$, $S_l > 0$, $R_l > 0$, $l = 0, 1$, S_{12}, $n_i \times n_i$ *matrices* $Q_i > 0$, $i = 1, 2$, *such that (9.16), (9.30), (9.33), (9.34) and*

$$
\eta_m S_0 + \lambda^{\eta_m}(\tau_M - \eta_m)S_1 \le \sigma P, \tag{9.40}
$$

$$
\lambda^{\tau_M} Q \le \bar{\sigma} P, \tag{9.41}
$$

$$
\begin{bmatrix} P\rho_{TOD}^{-1} & (K^i C^i)_j^T \\ * & \frac{1}{4}\beta \bar{u}_j^2 \end{bmatrix} \ge 0, \ j = 1, \dots, n_u, \tag{9.42}
$$

$$
\tilde{F}_0^T P \tilde{F}_0 + \tilde{\Sigma} + \tilde{F}_{01}^T W \tilde{F}_{01} - \lambda^{\eta_m} \tilde{F}_{12}^T R_0 \tilde{F}_{12} - \lambda^{\tau_M} \tilde{F}^T \hat{\Omega} \tilde{F} < 0, \tag{9.43}
$$

are feasible, where the notations W *and* $\hat{\Omega}$ *are given by (9.35) and (9.33), respectively, and where*

$$
\begin{aligned}
\tilde{F}_0 &= [A\ 0\ 0\ 0], \ \tilde{F}_{01} = [A - I\ 0\ 0\ 0], \ \tilde{F}_{12} = [I\ \ -I\ 0\ 0], \\
\tilde{F} &= \begin{bmatrix} 0 & I & -I & 0 \\ 0 & 0 & I & -I \end{bmatrix}, \\
\tilde{\Sigma} &= \mathrm{diag}\{S_0 - \lambda P + (1 - \mu)P, -\lambda^{\eta_m}(S_0 - S_1), 0, -\lambda^{\tau_M} S_1\}, \\
\rho_{TOD} &= \lambda^{\eta_m}(1 + \sigma) + (\mu^{\eta_M} - 1) + \bar{\sigma}, \ i = 1, 2.
\end{aligned} \tag{9.44}
$$

Then for all initial conditions x_0 *belonging to* $\mathscr{X}_\beta$, *the closed-loop system (9.22)–(9.24) is exponentially stable with respect to* x. *Moreover, if the above inequalities hold with* $\lambda = 1$, *then they are feasible for* $\lambda = 1 - \varepsilon$, *where* $\varepsilon > 0$ *is small enough.*

Proof Suppose that $x(t) \in \mathscr{P}(K^1, \bar{u}) \cap \mathscr{P}(K^2, \bar{u})$. As shown in Lemma 9.2, (9.30), (9.33) and (9.34) lead to (9.28) for $t = t_0, t_0 + 1, \dots$. Following the proof of Theorem 9.1, for $t = 0, 1, \dots, t_0 - 1$, (9.16) and (9.43) with (9.33) guarantee (9.36a) and (9.36b), respectively.

Next noting that (9.30) and (9.41), we have

$$
x_0^T C^{iT} Q_i C^i x_0 \le \lambda^{\tau_M} x_0^T Q x_0 \le \bar{\sigma} x_0^T P x_0, \tag{9.45}
$$

which implies that $|\sqrt{Q_i} e_i(t_0)|^2 = |-\sqrt{Q_i} C^i x_0|^2 < \bar{\sigma} x_0^T P x_0$, $i = 1, 2$.

Therefore, taking into (9.28), (9.37), (9.40) and (9.45), we obtain

$$\begin{aligned}x^T(t)Px(t) &\le V_{TOD}(t) \le \lambda^{t-t_0}V_e(t_0)\\ &\le \lambda^{t-t_0}\left\{V_{TOD}(t_0)+\min_{i=1,2}\{e_i^T(t_0)Q_ie_i(t_0)\}\right\}\\ &\le x_0^T[\lambda^{\eta_m}\Xi_{TOD}+(\mu^{\eta_M}-1)P+\bar{\sigma}P]x_0\\ &\le x_0^T[\lambda^{\eta_m}(1+\sigma)+(\mu^{\eta_M}-1)+\bar{\sigma}]Px_0\\ &= \rho_{TOD}x_0^TPx_0, \quad t=t_0,t_0+1,\ldots.\end{aligned}$$

Hence, for all $x(t)$, $x^T(t)Px(t) \le \rho_{TOD}\beta^{-1}$ implies $x^T(t)(K^iC^i)_j^T(K^iC^i)_jx(t) \le \frac{1}{4}\bar{u}_j^2$, if

$$x^T(t)(K^iC^i)_j^T(K^iC^i)_jx(t) \le \frac{1}{4}\beta\rho_{TOD}^{-1}x^T(t)Px(t)\bar{u}_j^2, \ j=1,\ldots,n_u, \ i=1,2.$$

The latter inequality is guaranteed if $\frac{1}{4}\beta\rho_{TOD}^{-1}P\bar{u}_j^2-(K^iC^i)_j^T(K^iC^i)_j \ge 0$, and thus, by Schur complements if (9.42) is feasible. Hence, the solutions of the hybrid system (9.22)–(9.24) initialized with (9.25) and $e(t_0)\in\mathbb{R}^{n_y}$ converge to the origin exponentially with respect to x, provided that $x_0\in\mathscr{X}_\beta$.

Remark 9.2 Note that for the stability analysis of discrete-time systems with time-varying delay in the state, a switched system transformation approach can be used in addition to Lyapunov–Krasovskii method. See more details in [90].

Remark 9.3 Note that

$$x_0^TPx_0 \le \lambda_{\max}(P)|x_0|^2 < \beta^{-1}. \tag{9.46}$$

Hence, the following initial region $|x_0|^2 < \beta^{-1}/\lambda_{\max}(P)$ is inside of $\mathscr{X}_\beta$. In order to maximize the initial ball, we can add the condition

$$P-\gamma I < 0, \tag{9.47}$$

to Theorems 9.1 and 9.2, where $\gamma > 0$ is minimized.

9.3 Example: Discrete-Time Cart-Pendulum

Consider the following linearized model of the inverted pendulum on a cart [133]:

$$\begin{bmatrix}\dot{x}\\ \ddot{x}\\ \dot{\theta}\\ \ddot{\theta}\end{bmatrix}=\begin{bmatrix}0&1&0&0\\0&0&\frac{-mg}{M}&0\\0&0&0&1\\0&0&\frac{(M+m)g}{Ml}&0\end{bmatrix}\begin{bmatrix}x\\ \dot{x}\\ \theta\\ \dot{\theta}\end{bmatrix}+\begin{bmatrix}0\\ \frac{a}{M}\\ 0\\ \frac{-a}{Ml}\end{bmatrix}u(t) \tag{9.48}$$

with $M=3.9249$kg, $m=0.2047$kg, $l=0.2302$m, $g=9.81$N/kg, $a=25.3$N/V and $\bar{u}=50$. In the model, x and θ represent the cart position coordinate and the pendulum angle from vertical, respectively. Such a model is discretized with a sampling time $\mathrm{T_s} = 0.001s$:

$$\begin{bmatrix} x(t+1) \\ \Delta x(t+1) \\ \theta(t+1) \\ \Delta\theta(t+1) \end{bmatrix} = \begin{bmatrix} 1 & 0.001 & 0 & 0 \\ 0 & 1 & -0.0005 & 0 \\ 0 & 0 & 1.00 & 0.001 \\ 0 & 0 & 0.0448 & 1 \end{bmatrix} \begin{bmatrix} x(t) \\ \Delta x(t) \\ \theta(t) \\ \Delta\theta(t) \end{bmatrix} + \begin{bmatrix} 0 \\ 0.0064 \\ 0 \\ -0.0280 \end{bmatrix} u(t),\ t \in \mathbb{Z}^+ \tag{9.49}$$

with $\bar{u} = 50$. The pendulum can be stabilized by a state feedback

$$u(t) = K \begin{bmatrix} x\ \Delta x\ \theta\ \Delta\theta \end{bmatrix}^T$$

with the gain $K = [K^1\ \ K^2]$

$$K^1 = \begin{bmatrix} 5.825\ 5.883 \end{bmatrix},\ \ K^2 = \begin{bmatrix} 24.941\ 5.140 \end{bmatrix}, \tag{9.50}$$

which leads to the closed-loop system eigenvalues

$$\{0.8997, 0.9980 + 0.0020i, 0.9980 - 0.0020i, 0.9980\}.$$

Suppose that the variables θ, $\Delta\theta$ and x, Δx are not accessible simultaneously. We consider measurements $y^i(t) = C^i x(t)$, $t \in \mathbb{Z}^+$, where

$$C^1 = \begin{bmatrix} 1\,0\,0\,0 \\ 0\,1\,0\,0 \end{bmatrix},\ C^2 = \begin{bmatrix} 0\,0\,1\,0 \\ 0\,0\,0\,1 \end{bmatrix}. \tag{9.51}$$

Choose $\lambda = 1$, $\beta = 1$, $\sigma = 1.0 \times 10^{-2}$ and $\eta_m = 1$, $\eta_M = 2$, $\tau_M = 3$. Applying Theorem 9.2 with $\mu = 1.02$, $\bar{\sigma} = 1.1$ and Remark 9.3, we find that the closed-loop system (9.22)–(9.24) is exponentially stable with respect to x starting from the initial ball $|x_0| < 0.3342$ by the presented TOD protocol. Applying Theorem 9.1 with $\mu = 1.02$ and Remark 9.3, we find that the closed-loop system (9.6), (9.10) is asymptotically stable and the largest ball of admissible initial conditions is $|x_0| < 0.4964$.

Then for different μ, by Theorems 9.1, 9.2 with $\lambda = 1$ and Remark 9.3, we give the corresponding largest ball of admissible initial conditions (see Table 9.1).

Table 9.1 Largest ball of admissible initial conditions for different μ

μ	1.02	1.2	1.5	2	3
$\|x_0\|$(Theorem 9.1, Round-Robin)	0.4964	0.3557	0.2304	0.1297	0.1339
$\|x_0\|$(Theorem 9.2, TOD)	0.3342	0.3069	0.2674	0.2168	0.1541

9.4 Notes

In this chapter, a time-delay approach was proposed for the stability analysis of discrete-time NCSs in the presence of actuator saturation under Round-Robin or under TOD protocol. A Lyapunov-based method was provided for finding the domain of attraction under both scheduling protocols. The conditions are given in terms of LMIs. Polytopic uncertainties in the system model can be easily included in the analysis. Numerical example illustrates the efficiency of our method.

Chapter 10
Quantized Control of Discrete-Time Systems Under Round-Robin Protocol

In Chap. 8, an LMI-based time-triggered zooming algorithm has been proposed via a time-delay approach for NCSs with dynamic quantization and variable communication delays. The scheduling protocols were not taken into account. This observation and the need for the scheduling in wireless NCSs motivate us to develop a time-delay approach to linear NCSs under scheduling protocol and dynamic quantization. In this chapter, we consider the stability analysis of discrete-time NCSs with multiple sensor nodes, where the system involves dynamic quantization, large communication delays, variable sampling intervals and Round-Robin scheduling protocol. The closed-loop quantized system is modeled as a switched system with multiple and ordered time-varying delays and bounded disturbances. In the presence of the Round-Robin protocol, a time-triggered zooming algorithm, which is implemented at the sensors, is proposed, and it is shown to lead to an exponentially stable closed-loop system. After each zooming-in instant, we suggest waiting for all the N (N is the number of sensor nodes) latest transmitted measurements to arrive at the controller side and then sending them together to the actuator side. Following [130] and Chap. 8, we propose a direct Lyapunov approach to the initialization of the zoom variable. The zooming algorithm given in this chapter for Round-Robin protocol can be extended to the case of TOD and stochastic protocols.

10.1 NCS Model and Problem Formulation

In this section, we present the considered discrete-time NCS model and some preliminary results on the problem to be addressed in this chapter.

K. Liu et al., *Networked Control Under Communication Constraints*, Advances in Delays and Dynamics 11, https://doi.org/10.1007/978-981-15-4230-5_10

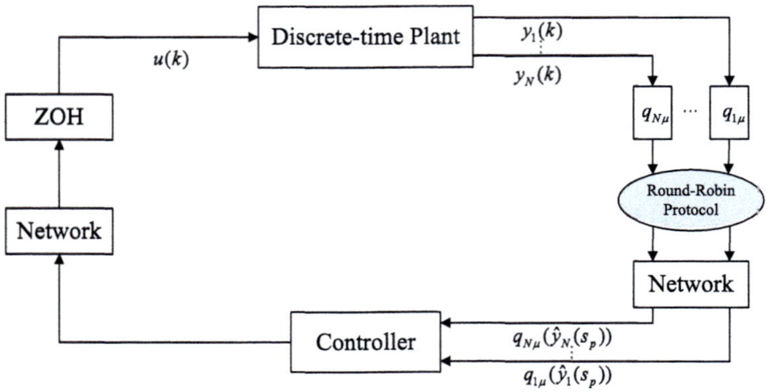

Fig. 10.1 Networked control systems with quantizers and Round-Robin protocol

10.1.1 Quantized NCS Under Round-Robin Protocol

The quantized NCS is depicted schematically in Fig. 10.1. It consists of a linear discrete-time plant, N distributed sensors and quantizers, a controller node and an actuator node, which are all connected via communication networks. The discrete-time plant is given by

$$x(k+1) = Ax(k) + Bu(k), \quad k \in \mathbb{Z}^+, \tag{10.1}$$

where $x(k) \in \mathbb{R}^n$ denotes the state of the plant and $u(k) \in \mathbb{R}^{n_u}$ the control input. The matrices A and B may be certain or uncertain. The initial condition is given by $x(0) = x_0$. It is assumed that x_0 may be unknown, but satisfies the bound $|x_0| < X_0$, where $X_0 > 0$ is known. The assumption on boundedness of the initial state is common, e.g., for interval observer design.

The measurement outputs of the plant are described by $y_i(k) = C_i x(k) \in \mathbb{R}^{n_i}$, $i = 1, \ldots, N$, $\sum_{i=1}^{N} n_i = n_y$. We denote

$$C = \left[C_1^T \cdots C_N^T \right]^T, \quad y(k) = \left[y_1^T(k) \cdots y_N^T(k) \right]^T \in \mathbb{R}^{n_y}.$$

Following [65], we consider the quantization effect from the sensors to the controller.

Let $z_i(k) \in \mathbb{R}^{n_i}$, $i = 1, \ldots, N$, be the vectors to be quantized. The quantizers are piecewise-constant functions $q_i(z_i(k))\colon \mathbb{R}^{n_i} \to \mathbb{D}_i$, where $\mathbb{D}_i$ is a finite subset of $\mathbb{R}^{n_i}$, $i = 1, \ldots, N$. It is assumed that there exist real numbers $M_i > \Delta_i > 0$, $i = 1, \ldots, N$, such that the following two conditions hold:

(a) If $|z_i(k)| \leq M_i$, then $|q_i(z_i(k)) - z_i(k)| \leq \Delta_i$;
(b) If $|z_i(k)| > M_i$, then $|q_i(z_i(k))| > M_i - \Delta_i$,

where Δ_i and M_i are the quantization error bounds and ranges, respectively. Condition (a) gives a bound on the quantization error when the quantizer does not saturate, and condition (b) provides a way to detect saturation.

We consider the quantized measurements as in [123]:

$$q_{i\mu}(z_i(k)) := \mu(k)q_i\Big(\frac{z_i(k)}{\mu(k)}\Big), \ i = 1, \ldots, N, \tag{10.2}$$

where $\mu(k) > 0$ is the zoom variable. The range of the quantizer $q_{i\mu}$, is $\mu(k)M_i$ and the quantization error is $\mu(k)\Delta_i$, $i = 1, \ldots, N$. The zoom variable $\mu(k)$ changes dynamically in order to achieve exponential stability.

Let s_p represent the unbounded and monotonously increasing sequence of sampling instants, i.e.,

$$0 = s_0 < s_1 < \cdots < s_p < \cdots, \quad p \in \mathbb{Z}^+, \ \lim_{p\to\infty} s_p = \infty, \ s_{p+1} - s_p \le \text{MATI}, \tag{10.3}$$

where $\{s_p\}$ is a subsequence of $\mathbb{Z}^+$. Denote by

$$q_\mu(\hat{y}(s_p)) = \big[q_{1\mu}^T(\hat{y}_1(s_p)) \cdots q_{N\mu}^T(\hat{y}_N(s_p))\big]^T \in \mathbb{R}^{n_y}$$

the output information submitted to the scheduling protocol. At each sampling instant s_p, one of the outputs $y_i(s_p) \in \mathbb{R}^{n_i}$ is quantized and transmitted over the network, that is, one of the $q_{i\mu}(\hat{y}_i(s_p))$ values is updated with the recent quantized output $q_{i\mu}(y_i(s_p))$. Let $i_p^* \in \mathscr{I} = \{1, \ldots, N\}$ denote the active output node at the sampling instant s_p, which will be chosen according to the scheduling protocol below.

Consider Round-Robin scheduling for the choice of the active quantized output node: $q_{i\mu}(y_i(k)) = q_{i\mu}(C_i x(k))$, $k \in \mathbb{Z}^+$, is transmitted only at the sampling instant $k = s_{N\ell+i-1}$, $\ell \in \mathbb{Z}^+$, $i = 1, \ldots, N$. After each transmission and reception, the values in $q_{i\mu}(y_i(k))$ are updated with the newly received values, while the values of $q_{j\mu}(y_j(k))$ for $j \neq i$ remain the same, as no additional information is received. This leads to the constrained data exchange expressed as

$$q_{i\mu}(\hat{y}_i(s_p)) = \begin{cases} q_{i\mu}(y_i(s_p)) = q_{i\mu}(C_i x(s_p)), & p = N\ell + i - 1, \\ q_{i\mu}(\hat{y}_i(s_{p-1})), & p \neq N\ell + i - 1, \end{cases} \quad \ell \in \mathbb{Z}^+. \tag{10.4}$$

It is assumed that the data packet loss does not occur. Denote by t_p the updating time instant of the ZOH. Suppose that the updating data at the instant t_p on the actuator side has experienced a variable transmission delay $\eta_p = t_p - s_p$. As in [161], the delays may be either smaller or larger than the sampling interval provided that the transmission order of data packets is maintained. Assume that the network-induced delay η_p and the time span between the updating and the current sampling instants are bounded:

$$t_{p+1} - 1 - t_p + \eta_p \le \tau_M^N, \ 0 \le \eta_m \le \eta_p \le \eta_M, \ p \in \mathbb{Z}^+, \tag{10.5}$$

where τ_M^N, η_m and η_M are known non-negative integers. Then, we have

$$\begin{aligned} (t_{p+1}-1)-s_p &= s_{p+1}-s_p+\eta_{p+1}-1 \leq \text{MATI}+\eta_M-1=\tau_M^N, \\ (t_{p+1}-1)-s_{p-N+j} &= s_{p+1}-s_{p-N+j}+\eta_{p+1}-1 \\ &\leq (N-j+1)\,\text{MATI}+\eta_M-1 \\ &= (N-j+1)\tau_M^N-(N-j)\eta_M+N-j \triangleq \tau_M^j,\ j=1,\ldots,N-1, \\ t_{p+1}-t_p &\leq \tau_M^N-\eta_m+1. \end{aligned} \tag{10.6}$$

10.1.2 Switched System Model with Ordered Time-Varying Delays

Suppose that the controller and the actuator are event-driven. The most recent output information on the controller side is denoted by $q_\mu(\hat{y}(s_p))$. Assume that there exists a matrix $K=[K_1\ \ldots\ K_N]$, $K_i\in\mathbb{R}^{m\times n_i}$ such that $A+BKC$ is Schur. Consider the static output feedback controller

$$u(k)=Kq_\mu(\hat{y}(s_p)),\ k\in[t_p,t_{p+1}-1],\ k\in\mathbb{Z}^+. \tag{10.7}$$

Due to (10.4), the controller (10.7) can be represented as

$$\begin{aligned} u(k) &= K_{i_p^*}q_{i_p^*\mu}(y_{i_p^*}(t_p-\eta_p)) \\ &\quad+\sum_{i=1,i\neq i_p^*}^{N} K_i q_{i\mu}(\hat{y}_i(t_{p-1}-\eta_{p-1})),\ k\in[t_p,t_{p+1}-1], \end{aligned} \tag{10.8}$$

where i_p^* is the index of the active node at s_p and η_p is the communication delay. The closed-loop system with Round-Robin protocol is modeled as a switched system:

$$\begin{aligned} x(k+1) &= Ax(k)+\sum_{j=1}^{N}A_{\theta(i,j)}x(t_{p-N+j}-\eta_{p-N+j}) \\ &\quad+\sum_{j=1}^{N}B_{\theta(i,j)}\omega_{\theta(i,j)}(k),\ k\in[t_p,t_{p+1}-1],\ i=1,\ldots,N, \end{aligned} \tag{10.9}$$

where $A_{\theta(i,j)}=BK_{\theta(i,j)}C_{\theta(i,j)}$, $B_{\theta(i,j)}=BK_{\theta(i,j)}$,

$$p=\begin{cases} N\ell+i-1, & \text{for } i\in\mathscr{I}\backslash\{N\},\ \ell\in\mathbb{N} \\ N\ell-1, & \text{for } i=N,\ \ell\in\mathbb{N} \end{cases}$$

$$\theta(i,j)=\begin{cases} i+j, & \text{if } i+j\leq N, \\ i+j-N, & \text{if } i+j>N,\ j=1,\ldots,N, \end{cases}$$

and

$$\omega_{\theta(i,j)}(k) = q_{\theta(i,j)\mu}(y_{\theta(i,j)}(s_{p-N+j})) - y_{\theta(i,j)}(s_{p-N+j}),\ i \in \mathscr{I},\ j = 1, \ldots, N,$$

denote the quantization errors. If $|y_{\theta(i,j)}(s_{p-N+j})| \le \mu(k) M_{\theta(i,j)}$, then we have for $k \in [t_p, t_{p+1} - 1]$

$$|\omega_{\theta(i,j)}(k)| \le \mu(k) \Delta_{\theta(i,j)},\ i \in \mathscr{I},\ j = 1, \ldots, N.$$

We represent $t_{p-N+j} - \eta_{p-N+j} = k - \tau_j(k),\ j = 1, \ldots, N$, where

$$\begin{array}{l} \tau_\vartheta(k) < \tau_{\vartheta-1}(k),\ \vartheta = 2, \ldots, N, \\ \tau_\vartheta(k) = k - t_{p-N+\vartheta} + \eta_{p-N+\vartheta}, \\ \tau_{\vartheta-1}(k) = k - t_{p-N+\vartheta-1} + \eta_{p-N+\vartheta-1}, \\ \tau_j(k) \in [\eta_m, \tau_M^j],\ \ k \in [t_p, t_{p+1} - 1],\ j = 1, \ldots, N. \end{array} \tag{10.10}$$

Therefore, (10.9) can be considered as a system with N time-varying interval delays, where $\tau_\vartheta(k) < \tau_{\vartheta-1}(k),\ \vartheta = 2, \ldots, N$.

This chapter is devoted to find an LMI-based time-triggered zooming algorithm (i.e., to choose a suitable time-varying parameter $\mu(k)$) for the exponential stability of the switched system (10.9). To do so, we first present a lemma for the ISS of system (10.9) with static quantization (i.e., $\mu(k) \equiv \mu$). This lemma plays a key role in achieving the main results.

10.1.3 ISS Under Round-Robin Protocol and Static Quantization

Definition 10.1 The switched system (10.9) is said to be ISS if there exist constants $b > 0$, $0 < \kappa < 1$ and $b' > 0$ such that, for initial condition $x_{t_{N-1}} \in \underbrace{\mathbb{R}^n \times \cdots \times \mathbb{R}^n}_{\tau_M^1+1 \text{ times}}$ and for disturbances ω_i, $i = 1, \ldots, N$, the solutions of the switched system (10.9) satisfy

$$|x(k)|^2 \le b\kappa^{2(k-t_{N-1})} \|x_{t_{N-1}}\|_c^2 + b' \max\{|\omega(t_{N-1})|^2, \ldots, |\omega(k)|^2\},\ k \ge t_{t_{N-1}},$$

where $\|x_{t_{N-1}}\|_c = \sup_{t_{N-1}-\tau_M^1 \le s \le t_{N-1}} |x(s)|$ and $\omega = \text{col}\{\omega_1, \ldots, \omega_N\}$.

Consider first the static quantizers with a constant zoom variable $\mu(k) \equiv \mu$. We apply the following discrete-time Lyapunov functional to system (10.9) with time-varying delay from the maximum delay interval $[\eta_m, \tau_M^1]$ [50]:

$$
\begin{aligned}
V(x_k) = x^T(k)Px(k) + \sum_{s=k-\eta_m}^{k-1} \lambda^{k-s-1} x^T(s)S_0x(s) \\
+\eta_m \sum_{j=-\eta_m}^{-1} \sum_{s=k+j}^{k-1} \lambda^{k-s-1}\eta^T(s)R_0\eta(s) + \sum_{s=k-\tau_M^1}^{k-\eta_m-1} \lambda^{k-s-1}x^T(s)S_1x(s) \\
+(\tau_M^1 - \eta_m) \sum_{j=-\tau_M^1}^{-\eta_m-1} \sum_{s=k+j}^{k-1} \lambda^{k-s-1}\eta^T(s)R_1\eta(s),
\end{aligned}
\tag{10.11}
$$

where $\eta(k) = x(k+1) - x(k)$, $P > 0$, $S_i > 0$, $R_i > 0$, $i = 0, 1$, $0 < \lambda < 1$, $x_k(j) \triangleq x(k+j)$, $j = -\tau_M^1, \ldots, -1, 0$, and $x(k) = x_0$, $k = -\tau_M^1, \ldots, -1, 0$. Following [50], we find the conditions such that

$$
V(x_{k+1}) - \lambda V(x_k) - \sum_{i=1}^{N} b_i|\omega_i(k)|^2 \le 0,\ k \ge t_{N-1}, \tag{10.12}
$$

holds, where $0 < \lambda < 1$, $b_i > 0$, $i = 1, \ldots, N$. Then, we arrive at the following conditions to guarantee (10.12) and thus, for ISS of the switched system (10.9).

Lemma 10.1 *Given scalar $0 < \lambda < 1$, positive integers $0 \le \eta_m < \tau_M^N$, and K_i, $i = 1, \ldots, N$, assume that there exist scalars $b_i > 0$, $i = 1, \ldots, N$, $n \times n$ matrices $P > 0$, $S_i > 0$, $R_i > 0$, $i = 0, 1$, $G^i_{\ell,\vartheta}$ $i = 1, \ldots, N$, $\ell = 1, \ldots, N$, $\vartheta = 2, \ldots, N+1$, $\ell < \vartheta$, such that the following LMIs*

$$
\Omega_i = \begin{bmatrix} R_1 & * \\ (G^i_{\ell,\vartheta})^T & R_1 \end{bmatrix} \ge 0, \tag{10.13}
$$

$$
\begin{bmatrix} \Psi & * & * \\ PF_0^i & -P & * \\ H(F_0^i - F_1) & 0 & -H \end{bmatrix} < 0, \tag{10.14}
$$

are feasible, where

$$
\begin{aligned}
F_0^i &= [A\ 0_{n\times n}\ A_{\theta(i,N)} \ldots A_{\theta(i,1)}\ 0_{n\times n}\ B_{\theta(i,N)} \ldots B_{\theta(i,1)}],\ F_1 = [I_n\ 0_{n\times((N+2)n+n_y)}], \\
F_2 &= [0_{n\times n}\ I_n\ 0_{n\times((N+1)n+n_y)}],\ \ldots,\ F_{N+3} = [0_{n\times((N+2)n)}\ I_n\ \ 0_{n\times n_y}], \\
\Sigma &= \mathrm{diag}\{S_0 - \lambda P, -\lambda^{\eta_m}(S_0 - S_1), 0_{(Nn)\times(Nn)}, -\lambda^{\tau_M^1} S_1, -b_{\theta(i,N)} I_{\theta(i,N)}, \\
&\quad \ldots, -b_{\theta(i,1)} I_{\theta(i,1)}\}, \\
\Psi &= \Sigma - \lambda^{\eta_m}(F_1 - F_2)^T R_0 (F_1 - F_2) - \lambda^{\tau_M^1} \sum_{i=2}^{N+2} (F_i - F_{i+1})^T R_1 (F_i - F_{i+1}) \\
&\quad -2\lambda^{\tau_M^1} \sum_{j=2}^{N+1} (F_j - F_{j+1})^T \sum_{s=j+1}^{N+2} G^i_{j-1,s-1}(F_s - F_{s+1}), \\
H &= \eta_m^2 R_0 + (\tau_M^1 - \eta_m)^2 R_1,\ i = 1, \ldots, N.
\end{aligned}
\tag{10.15}
$$

Let $\mu > 0$ be a constant and $|\omega_i(k)| \leq \mu \Delta_i$, $i = 1, \ldots, N$. Then, the solutions of the switched system (10.9) with the initial conditions $x_{t_{N-1}} \in \underbrace{\mathbb{R}^n \times \cdots \times \mathbb{R}^n}_{\tau_M^1+1 \text{ times}}$ satisfy the following inequalities:

$$V(x_k) \leq \lambda^{k-t_{N-1}} V(x_{t_{N-1}}) + \frac{\mu^2}{1-\lambda} \sum_{i=1}^{N} b_i \Delta_i^2, \ k \geq t_{N-1}. \tag{10.16}$$

Proof The proof is given in the appendix.

In order to derive a bound on $V(t_{N-1})$ in terms of x_0 in a simple way, we suggest waiting for all the N latest transmitted measurements $q_{1\mu}(y_1(s_0)), q_{2\mu}(y_2(s_1)), \ldots, q_{N\mu}(y_N(s_{N-1}))$ on the controller side and then sending them together to the actuator side. This is a reasonable approach which can be easily implemented. Then for $k = 0, 1, \ldots, t_{N-1} - 1$, (10.1) is given by

$$x(k+1) = Ax(k), \ k = 0, 1, \ldots, t_{N-1} - 1. \tag{10.17}$$

Remark 10.1 A common Lyapunov functional (10.11) has been applied to the switched system (10.9) to derive the sufficient conditions for the ISS. It should be pointed out that the multiple Lyapunov functional method and the dwell time approach can be utilized to find a suitable switching signal to improve the performance [87, 215].

10.2 Dynamic Quantization of NCSs Under Round-Robin Protocol

In the sequel, based on ISS of system (10.9), we give the results on dynamic quantization of NCSs in the presence of Round-Robin protocol. By defining the initial and level sets in Sect. 10.2.1, in Sect. 10.2.2 we provide an LMI-based time-triggered zooming algorithm for the exponential stability of the switched system (10.9). In Sect. 10.2.3, we propose a novel Lyapunov-based method to initialize the zoom parameter. Under Round-Robin protocol, the digit "1" is transmitted in the protocol along with the measurements at the zooming-in sampling instants (otherwise, the digit "0" is transmitted). Thus, on the controller side it is known whether the zoom variable μ of the received measurement is updated or not. Once the value of μ is updated, all N latest transmitted measurements are waited on the controller side and then sent together to the actuator side.

10.2.1 Initial and Level Sets

Given positive numbers σ and ρ, the initial and level sets are defined as

$$\begin{aligned}\mathscr{S}_\sigma = \{x_{t_{N-1}} \in \underbrace{\mathbb{R}^n \times \cdots \times \mathbb{R}^n}_{\tau_M^1+1 \text{ times}} : V(x_{t_{N-1}}) < \sigma,\\ x^T(k)Px(k) < \sigma,\ k \in [t_{N-1} - \eta_M, t_{N-1}]\}\end{aligned} \tag{10.18}$$

and

$$\mathscr{X}_{k^*,\rho} = \{x_k \in \underbrace{\mathbb{R}^n \times \cdots \times \mathbb{R}^n}_{\tau_M^1+1 \text{ times}} : V(x_k) < \rho, k = k^*, k^*+1, \ldots\},$$

respectively. Given positive numbers μ, M_0, $\beta < 1$ and $\nu < 1$, the following lemma ensures that all the solutions of (10.9) with $x_{t_{N-1}} \in \mathscr{S}_{\mu^2 M_0^2}$ stay inside the region $\mathscr{X}_{t_{N-1},(1+\beta\nu^2)\mu^2 M_0^2}$ for all $k \geq t_{N-1}$, and enter a smaller region $\mathscr{X}_{t_{N-1}+\mathrm{T},\nu^2\mu^2 M_0^2}$ in a finite time T.

Lemma 10.2 *Given $M_j > 0,\ j = 0, 1, \ldots, N,\ \Delta_i > 0,\ i = 1, \ldots, N,$ positive integers $0 \leq \eta_m < \tau_M^N$ and tuning parameters $0 < \lambda < 1,\ 0 < \nu < 1$, assume that there exist scalars $0 < \beta < 1,\ b_i,\ i = 1, \ldots, N,\ n \times n$ matrices $P > 0,\ S_i > 0,$ $R_i > 0,\ i = 0, 1, G_{\ell,j}^i\ i = 1, \ldots, N, \ell = 1, \ldots, N, j = 2, \ldots, N+1, \ell < j$, such that LMIs (10.13)–(10.14) and*

$$(1 + \beta\nu^2)M_0^2 C_i^T C_i < PM_i^2,\ i = 1, \ldots, N, \tag{10.19}$$

$$\frac{1}{1-\lambda}\sum_{i=1}^{N} b_i \Delta_i^2 < \beta\nu^2 M_0^2 \tag{10.20}$$

hold. Let $\mu > 0$ be a constant. Then, the solutions of (10.9) that start in the region $\mathscr{S}_{\mu^2 M_0^2}$

(i) Satisfy $|C_i x(t_p - \eta_p)| = |y_i(t_p - \eta_p)| < \mu M_i,\ p \in \mathbb{Z}^+$, (implying $|\omega_i(k)| \leq \mu\Delta_i$ for all $i \in \mathscr{I}$ and $k = t_{N-1}, t_{N-1}+1, \ldots$);
(ii) Remain in the set $\mathscr{X}_{t_{N-1},(1+\beta\nu^2)\mu^2 M_0^2}$;
(iii) Enter a smaller set $\mathscr{X}_{t_{N-1}+\mathrm{T},\nu^2\mu^2 M_0^2}$ in a finite time $\mathrm{T} = \lceil \tilde{\mathrm{T}} \rceil$, where $\tilde{\mathrm{T}}$ is the solution of

$$\lambda^{\tilde{\mathrm{T}}} = (1-\beta)\nu^2. \tag{10.21}$$

The proof of Lemma 10.2 follows from Lemma 8.2. The second inequality in (10.18) allows us to guarantee the bounds on $y(s_p)$, $s_p < t_{N-1}$ by verifying (10.19).

Remark 10.2 The functional $V(x_k)$ is a standard Lyapunov functional for delay-dependent analysis. The LMIs of Lemma 10.2 are feasible for small enough delay bound τ_M^N, large enough quantization ranges $M_1, \ldots, M_N$ and small enough quantization errors $\Delta_1, \ldots, \Delta_N$. Indeed, the LMIs (10.13) and (10.14) are feasible for

$\tau_M^N = 0$ (i.e., in the absence of delay) since $A + BKC$ is Schur. Hence, (10.13) and (10.14) are feasible for small enough τ_M^N. The LMIs (10.19) and (10.20) are feasible for large enough quantization ranges and small enough quantization intervals. Moreover, the initial values of λ and ν can be set to be 1. It is noted that the conditions are sufficient only and always may be improved.

10.2.2 Dynamic Quantization and Zooming Algorithm

In this section, we consider dynamic quantizers with the zoom variable μ. The zooming is performed on the sensor level. Therefore, in the closed-loop system $\mu = \mu(s_p)$ is constant on $[t_p, t_{p+1} - 1]$.

Given $\mu_0 > 0$, let $\mu = \mu_0$, $x_{t_{N-1}} \in \mathscr{S}_{\mu^2 M_0^2} = \mathscr{S}_{\mu_0^2 M_0^2}$. We will show how to choose μ_0 in Theorem 10.1 below. Assume that the LMIs of Lemma 10.2 are feasible. In the sequel, we propose a zooming-in algorithm in Fig. 10.2, where μ is decreased and

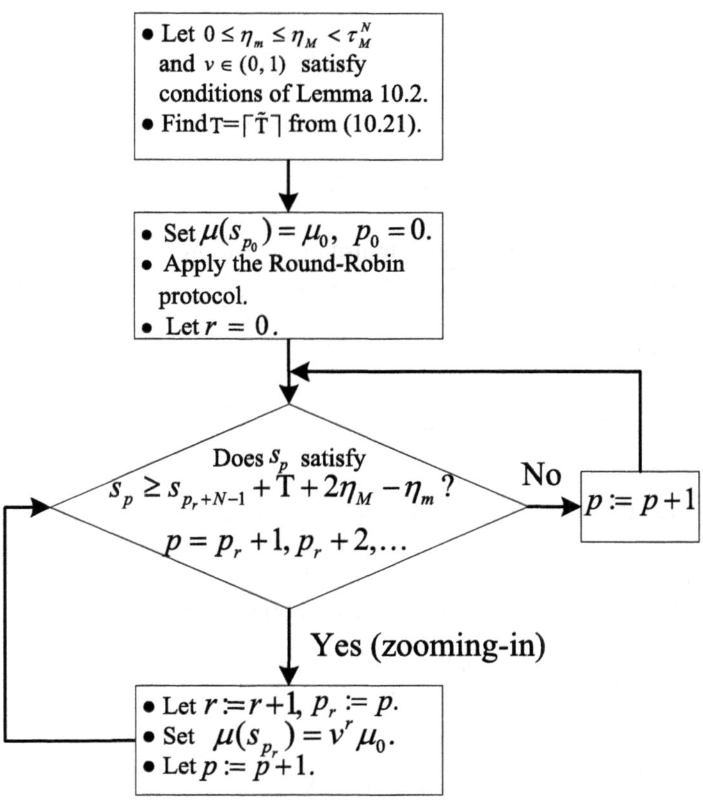

Fig. 10.2 "Zooming-in" algorithm for dynamic quantization and Round-Robin protocol

thus, the resulting quantization error is reduced to drive the state of (10.9) to the origin exponentially.

Definition 10.2 If there exist constants $b > 0$ and $0 < \kappa < 1$ such that

$$|x(k)|^2 \leq b\kappa^{2(k-t_{N-1})}\mu_0^2 M_0^2,\ \forall k \geq t_{N-1},\ k \in \mathbb{N},$$

for the solutions of the system (10.9) initialized with $x_{t_{N-1}} \in \mathcal{S}_{\mu_0^2 M_0^2}$, then the system (10.9) with $|\omega_i(k)| \leq \mu \Delta_i$, $i = 1, \ldots, N$, is said to be exponentially stable with the decay rate κ for some choice of the zoom variable $\mu > 0$.

Proposition 10.1 *Assume that the LMIs of Lemma 10.2 are feasible. Given $\mu_0 > 0$, let $\mu = \mu_0$, $x_{t_{N-1}} \in \mathcal{S}_{\mu_0^2 M_0^2}$. Then under the algorithm in Fig. 10.2, the system (10.9) is exponentially stable with a decay rate $\kappa = \nu^{\frac{1}{\bar{\tau}_M}}$, where*

$$\bar{\tau}_M = \mathrm{T} + N\tau_M^N + 2\eta_M - N\eta_m - \eta_m + N. \tag{10.22}$$

Proof Set $r = 0$. Since

$$t_{p_1} - \eta_M = s_{p_1} + \eta_{p_1} - \eta_M \geq t_{N-1} + \mathrm{T} + \eta_{p_1} - \eta_m \geq t_{N-1} + \mathrm{T},$$

the application of Lemma 10.2 with $\mu = \mu_0$ leads to

$$x^T(k)Px(k) \leq V(x_k) < \nu^2\mu_0^2 M_0^2,\ \forall k \geq t_{p_1} - \eta_M,\ k \in \mathbb{N}.$$

Set $r = 1$. We wait for all N latest transmitted measurements to arrive into the reduced domain with $x^T(s_p)C_i^T C_i x(s_p) < \mu_0^2\nu^2 M_i^2$, $i = 1, \ldots, N$, for $p \geq p_1$, where $|\omega_i(k)| \leq \mu_0\nu\Delta_i$, $k \geq t_{p_1}$, $k \in \mathbb{N}$. After sampling instant s_{p_1+N-1}, the resulting closed-loop system has the initial condition

$$x_{t_{p_1+N-1}} \in \underbrace{\mathbb{R}^n \times \cdots \times \mathbb{R}^n}_{\tau_M^1+1 \text{ times}} : V(x_{t_{p_1+N-1}}) < \nu^2\mu_0^2 M_0^2. \tag{10.23}$$

Then, Lemma 10.2 is applied with $\mu = \mu_0\nu$, where t_{N-1} and η_{N-1} are changed by t_{p_1+N-1} and η_{p_1+N-1}, respectively. Thus, the solutions of (10.9) initiated by (10.23) remain in a region $\mathcal{X}_{t_{p_1+N-1},(1+\beta\nu^2)\nu^2\mu_0^2 M_0^2}$ for all $k \geq t_{p_1+N-1}$, $k \in \mathbb{N}$. Since $s_{p_1+N-1} = t_{p_1+N-1} - \eta_{p_1+N-1} \geq t_{p_1+N-1} - \eta_M$, from (10.19) it follows that

$$x^T(s_p)C_i^T C_i x(s_p) < \frac{x^T(s_p)Px(s_p) \cdot \nu^2\mu_0^2 M_i^2}{(1+\beta\nu^2)\nu^2\mu_0^2 M_0^2} < \nu^2\mu_0^2 M_i^2,\ i = 1, \ldots, N, \forall p \geq p_1 + N - 1,$$

and thus, $|\omega_i(k)| \leq \nu\mu_0\Delta_i$, $i = 1, \ldots, N$, $k \geq t_{p_1+N-1} = s_{p_1+N-1} + \eta_{p_1+N-1}$. Therefore, for $k \geq t_{p_2} - \eta_M \geq t_{p_1+N-1} + \mathrm{T}$, we obtain

$$
\begin{aligned}
V(x_k) &\le \lambda^{k-t_{p_1+N-1}} V(x_{t_{p_1+N-1}}) + \frac{\nu^2\mu_0^2}{1-\lambda}\sum_{i=1}^{N} b_i \Delta_i^2 \\
&\le \lambda^{\mathrm{T}} V(x_{t_{p_1+N-1}}) + \frac{\nu^2\mu_0^2}{1-\lambda}\sum_{i=1}^{N} b_i \Delta_i^2 \\
&< (1-\beta)\nu^2 \cdot \nu^2\mu_0^2 M_0^2 + \beta\nu^2\mu_0^2 M_0^2 \cdot \nu^2 \\
&= \nu^4 \mu_0^2 M_0^2.
\end{aligned}
$$

Similarly, for $r = 2, 3, \ldots$ we have $V(x_k) < \nu^{2r}\mu_0^2 M_0^2$ for all $k \in [t_{p_r} - \eta_M, t_{p_{r+1}} - \eta_M - 1]$. Noting that

$$
\begin{aligned}
r\mathrm{T} + (r-1)(2\eta_M - \eta_m + N - 1) + t_{N-1} \le t_{p_r} - \eta_M \le k \\
\le t_{p_{r+1}} - \eta_M - 1 < (r+1)\bar{\tau}_M - 1 + t_{N-1},
\end{aligned}
$$

we obtain for $k \in [t_{p_r} - \eta_M, t_{p_{r+1}} - \eta_M - 1],\ r \in \mathbb{N}$,

$$
\begin{aligned}
V(x_k) &< \nu^{2r}\mu_0^2 M_0^2 < \nu^{2\left(\frac{k-t_{N-1}}{\bar{\tau}_M} - \frac{\bar{\tau}_M - 1}{\bar{\tau}_M}\right)} \mu_0^2 M_0^2 \\
&= \nu^{-\frac{2(\bar{\tau}_M - 1)}{\bar{\tau}_M}} \left(\nu^{\frac{1}{\bar{\tau}_M}}\right)^{2(k-t_{N-1})} \mu_0^2 M_0^2.
\end{aligned}
$$

Then, the following holds for $k \ge t_{N-1}$

$$
|x(k)|^2 \le \nu^{-\frac{2(\bar{\tau}_M - 1)}{\bar{\tau}_M}} [\lambda_{\min}(P)]^{-1} \left(\nu^{\frac{1}{\bar{\tau}_M}}\right)^{2(k-t_{N-1})} \mu_0^2 M_0^2.
$$

Remark 10.3 In the above analysis, it is assumed that the data packet dropout does not occur. However, for small delays $\eta_p < s_{p+1} - s_p$, if the number of successive packet dropouts is upper bounded by $\bar{d}$, in the presence of Round-Robin scheduling we could accommodate the packet dropouts by modeling them as prolongations of the transmission interval and replace T by $\mathrm{T} + 2\bar{d} \cdot \mathrm{MATI}$ in the algorithm.

10.2.3 Initialization of the Zoom Variable

The algorithm of the previous section is given in terms of the initial set $\mathscr{S}_{\mu_0^2 M_0^2}$ that involves the bound on $V(x_{t_{N-1}})$. In this section, we find the ball of initial conditions $x(0) = x_0$, starting from which the solutions of (10.9) and (10.17) remain in the initial set $\mathscr{S}_{\mu_0^2 M_0^2}$. From (10.5) and the bound

$$
s_{N-2} = s_{N-2} - s_{N-3} + s_{N-3} - \cdots + s_1 - s_0 \le (N-2)(\tau_M^N - \eta_m + 1),
$$

it holds that

$$
t_{N-1} \le s_{N-2} + \tau_M^N + 1 \le (N-2)(\tau_M^N - \eta_m + 1) + \tau_M^N + 1 \triangleq \hat{\tau}_M. \tag{10.24}
$$

Then following [130], we derive a bound on $V(x_{t_{N-1}})$ in terms of x_0 in the next lemma.

Lemma 10.3 ([130]) *Consider the Lyapunov functional* $V(x_k)$ *given by (10.11) and denote* $V_0(k) = x^T(k)Px(k)$. *Under the constant initial condition* $x(k) = x_0$, $k < 0$, *if there exist* $0 < \lambda < 1$ *and* $c > 1$ *such that the following inequalities*

$$V_0(k+1) - cV_0(k) \leq 0, \tag{10.25a}$$
$$V(x_{k+1}) - \lambda V(x_k) - (c-1)V_0(k) \leq 0, \tag{10.25b}$$

hold for $k = 0, 1, \ldots, t_{N-1} - 1$ *along (10.17), then we have*

$$\begin{aligned} V_0(k) &\leq \lambda_{\max}(c^{\hat{\tau}_M} P)|x_0|^2, \ k = 0, 1, \ldots, t_{N-1}, \\ V(x_{t_{N-1}}) &\leq \lambda_{\max}(c^{\hat{\tau}_M} P + \Omega)|x_0|^2, \end{aligned} \tag{10.26}$$

where $\hat{\tau}_M$ *is given by (10.24) and*

$$\Omega = \eta_m S_0 + \lambda^{\eta_m}(\tau_M^1 - \eta_m)S_1. \tag{10.27}$$

As a consequence, we achieve the main result.

Theorem 10.1 *Given* $M_j > 0$, $j = 0, 1, \ldots, N$, $\Delta_i > 0$, $i = 1, \ldots, N$, *positive integers* $0 \leq \eta_m \leq \eta_M < \tau_M^N$ *and tuning parameters* $0 < \lambda < 1$, $0 < \nu < 1$, $c > 1$, *assume that there exist scalars* $0 < \beta < 1$, b_i, $i = 1, \ldots, N$, $n \times n$ *matrices* $P > 0$, $S_i > 0$, $R_i > 0$, $i = 0, 1$, $G_{\ell,\vartheta}^i$, $i = 1, \ldots, N$, $\ell = 1, \ldots, N$, $\vartheta = 2, \ldots, N+1$, $\ell < \vartheta$, *such that the LMIs (10.13)–(10.14), (10.19)–(10.20) and the following LMIs*

$$\begin{bmatrix} -cP & * \\ PA & -P \end{bmatrix} < 0, \tag{10.28}$$

$$\begin{bmatrix} \tilde{\Psi} & * & * \\ P\tilde{F}_0 & -P & * \\ H(\tilde{F}_0 - \tilde{F}_1) & 0 & -H \end{bmatrix} < 0, \tag{10.29}$$

are feasible, where

$$\begin{aligned} \tilde{F}_0 &= [A \ 0_{n\times((N+2)n)}], \ \tilde{F}_1 = [I_n \ 0_{n\times((N+2)n)}], \\ \tilde{F}_2 &= [0_{n\times n} \ I_n \ 0_{n\times((N+1)n)}], \ldots, \ \tilde{F}_{N+3} = [0_{n\times((N+2)n)} \ I_n], \\ \tilde{\Sigma} &= \mathrm{diag}\{S_0 - \lambda P - (c-1)P, -\lambda^{\eta_m}(S_0 - S_1), 0_{(Nn)\times(Nn)}, -\lambda^{\tau_M^1} S_1\}, \\ \tilde{\Psi} &= \tilde{\Sigma} - \lambda^{\eta_m}(\tilde{F}_1 - \tilde{F}_2)^T R_0(\tilde{F}_1 - \tilde{F}_2) - \lambda^{\tau_M^1} \sum_{i=2}^{N+2} (\tilde{F}_i - \tilde{F}_{i+1})^T R_1(\tilde{F}_i - \tilde{F}_{i+1}) \\ &\quad -2\lambda^{\tau_M^1} \sum_{j=2}^{N+1} (\tilde{F}_j - \tilde{F}_{j+1})^T \sum_{s=j+1}^{N+2} G_{j-1,s-1}^1(\tilde{F}_s - \tilde{F}_{s+1}), \end{aligned} \tag{10.30}$$

and the notation H is given by (10.15). If the initial condition satisfies the inequality $|x_0| < X_0$, where $X_0 > 0$ is known, then the zooming-in algorithm of Sect. 10.2.2 starting with $\mu(s_0) = \mu_0$ with μ_0 given by

$$\mu_0^2 = \frac{\lambda_{\max}(c^{\hat{\tau}_M} P + \Omega)}{M_0^2} X_0^2 \tag{10.31}$$

exponentially stabilizes system (10.9) and (10.17), where $\hat{\tau}_M$ and Ω are given by (10.24) and (10.27), respectively.

Proof From [130], it follows that the matrix inequalities (10.13), (10.28) and (10.29) guarantee (10.25) along (10.17) for $k = 0, 1, \ldots, t_{N-1} - 1$. Therefore, if the initial condition satisfies the inequality $|x_0| < X_0$, then we have

$$\begin{aligned} \max\{V_0(k), V(x_{t_{N-1}})\} &\leq \lambda_{\max}(c^{\hat{\tau}_M} P + \Omega) X_0^2 \\ &= \mu_0^2 M_0^2, \ k = 0, 1, \ldots, t_{N-1}, \end{aligned}$$

meaning that $x_{t_{N-1}} \in \mathscr{S}_{\mu_0^2 M_0^2}$. The result then follows from Proposition 10.1.

Remark 10.4 Note that given a bound $X_0 > 0$ on the initial state conditions and the values of the quantizer range $M_i > 0$ and the error $\Delta_i > 0$, $i = 1, \ldots, N$, Eq. (10.31) defines the initial value of the zoom variable, starting from which the exponential stability is guaranteed by using zooming-in only. If the initial value of the zoom variable is given by μ_0, then the zooming-in algorithm of Sect. 10.2.2 starting with $\mu(s_0) = \mu_0$ exponentially stabilizes all the solutions of (10.9) and (10.17) starting from the initial ball

$$|x_0| < X_0, \quad X_0 = \frac{\mu_0 M_0}{\sqrt{\lambda_{max}(c^{\hat{\tau}_M} P + \Omega)}}, \tag{10.32}$$

where $\hat{\tau}_M$ and Ω are given by (10.24) and (10.27), respectively. In order to maximize the initial ball (10.32), the condition $c^{\hat{\tau}_M} P + \Omega - \gamma I < 0$ can be added to the conditions of Theorem 10.1, where $\gamma > 0$ is to be minimized.

Remark 10.5 The conditions of Theorem 10.1 possess $(N + 1)$ of $2n \times 2n$, one of $(N + 5)n \times (N + 5)n$, N of $((N + 5)n + n_y) \times ((N + 5)n + n_y)$ LMIs, and have the number $\frac{N^2(N+1)+5}{2} n^2 + 2.5n + N + 1$ of decision variables. The huge numerical complexity is caused by the switched closed-loop system (10.9) composed of N subsystems.

Remark 10.6 The LMIs of Theorem 10.1 are affine in the system matrices. Therefore, in the case of system matrices from the uncertain time-varying polytope

$$\Theta = \sum_{j=1}^{M} g_j(k)\Theta_j,\ 0 \le g_j(k) \le 1,$$
$$\sum_{j=1}^{M} g_j(k) = 1,\ \Theta_j = \begin{bmatrix} A^{(j)} & B^{(j)} \end{bmatrix},$$

where $g_j(k),\ j = 1, \ldots, M$, are uncertain time-varying parameters and the system matrices $A^{(j)}$ and $B^{(j)},\ j = 1, \ldots, M$, are known with appropriate dimensions, one has to solve these LMIs simultaneously for all the M vertices Θ_j, applying the same decision matrices.

10.3 Illustrative Examples

10.3.1 Inverted Pendulum

Example 10.1 The inverted pendulum system is widely used as a benchmark for testing control algorithm. The dynamics of the inverted pendulum on a cart shown in Fig. 10.3 can be described in the following as in e.g., [255]:

$$\begin{bmatrix} \dot{x} \\ \ddot{x} \\ \dot{\theta} \\ \ddot{\theta} \end{bmatrix} = \begin{bmatrix} 0 & 1 & 0 & 0 \\ 0 & -\frac{(a+ml^2)b}{a(M+m)+Mml^2} & -\frac{m^2gl^2}{a(M+m)+Mml^2} & 0 \\ 0 & 0 & 0 & 1 \\ 0 & -\frac{mlb}{a(M+m)+Mml^2} & -\frac{mgl(M+m)}{a(M+m)+Mml^2} & 0 \end{bmatrix} \begin{bmatrix} x \\ \dot{x} \\ \theta \\ \dot{\theta} \end{bmatrix} + \begin{bmatrix} 0 \\ \frac{a+ml^2}{a(M+m)+Mml^2} \\ 0 \\ \frac{ml}{a(M+m)+Mml^2} \end{bmatrix} u, \tag{10.33}$$

where the parameters a and b represent the friction of the cart and the inertia of the pendulum, respectively, and are chosen as $a = 0.0034\,\mathrm{kg}\cdot\mathrm{m}^2$ and $b = 0.1\,\mathrm{N/m/s}$; the mass of the cart M and the mass of the pendulum m are given by $M = 1.096\,\mathrm{kg}$ and $m = 0.109\,\mathrm{kg}$, respectively; the length of the pendulum l is chosen as $0.25\,\mathrm{m}$, and $g = 9.8\,\mathrm{m/s^2}$ is the gravity acceleration. In the model, x and θ represent the cart position coordinate and the pendulum angle from vertical, respectively. By choosing a sampling time $\mathrm{T_s} = 0.01\,\mathrm{s}$, we obtain the following discrete-time system model:

$$x(k+1) = \begin{bmatrix} 1 & 0.01 & 0 & 0 \\ 0 & 0.9991 & 0.0063 & 0 \\ 0 & 0 & 1.0014 & 0.01 \\ 0 & -0.0024 & 0.2784 & 1.0014 \end{bmatrix} x(k) + \begin{bmatrix} 0 \\ 0.0088 \\ 0.0001 \\ 0.0236 \end{bmatrix} u(k),\ k \in \mathbb{Z}^+. \tag{10.34}$$

The pendulum can be stabilized by a state feedback $u(k) = Kx(k)$ with the gain $K = [K_1\ \ K_2]$

$$K_1 = \begin{bmatrix} 0.9163 & 2.0169 \end{bmatrix},\ K_2 = \begin{bmatrix} -27.4850 & -5.3437 \end{bmatrix},$$

Fig. 10.3 Inverted pendulum system

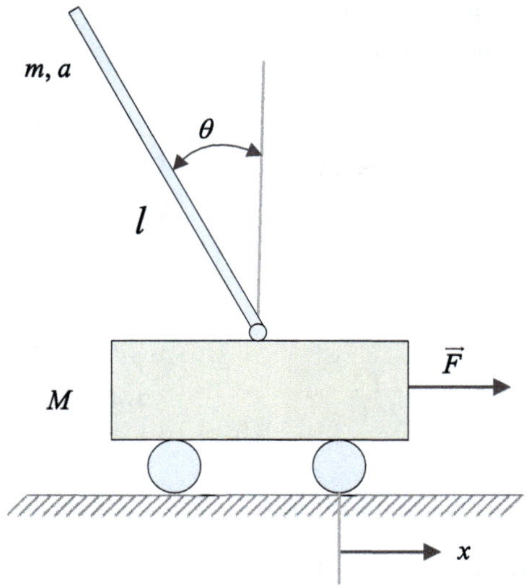

which leads to the closed-loop system having eigenvalues

$$\{0.9419, 0.9865 + 0.0035i,\ 0.9865 - 0.0035i,\ 0.9813\}.$$

Suppose that the spatially distributed components of the state of the cart-pendulum system (10.34) are not accessible simultaneously.

Consider $N = 2$ and

$$C_1 = \begin{bmatrix} 1\,0\,0\,0 \\ 0\,1\,0\,0 \end{bmatrix},\ C_2 = \begin{bmatrix} 0\,0\,1\,0 \\ 0\,0\,0\,1 \end{bmatrix}. \tag{10.35}$$

The quantizer is chosen as

$$q_\mu(y^i) = \begin{cases} 100\mu\ \mathrm{sgn}(y^i), & \text{if } |y^i| > 100\mu, \\ \mu\left\lfloor \frac{y^i}{\mu} + 0.01 \right\rfloor, & \text{if } |y^i| \le 100\mu, \end{cases}$$

where y^i is the ith component of y, $i = 1, \ldots, 4$. Therefore, we can take $M_1 = M_2 = 100$, $\Delta_1 = \Delta_2 = 0.01$. Choose $\mu_0 = 1$, $M_0 = 100$, $\nu = 0.9$, $\lambda = 0.984$, $c = 1.37$, $\tau_M^N = 4$, $\eta_m = 0$, $\eta_M = 2$. Then from (10.5), it follows that the network-induced delays η_p and the sampling intervals are bounded by $0 \le \eta_p \le 2$ and $1 \le s_{p+1} - s_p \le 3$, $p \in \mathbb{Z}^+$, respectively. It is observed that we allow the network-induced delays are larger than the sampling intervals.

The initial state is assumed to be $x_0 = [0.5\ 0.3\ \ -0.2\ \ -0.9]^T$. In the simulation, the network-induced delays are generated randomly according to the aforementioned assumption and shown in Fig. 10.4. By Theorem 10.1 we find $\mathrm{T} = \lceil -\frac{\ln(1-\beta)+2\ln\nu}{\ln\lambda} \rceil = 15$ from (10.21). Then, the zooming-in algorithm of Sect. 10.2.2

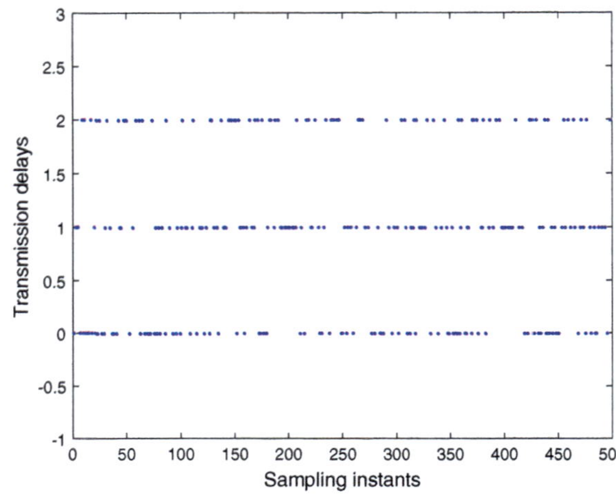

Fig. 10.4 Example 10.1: transmission delays

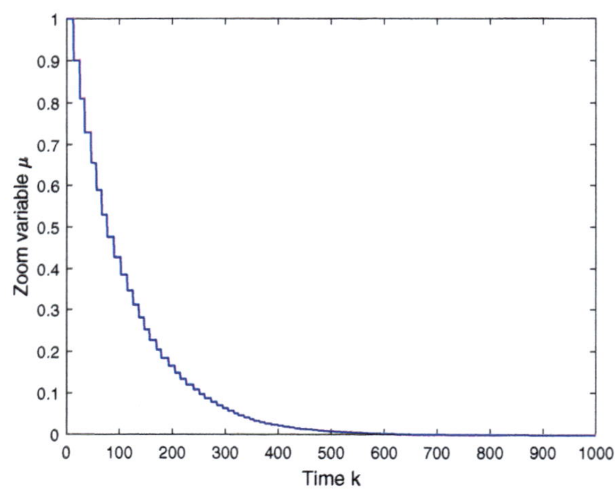

Fig. 10.5 Example 10.1: evolution of the zoom variable μ in the zooming-in algorithm

and Round-Robin protocol with T $= 15$ and $\nu = 0.9$ exponentially stabilizes all the solutions of (10.9) and (10.17) starting from the initial ball $|x_0| < 1.2376$. Figure 10.5 shows the evolution of the zoom variable $\mu(k)$.

Moreover, it is found that the system is exponentially stable with a decay rate $\kappa = \nu^{\frac{1}{\bar{\tau}_M}} = 0.9964$, where (following (10.22)) $\bar{\tau}_M = \mathrm{T} + 2\tau_M^N + 2\eta_M - 3\eta_m + 2 = 29$. The evolution of the control input and the state is depicted in Figs. 10.6 and 10.7, respectively.

For the case of $N = 1$, i.e., the scheduling is not taken into account and $y(k) = [C_1^T \; C_2^T]^T x(k)$, we achieve a slightly better $\kappa = 0.9950$ for essentially larger initial ball $|x_0| < 11.1951$.

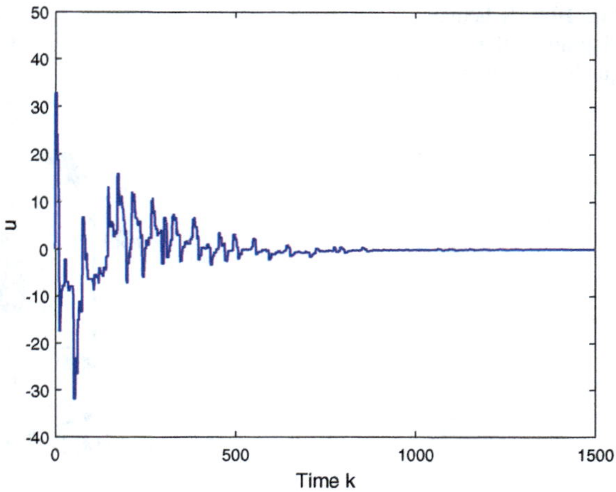

Fig. 10.6 Example 10.1: evolution of the control input

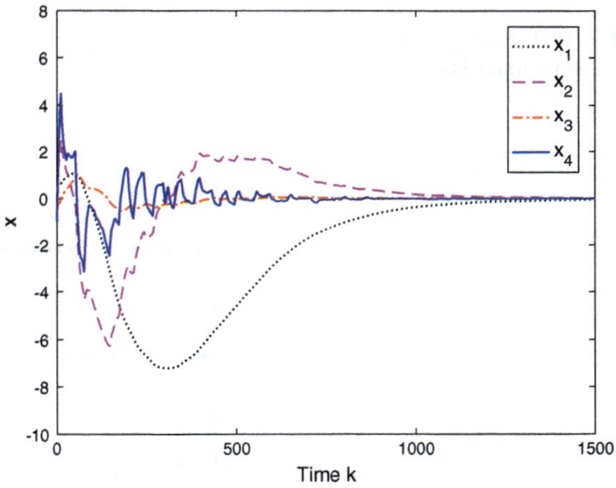

Fig. 10.7 Example 10.1: trajectory of the closed-loop system

10.3.2 Quadruple-Tank Process

Example 10.2 We also illustrate the efficiency of the given conditions on the example of the quadruple-tank process [106] described in Fig. 10.8. The linear discrete-time model obtained in [223] is given by:

$$x(k+1) = \begin{bmatrix} 0.975 & 0 & 0.042 & 0 \\ 0 & 0.977 & 0 & 0.044 \\ 0 & 0 & 0.958 & 0 \\ 0 & 0 & 0 & 0.956 \end{bmatrix} x(k) + \begin{bmatrix} 0.0515 & 0.0016 \\ 0.0019 & 0.0447 \\ 0 & 0.0737 \\ 0.0850 & 0 \end{bmatrix} u(k),\ k \in \mathbb{Z}^+. \tag{10.36}$$

Here, the open-loop system is exponentially stable with a decay rate $\kappa = 0.9770$.

Consider $N = 2$ and choose a controller gain $K = [K_1 \;\; K_2]$, where

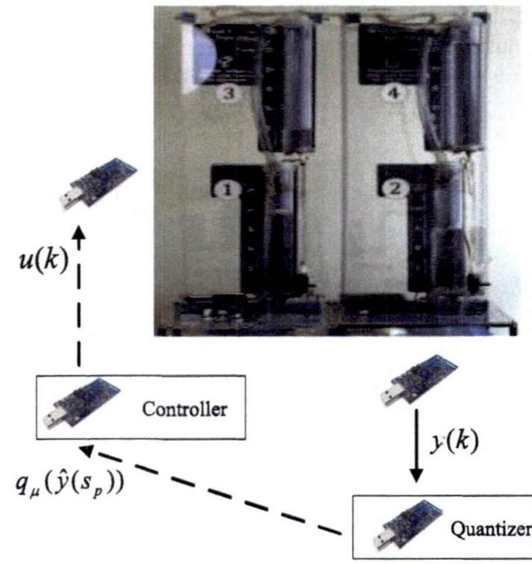

Fig. 10.8 Schematic diagram of the quadruple-tank process

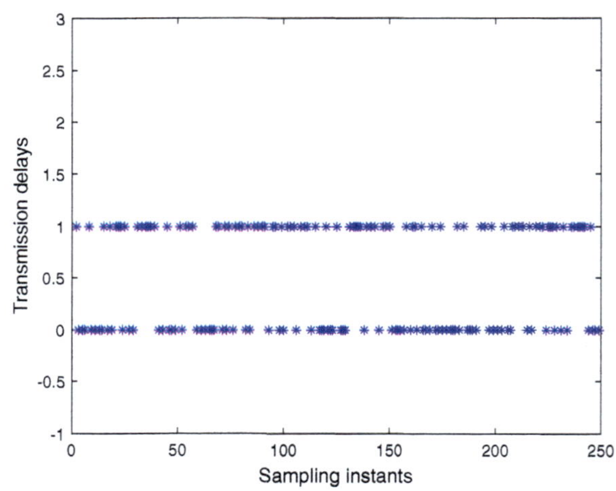

Fig. 10.9 Example 10.2: transmission delays

$$K_1 = \begin{bmatrix} 0.0449 & -0.3007 \\ -0.3080 & 0.0469 \end{bmatrix}, \quad K_2 = \begin{bmatrix} 0.1651 & -0.5644 \\ -0.6275 & 0.1681 \end{bmatrix}.$$

The measurement outputs are $y_i(k) = C_i x(k)$ with C_i, $i = 1, 2$, given by (10.35). Suppose that the components of the state of system (10.36) are not accessible simultaneously. The quantizer is chosen as

$$q_\mu(y^i) = \begin{cases} 150\mu \operatorname{sgn}(y^i), & \text{if } |y^i| > 150\mu, \\ \mu \left\lfloor \frac{y^i}{\mu} + 0.001 \right\rfloor, & \text{if } |y^i| \leq 150\mu, \end{cases}$$

where y^i is the ith component of y, $i = 1, \ldots, 4$. Therefore, we can take $M_1 = M_2 = 150$, $\Delta_1 = \Delta_2 = 0.001$.

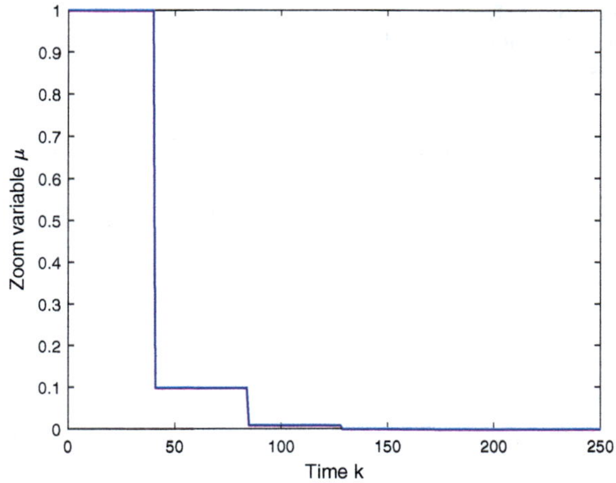

Fig. 10.10 Example 10.2: evolution of the zoom variable μ in the zooming-in algorithm

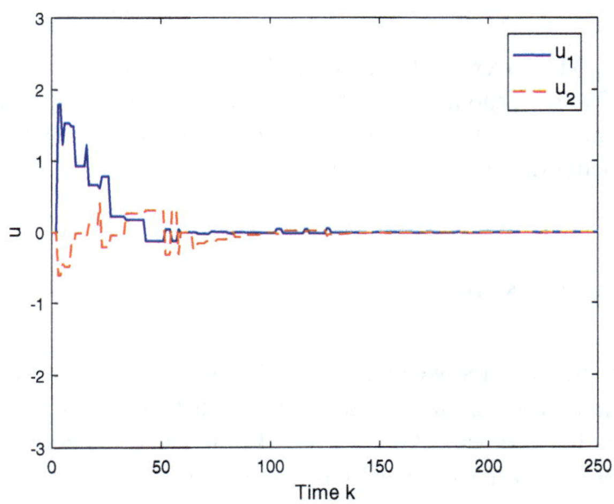

Fig. 10.11 Example 10.2: evolution of the control input

Choose $\tau_M^N = 2,\ \eta_m = 0,\ \eta_M = 1$. From (10.5), it follows that the network-induced delays η_p and the sampling intervals are bounded by $0 \le \eta_p \le 1$ and $1 \le s_{p+1} - s_p \le 2,\ p \in \mathbb{Z}^+$, respectively. The network-induced delays are depicted in Fig. 10.9.

Then, we find that given $\mu_0 = 1,\ M_0 = 100,\ \nu = 0.1,\ \lambda = 0.926,\ c = 1.10$, the zooming-in algorithm of Sect. 10.2.2 and Round-Robin protocol with T = 60 exponentially stabilizes all the solutions of (10.9) and (10.17) starting from the initial ball $|x_0| < 23.7748$ with a decay rate $\kappa = 0.9667$. The decay rate for closed-loop system is improved compared to the one for the open-loop system. The evolution of the zoom variable $\mu(k)$ is presented in Fig. 10.10. The evolution of the control input and the state with the initial state $x_0 = [5\ \ 2\ \ -2\ \ -4]^T$ is given in Figs. 10.11 and 10.12, respectively.

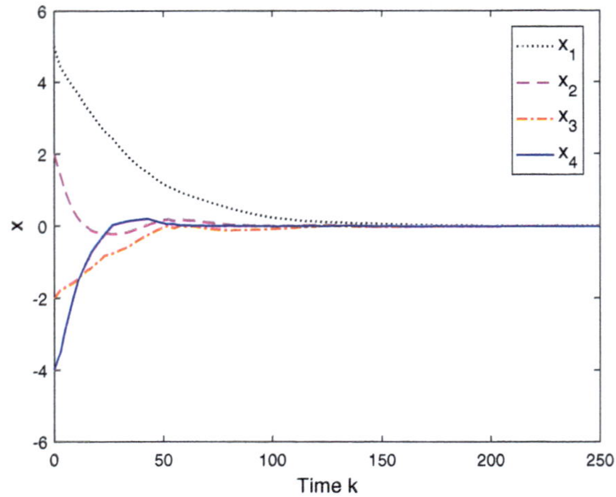

Fig. 10.12 Example 10.2: trajectory of the closed-loop system

Moreover, for the case of $N = 1$, it is shown that the zooming-in algorithm with $\mathrm{T} = 60$ exponentially stabilizes all the solutions of the closed-loop system (10.9) and (10.17) ($N = 1$) starting from a larger initial ball $|x_0| < 91.4413$ with a slightly better decay rate $\kappa = 0.9647$.

10.4 Notes

In this chapter, we investigated linear discrete-time NCSs that are subject to dynamic quantization, variable communication delays, variable sampling intervals and Round-Robin protocol. An LMI-based time-triggered zooming algorithm, which includes proper initialization of the zoom parameter, was proposed for exponential stability of the switched closed-loop system.

Chapter 11
Stability Conditions for Discrete-Time Systems Under Dynamic Protocols

In Chap. 9 the time-delay approach has been applied to the stability analysis of discrete-time NCSs with actuator constraints and two sensor nodes under TOD scheduling protocol in the presence of large communication delays. Only partial stability of the resulting hybrid-delayed system has been guaranteed. The extension from two to multiple sensor nodes is far from being straightforward. On the one hand, the time-dependent Lyapunov functional of Chap. 9 is not applicable any more. On the other hand, it is important to guarantee the stability of the resulting closed-loop system with respect to the full state and not only to the partial state.

In this chapter, we focus on the stability problem of discrete-time NCSs with multiple sensor nodes under dynamic scheduling protocols, aiming at presenting an improved stability criterion to find MATI and MAD such that the resulting closed-loop system is exponentially stable with respect to the full state. This chapter presents the discrete-time counterpart of the continuous-time results of Chap. 5 for TOD protocol and Chap. 6 for stochastic protocol with an improvement achieved by the developed reciprocally convex combination inequality proposed in [253] and the discrete-time counterpart of augmented Lyapunov functional provided in [146] for stability analysis of continuous-time systems with time-varying delays. The preliminary results on the stabilization of networked discrete-time system with multiple sensor nodes under TOD protocol were provided in [143], where the condition was obtained by Jensen's inequality and the reciprocally convex combination approach [173].

11.1 Discrete-Time NCS Model and Preliminaries

In this section, we first demonstrate the discrete-time description of the NCS model and then introduce the dynamic scheduling protocols to be adopted in the chapter.

K. Liu et al., *Networked Control Under Communication Constraints*, Advances in Delays and Dynamics 11, https://doi.org/10.1007/978-981-15-4230-5_11

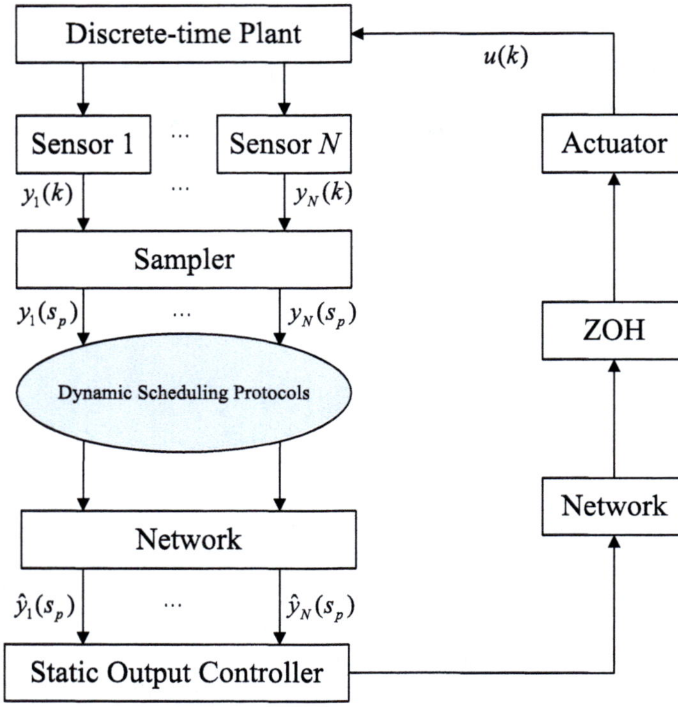

Fig. 11.1 Architecture of networked control systems under dynamic scheduling protocols

11.1.1 Description of System Data

Consider the networked control scheme shown in Fig. 11.1, where a linear discrete-time plant, N distributed sensors, a controller node and an actuator node are all connected via communication networks. The linear time-invariant discrete-time plant is given by

$$\begin{aligned} x(k+1) &= Ax(k) + Bu(k), \quad k \in \mathbb{Z}^+, \\ y_i(k) &= C_i x(k), \quad i = 1, \ldots, N, \end{aligned} \tag{11.1}$$

where $x(k) \in \mathbb{R}^n$ denotes the state of the plant, $u(k) \in \mathbb{R}^{n_u}$ denotes the control input, $y_i(k) \in \mathbb{R}^{n_i}$ $(i = 1, \ldots, N)$ denotes the measurement outputs of the plant, and A, B and C_i, $i = 1, \ldots, N$ are system matrices with appropriate dimensions. These matrices can be uncertain with polytopic-type uncertainty. The initial condition is given by $x(k) = x_0$. We denote

$$C = \left[\, C_1^T \cdots C_N^T \,\right]^T, \quad y(k) = \left[\, y_1^T(k) \cdots y_N^T(k) \,\right]^T \in \mathbb{R}^{n_y}$$

and $\sum_{i=1}^N n_i = n_y$.

The sequence of sampling instants $0 = s_0, s_1, s_2, \ldots$ is strictly increasing in the sense that $s_{p+1} - s_p \leq \text{MATI}$, where $\{s_p\}$ is a subsequence of $\mathbb{Z}^+$. Denote by t_p the updating time instant of the ZOH. Suppose that the updating data at the instant t_p on the actuator side has experienced an uncertain transmission delay $h_p = t_p - s_p$ as it is transmitted through the network (both from the sensor to the controller and from the controller to the actuator). The delays may be either smaller or larger than the sampling interval provided that the transmission order of the data packets is maintained [161]. Assume that the network-induced delay h_p is bounded with the interval $[h_m,\ h_M]$, where h_m and h_M are known as nonnegative integers.

Denote by $\hat{y}(s_p) = [\hat{y}_1^T(s_p) \cdots \hat{y}_N^T(s_p)]^T \in \mathbb{R}^{n_y}$ the output information submitted to the dynamic scheduling protocol. At each sampling instant s_p, one of the outputs $y_i(s_p) \in \mathbb{R}^{n_i}$ is transmitted over the network; that is, one of the $\hat{y}_i(s_p)$ values is updated with the recent state $y_i(s_p)$. Let $i_p^* \in \mathscr{I}_N = \{1, \ldots, N\}$ denote the active output node at the sampling instant s_p. Then, we have

$$\hat{y}_i(s_p) = \begin{cases} y_i(s_p), & i = i_p^*, \\ \hat{y}_i(s_{p-1}), & i \neq i_p^*. \end{cases} \tag{11.2}$$

Following (5.5), we denote $e(k)$ by the error between the system output $y(s_p)$ and the latest available information $\hat{y}(s_{p-1})$:

$$\begin{aligned} &e(k) = \text{col}\{e_1(k), \ldots, e_N(k)\} \equiv \hat{y}(s_{p-1}) - y(s_p), \\ &k \in [t_p, t_{p+1} - 1],\ k \in \mathbb{Z}^+,\ \hat{y}(s_{-1}) \triangleq 0,\ e(k) \in \mathbb{R}^{n_y}. \end{aligned} \tag{11.3}$$

The choice of i_p^* will depend on the transmission error and will be chosen according to the scheduling protocols which are defined below.

11.1.2 Dynamic Scheduling Protocols

11.1.2.1 TOD Protocol

In TOD protocol, the output node $i_p^* \in \mathscr{I}_N$ with the greatest weighted error will be granted the access to the network.

Let $Q_i > 0$, $i = 1, \ldots, N$ be some weighting matrices that will be computed in Theorem 11.2. At the sampling instant s_p, the weighted TOD protocol is a protocol in which the active output node with the index i_p^* is defined as any index that satisfies

$$\left|\sqrt{Q}_{i_p^*} e_{i_p^*}(k)\right|^2 \geq \left|\sqrt{Q}_i e_i(k)\right|^2,\ k \in [t_p, t_{p+1} - 1],\ i \in \mathscr{I}_N \backslash \{i_p^*\}. \tag{11.4}$$

A possible selection of i_p^* is given by

$$i_p^* = \min\left\{\arg \max_{i \in \{1, \ldots, N\}} |\sqrt{Q}_i \left(\hat{y}_i(s_{p-1}) - y_i(s_p)\right)|^2\right\}.$$

11.1.2.2 Iid Protocol

The selection of i_p^* is assumed to be iid with the probabilities given by

$$\text{Prob}\{i_p^* = i\} = \beta_i, \; i \in \mathcal{I}_N, \tag{11.5}$$

where $\beta_i \in \mathcal{I}_N$ are nonnegative scalars and $\sum_{i=1}^{N} \beta_i = 1$. Here, β_j, $j = 1, \ldots, N$, are the probabilities of the state $x_j(s_p)$ to be transmitted at s_p.

Remark 11.1 The iid protocol, as one of the dynamic protocols, is adopted to schedule which sensor node can access to the communication medium at each sampling instant. It has been applied to describe probabilistic measurements missing [235], stochastic sampling intervals [67], stochastic interval time-delays [249] and the probability of a switched system staying in each subsystem [224, 225].

11.2 A Discrete-Time Hybrid System Model

Consider (11.1) under the static output-feedback controller. In the sequel, we propose a hybrid system model for the closed-loop system of NCS provided above.

The controller and the actuator are supposed to be event-driven. The most recent output information on the controller side is denoted by $\hat{y}(s_p)$. Assume that there exists a matrix $K = [K_1 \; \cdots \; K_N]$, $K_i \in \mathbb{R}^{n_u \times n_i}$ such that $A + BKC$ is Schur. Consider the static output-feedback controller

$$u(k) = K\hat{y}(s_p), \; k \in [t_p, t_{p+1} - 1], \; k \in \mathbb{Z}^+. \tag{11.6}$$

From (11.2), it follows that the controller (11.6) can be rewritten as

$$u(k) = K_{i_p^*} y_{i_p^*}(s_p) + \sum_{i=1, i \neq i_p^*}^{N} K_i \hat{y}_i(s_{p-1}), \tag{11.7}$$

for $k \in [t_p, t_{p+1} - 1]$, where i_p^* is the index of the active node at s_p.

Therefore, from (11.1), (11.3) and (11.7), we obtain the following augmented closed-loop system for $k \in [t_p, t_{p+1} - 2]$, $k \in \mathbb{Z}^+$:

$$\begin{cases} x(k+1) = Ax(k) + A_1 x(s_p) + \sum_{i=1, i \neq i_p^*}^{N} B_i e_i(t_p), \\ e(k+1) = e(k) \end{cases} \tag{11.8}$$

with the delayed reset system for $k = t_{p+1} - 1$,

$$\begin{cases} x(t_{p+1}) = Ax(t_{p+1}-1) + A_1 x(s_p) + \sum_{i=1, i \neq i_p^*}^{N} B_i e_i(t_p), \\ e_i(t_{p+1}) = C_i[x(s_p) - x(s_{p+1})],\ i = i_p^*, \\ e_i(t_{p+1}) = e_i(t_p) + C_i[x(s_p) - x(s_{p+1})],\ i \neq i_p^*, \end{cases} \tag{11.9}$$

where the augmented state is col$\{x(k), e(k)\}$, $e(k)$ the error between the system output $y(s_p)$ and the latest available information $\hat{y}(s_{p-1})$ is defined in (11.3) and

$$A_1 = BKC,\ K = [K_1 \cdots K_N],\ B_i = BK_i,\ i = 1, \ldots, N.$$

For $k \in [t_p, t_{p+1} - 1]$, denote $h(k) = k - s_p$. Then, $h_m \le h(k) \le \text{MATI} + h_M - 1 \triangleq \tau_M$. Therefore, (11.8)–(11.9) can be considered as a discrete-time hybrid system with a time-varying interval delay.

The objective of the present chapter is to provide improved stability criteria to find MATI and MAD (i.e., h_M) such that the closed-loop system (11.8)–(11.9) under dynamic protocol (11.4) (or (11.5)) is exponentially stable (exponentially mean-square stable) with respect to the full state. To do so, in the sequel, we first establish a new stability criterion for discrete-time systems with time-varying delays by the discrete-time counterpart of augmented Lyapunov functional provided in [146] for stability analysis of continuous-time systems with time-varying delays. Then, the proposed approach is applied to guarantee the stability of the resulting discrete-time hybrid system model with respect to the full state under TOD and iid scheduling protocols, respectively.

11.3 Preliminary Results on Discrete-Time Systems with Time-Varying Delays

In this section, we propose a new stability condition for linear discrete-time system with time-varying delays. This class of system is governed by

$$\begin{cases} x(k+1) = Ax(k) + A_1 x(k - h(k)),\ k \ge 0,\ k \in \mathbb{Z}^+, \\ x(k) = \phi(k),\ \ k \in [-h_2, 0] \cap \mathbb{Z}, \end{cases} \tag{11.10}$$

where $x(k) \in \mathbb{R}^n$ is the state vector, $A \in \mathbb{R}^{n \times n}$, $A_1 \in \mathbb{R}^{n \times n}$ are constant matrices and $\phi(k)$ is an initial condition. The time-varying delay $h(k)$ is a positive integer satisfying $h_1 \le h(k) \le h_2,\ \ h_{12} \triangleq h_2 - h_1$, where h_1, h_2 are known as positive integers.

This study indeed represents the first stage toward the stability analysis of system (11.8)–(11.9) since it could represent the situation where $e(k) = 0$, $\forall k \ge 0,\ k \in \mathbb{Z}$. This preliminary result will then be extended to address the objective of this chapter, i.e., exponential stability (exponential mean-square stability) of (11.8)–(11.9) under dynamic protocol (11.4) (or (11.5)).

Let us first recall the discrete-time counterpart of the second-order Bessel–Legendre integral inequality [198] (i.e., the Bessel–Legendre inequality with the degree of Legendre polynomials equaling to two). The same inequality was also proposed in [93]. In this inequality, an improvement of Abel lemma-based inequality [262] or Wirtinger-type inequality [202] has been achieved.

Lemma 11.1 *For a given matrix $R \in \mathbb{S}^n_+$, integers a and b with $b > a$, any vector function $x\colon \mathbb{Z}[a, b] \to \mathbb{R}^n$, the inequality*

$$\sum_{s=a}^{b} \omega^T(s) R \omega(s) \geq \frac{1}{b-a+1} \Omega^T \operatorname{diag}\{R, 3R, 5R\} \Omega \tag{11.11}$$

holds, where $\omega(s) = x(s) - x(s-1)$ and

$$\Omega = \begin{bmatrix} x(b) - x(a-1) \\ x(b) + x(a-1) - \dfrac{2}{b-a+2} \displaystyle\sum_{s=a-1}^{b} x(s) \\ x(b) - x(a-1) - \dfrac{6}{b-a+2} \displaystyle\sum_{s=a-1}^{b} \delta_{a,b}(s) x(s) \end{bmatrix},$$

$$\delta_{a,b}(s) = 2\left(\frac{s-a}{b-a}\right) - 1.$$

In the sequel, based on Lemma 11.1 together with a newly developed delay-dependent reciprocally convex combination Lemma [253], a novel stability criterion is provided for discrete-time system (11.10) with time-varying delays. For the simplicity of presentation, in this section we denote by $\rho_i (i = 1, \ldots, 14)$ the block row vectors of the identity matrix I_{14n} and use the following notations:

$$\begin{aligned}
G_1 &= [\rho_1^T \ \ -\rho_1^T + (h_1+1)\rho_5^T \ \ -\rho_2^T - \rho_3^T + \rho_7^T + \rho_9^T - \rho_1^T + (h_1+1)\rho_6^T \ \ \hat{G}_1^T]^T, \\
\hat{G}_1 &= -(h_{12} - 1)\rho_2 + (h_{12} + 1)\rho_3 - \rho_{12} - \rho_{14}, \\
G(h) &= [0 \ \ 0 \ \ (h - h_1)\rho_7^T + (h_2 - h)\rho_9^T \ \ 0 \ \ \hat{G}(h)^T]^T, \\
\hat{G}(h) &= -2(h_2 - h)\rho_3 + (h_2 - h)(\rho_{11} + \rho_{14}) + (h - h_1)(\rho_{12} - \rho_{13}), \\
G_2 &= [\rho_1^T - \rho_2^T \ \ \rho_1^T + \rho_2^T - 2\rho_5^T \ \ \rho_1^T - \rho_2^T - 6\rho_6^T]^T, \\
G_3 &= [\rho_2^T - \rho_3^T \ \ \rho_2^T + \rho_3^T - 2\rho_7^T \ \ \rho_2^T - \rho_3^T - 6\rho_8^T]^T, \\
G_4 &= [\rho_3^T - \rho_4^T \ \ \rho_3^T + \rho_4^T - 2\rho_9^T \ \ \rho_3^T - \rho_4^T - 6\rho_{10}^T]^T, \\
\Gamma &= [G_3^T \ \ G_4^T]^T, \qquad \Sigma = A\rho_1 + A_1\rho_3, \\
G_0 &= [\Sigma^T \ \ -\rho_2^T + (h_1 + 1)\rho_5^T \ \ -\rho_3^T - \rho_4^T + \rho_7^T + \rho_9^T \hat{G}_{01}^T \ \ \hat{G}_{02}^T]^T, \\
\hat{G}_{01} &= \left(1 + \frac{4}{h_1 - 1}\right)\rho_2 - \frac{2(h_1 + 1)}{h_1}\rho_5 + (h_1 + 1)\rho_6, \\
\hat{G}_{02} &= (h_{12} + 3)(\rho_3 + \rho_4) - 2(\rho_{11} + \rho_{13}) - (\rho_{12} + \rho_{14}),
\end{aligned} \tag{11.12}$$

and

$$\begin{aligned}
\eta_0(k) &= [x^T(k)\; x^T(k-h_1)\; x^T(k-h(k))\; x^T(k-h_2)]^T, \\
\eta_1(k) &= \frac{1}{h_1+1}\left[\sum_{s=k-h_1}^{k} x^T(s) \;\; \sum_{s=k-h_1}^{k} \delta_1(k,s)x^T(s)\right]^T, \\
\eta_2(k) &= \frac{1}{h-h_1+1}\left[\sum_{s=k-h}^{k-h_1} x^T(s) \;\; \sum_{s=k-h}^{k-h_1} \delta_2(k,s)x^T(s)\right]^T, \\
\eta_3(k) &= \frac{1}{h_2-h+1}\left[\sum_{s=k-h_2}^{k-h} x^T(s) \;\; \sum_{s=k-h_2}^{k-h} \delta_3(k,s)x^T(s)\right]^T, \\
\eta_4(k) &= (h-h_1+1)\eta_2(k), \;\; \eta_5(k) = (h_2-h+1)\eta_3(k), \\
\eta_6(k) &= \left[\sum_{s=k-h_2}^{k-h_1} x^T(s) \;\; h_{12}\sum_{s=k-h_2}^{k-h_1} \delta_4(k,s)x^T(s)\right]^T,
\end{aligned} \tag{11.13}$$

where the functions δ_i, for $i = 1, \ldots, 4$, which refer to the functions $\delta_{a,b}$ given in Lemma 11.1, are given by

$$\begin{aligned}
\delta_1(k,s) &= 2\left(\frac{s-k+h_1-1}{h_1-1}\right)-1, \;\; \delta_2(k,s) = 2\left(\frac{s-k+h-1}{h-h_1-1}\right)-1, \\
\delta_3(k,s) &= 2\left(\frac{s-k+h_2-1}{h_2-h-1}\right)-1, \;\; \delta_4(k,s) = 2\left(\frac{s-k+h_2-1}{h_{12}-1}\right)-1.
\end{aligned} \tag{11.14}$$

The following theorem gives the sufficient conditions for the exponential stability of system (11.10).

Theorem 11.1 *The system* (11.10) *is exponentially stable with the decay rate* $\lambda \in (0,1)$ *for all time-varying delays* $h(k) \in [h_1, h_2]$ *if there exist matrices* $P \in \mathbb{S}_+^{5n}$, S_1, S_2, R_1, $R_2 \in \mathbb{S}_+^n$, N_1, $N_2 \in \mathbb{R}^{14n\times 2n}$, *and a matrix* $X \in \mathbb{R}^{3n\times 3n}$ *such that the matrix inequalities*

$$\begin{bmatrix} \tilde{\Phi}_i & \begin{bmatrix} (\Sigma-\rho_1)^T H \\ 0 \end{bmatrix} \\ * & -H \end{bmatrix} < 0, \; i = 1, 2, \tag{11.15}$$

hold, where

$$\begin{aligned}
\tilde{\Phi}_1 &= \begin{bmatrix} \Phi_0(h_1) - \lambda^{h_2}\Gamma^T\Psi(h_1)\Gamma & G_3^T X & \sqrt{1-\lambda}G^T(h)P \\ * & -\lambda^{-h_2}\tilde{R}_2 & 0 \\ * & * & -P \end{bmatrix}, \\
\tilde{\Phi}_2 &= \begin{bmatrix} \Phi_0(h_2) - \lambda^{h_2}\Gamma^T\Psi(h_2)\Gamma & G_4^T X^T & \sqrt{1-\lambda}G^T(h)P \\ * & -\lambda^{-h_2}\tilde{R}_2 & 0 \\ * & * & -P \end{bmatrix},
\end{aligned} \tag{11.16}$$

and for any θ *in* $\mathbb{R}$,

$$\begin{aligned}\Phi_0(\theta) &= G_0^T P G_0 - \lambda G_1^T P G_1 + \mathrm{He}(G^T(h)P(G_0 - \lambda G_1)\\ &\quad + N_1 g_1(\theta) + N_2 g_2(\theta)) + \hat{S} - G_2^T \tilde{R}_1 G_2,\\ \hat{S} &= \mathrm{diag}\{S_1, -\lambda^{h_1}(S_1 - S_2), 0_{n\times n}, -\lambda^{h_2} S_2, 0_{10n\times 10n}\},\\ \tilde{R}_i &= \mathrm{diag}\{R_i, 3R_i, 5R_i\}, \quad i = 1, 2,\\ H &= h_1^2 R_1 + h_{12}^2 R_2,\end{aligned} \tag{11.17}$$

and

$$\begin{aligned}\Psi(\theta) &= \begin{bmatrix} (2-(h-h_1)/h_{12})\tilde{R}_2 & X \\ X^T & (1+(h-h_1)/h_{12})\tilde{R}_2 \end{bmatrix},\\ g_1(\theta) &= (\theta - h_1 + 1)\begin{bmatrix}\rho_7\\ \rho_8\end{bmatrix} - \begin{bmatrix}\rho_{11}\\ \rho_{12}\end{bmatrix},\ g_2(\theta) = (h_2 - \theta + 1)\begin{bmatrix}\rho_9\\ \rho_{10}\end{bmatrix} - \begin{bmatrix}\rho_{13}\\ \rho_{14}\end{bmatrix}.\end{aligned} \tag{11.18}$$

Proof Consider the Lyapunov functional given by

$$V(k) = V_1(k) + V_2(k) + V_3(k), \tag{11.19}$$

where

$$\begin{aligned}V_1(k) &= \tilde{x}^T(k) P \tilde{x}(k),\\ V_2(k) &= \sum_{s=k-h_1}^{k-1} \lambda^{k-s-1} |\sqrt{S_1} x(s)|^2 + \sum_{s=k-h_2}^{k-h_1-1} \lambda^{k-s-1} |\sqrt{S_2} x(s)|^2,\\ V_3(k) &= h_1 \sum_{j=-h_1+1}^{0} \sum_{s=k+j}^{k} \lambda^{k-s} |\sqrt{R_1}\eta(s)|^2 + h_{12} \sum_{j=-h_2+1}^{-h_1} \sum_{s=k+j}^{k} \lambda^{k-s} |\sqrt{R_2}\eta(s)|^2,\end{aligned}$$

and where $\eta(k) = x(k) - x(k-1)$ and

$$\tilde{x}(k) = \begin{bmatrix} x(k) \\ \sum_{s=k-h_1}^{k-1} x(s) \\ \sum_{s=k-h_2}^{k-h_1-1} x(s) \\ \sum_{s=k-h_1}^{k-1} \delta_1(k,s) x(s) \\ (h_{12}-1) \sum_{s=k-h_2}^{k-h_1-1} \delta_4(k,s) x(s) \end{bmatrix}$$

with $\delta_1(k, s)$ and $\delta_4(k, s)$ given in (11.14).

The objective of the next development consists in finding an upper bound of the difference of $V(k+1) - \lambda V(k)$ along the trajectories of system (11.10) using the augmented vector

$$\zeta(k) = \text{col}\left\{\eta_0(k), \eta_1(k), \eta_2(k), \eta_3(k), \eta_4(k), \eta_5(k)\right\}, \tag{11.20}$$

where $\eta_i(k)$, for $i = 0, \ldots, 5$, are given in (11.13).

To express $\Delta V_1(k) = \tilde{x}^T(k+1)P\tilde{x}(k+1) - \lambda\tilde{x}^T(k)P\tilde{x}(k)$ in terms of the augmented vector $\zeta(k)$, we need to express $\tilde{x}(k+1)$ and $\tilde{x}(k)$ using $\zeta(k)$. On the one hand, we note that the first four components of $\tilde{x}(k)$ and $\tilde{x}(k+1)$ can be straightforwardly expressed as the components of $\zeta(k)$. Simple calculations show that

$$\begin{aligned}
x(k) &= \rho_1\zeta(k), \\
\sum_{s=k-h_1}^{k-1} x(s) &= -x(k) + \sum_{s=k-h_1}^{k} x(s) = [-\rho_1 + (h_1+1)\rho_5]\zeta(k), \\
\sum_{s=k-h_2}^{k-h_1-1} x(s) &= -x(k-h_1) - x(k-h) + \sum_{s=k-h}^{k-h_1} x(s) + \sum_{s=k-h_2}^{k-h} x(s) \\
&= [-\rho_2 - \rho_3 + (h-h_1+1)\rho_7 + (h_2-h+1)\rho_9]\zeta(k), \\
\sum_{s=k-h_1}^{k-1} \delta_1(k,s)x(s) &= -x(k) + \sum_{s=k-h_1}^{k} \delta_1(k,s)x(s) = [-\rho_1 + (h_1+1)\rho_6]\zeta(k)
\end{aligned}$$

and

$$\begin{aligned}
x(k+1) &= \Sigma\zeta(k), \\
\sum_{s=k+1-h_1}^{k} x(s) &= -x(k-h_1) + \sum_{s=k-h_1}^{k} x(s) = [-\rho_2 + (h_1+1)\rho_5]\zeta(k), \\
\sum_{s=k+1-h_2}^{k-h_1} x(s) &= -x(k-h) - x(k-h_2) + \sum_{s=k-h}^{k-h_1} x(s) + \sum_{s=k-h_2}^{k-h} x(s) \\
&= [-\rho_3 - \rho_4 + (h-h_1+1)\rho_7^T + (h_2-h+1)\rho_9^T]\zeta(k), \\
\sum_{s=k+1-h_1}^{k} \delta_1(k+1,s)x(s) &= \left(1 + \frac{4}{h_1-1}\right)x(k-h_1) - \frac{2}{h_1-1}\sum_{s=k-h_1}^{k} x(s) \\
&\quad + \sum_{s=k-h_1}^{k} \delta_1(k,s)x(s) \\
&= \left[\left(1 + \frac{4}{h_1-1}\right)\rho_2 - \frac{2(h_1+1)}{h_1-1}\rho_5 + (h_1+1)\rho_6\right]\zeta(k),
\end{aligned}$$

where the matrix Σ is given in (11.12). The last component of $\tilde{x}(k)$ and $\tilde{x}(k+1)$ requires a more dedicated development. To achieve this goal, we first note that

$$\begin{aligned}(h_{12}-1)\sum_{s=k-h_2}^{k-h_1-1}\delta_4(k,s)x(s) &= (h_{12}-1)\sum_{s=k-h}^{k-h_1}\delta_4(k,s)x(s)\\ &+(h_{12}-1)\sum_{s=k-h_2}^{k-h}\delta_4(k,s)x(s)-(h_{12}-1)x(k-h_1)\\ &-[2(h_2-h-1)-h_{12}+1]x(k-h).\end{aligned}\tag{11.21}$$

We need to find two expressions of $\delta_4(k,s)$, which depend on $\delta_2(k,s)$ and $\delta_3(k,s)$, respectively. Some calculations show

$$\begin{aligned}\delta_4(k,s) &= 2\Big(\frac{s-k+h_2-1}{h_{12}-1}\Big)-1\\ &= 2\frac{(s-k+h-1)+(h_2-h)}{h_{12}-1}-\frac{(h_2-h)+(h-h_1-1)}{h_{12}-1}\\ &= 2\frac{s-k+h-1}{h-h_1-1}\frac{h-h_1-1}{h_{12}-1}-\frac{h-h_1-1}{h_{12}-1}+\frac{h_2-h}{h_{12}-1}\\ &= \frac{h-h_1-1}{h_{12}-1}\delta_2(k,s)+\frac{h_2-h}{h_{12}-1},\end{aligned}\tag{11.22}$$

and, similarly,

$$\begin{aligned}\delta_4(k,s) &= 2\frac{h_2-h-1}{h_{12}-1}\frac{s-k+h_2-1}{h_2-h-1}-\frac{(h_2-h-1)+(h-h_1)}{h_{12}-1}\\ &= \frac{h_2-h-1}{h_{12}-1}\delta_3(k,s)-\frac{h-h_1}{h_{12}-1}.\end{aligned}\tag{11.23}$$

Reinjecting (11.22) and (11.23) into (11.21) leads to

$$\begin{aligned}(h_{12}-1)\sum_{s=k-h_2}^{k-h_1-1}\delta_4(k,s)x(s) &= (h-h_1-1)\sum_{s=k-h}^{k-h_1}\delta_2(k,s)x(s)+(h_2-h)\sum_{s=k-h}^{k-h_1}x(s)\\ &+(h_2-h-1)\sum_{s=k-h_2}^{k-h}\delta_3(k,s)x(s)\\ &-(h-h_1)\sum_{s=k-h_2}^{k-h}x(s)-(h_{12}-1)x(k-h_1)\\ &-[2(h_2-h-1)-h_{12}+1]x(k-h)\\ &= [0_{n\times 4n}\ I_n](G_1+G(h))\zeta(k).\end{aligned}$$

In the same way, the last component of $\tilde{x}(k+1)$ can be written as

$$
\begin{aligned}
&(h_{12}-1)\sum_{s=k+1-h_2}^{k-h_1}\delta_4(k+1,s)x(s) \\
&= (h_{12}-1)\sum_{s=k-h}^{k-h_1}\delta_4(k,s)x(s) + (h_{12}-1)\sum_{s=k-h_2}^{k-h}\delta_4(k,s)x(s) \\
&\quad - [2(h_2-h-1)-h_{12}-1]x(k-h) + (h_{12}+3)x(k-h_2) \\
&\quad - 2\sum_{s=k-h}^{k-h_1}x(s) - 2\sum_{s=k-h_2}^{k-h}x(s) \\
&= (h-h_1-1)\sum_{s=k-h}^{k-h_1}\delta_2(k,s)x(s) + (h_2-h-2)\sum_{s=k-h}^{k-h_1}x(s) \\
&\quad + (h_2-h-1)\sum_{s=k-h_2}^{k-h}\delta_3(k,s)x(s) - (h-h_1+2)\sum_{s=k-h_2}^{k-h}x(s) \\
&\quad - [2(h_2-h-1)-h_{12}-1]x(k-h) + (h_{12}+3)x(k-h_2) \\
&= [0_{n\times 4n}\ I_n](G_0+G(h))\zeta(k).
\end{aligned}
\tag{11.24}
$$

Hence, we finally obtain that $\tilde{x}(k) = (G_1 + G(h))\zeta(k)$ and $\tilde{x}(k+1) = (G_0 + G(h))\zeta(k)$. Thus, $\Delta V_1(k)$ can be written as

$$
\begin{aligned}
\Delta V_1(k) &= \tilde{x}^T(k+1)P\tilde{x}(k+1) - \lambda\tilde{x}^T(k)P\tilde{x}(k) \\
&= \zeta^T(k)\Big[G_0^T P G_0 - \lambda G_1^T P G_1 + (1-\lambda)G^T(h)PG(h) \\
&\qquad + \mathrm{He}(G^T(h)P(G_0-\lambda G_1))\Big]\zeta(k).
\end{aligned}
\tag{11.25}
$$

Moreover, from the definition of the augmented vector $\zeta(k)$, one can see that the last four components can be seen as linear combination of the other components of $\zeta(k)$ since the relations $\eta_4(k) = (h - h_1 + 1)\eta_2(k)$ and $\eta_5(k) = (h_2 - h + 1)\eta_3(k)$ hold. Then for any matrices N_1, N_2 in $\mathbb{R}^{14n\times 2n}$, it holds that

$$
2\zeta^T(k)[N_1 g_1(h) + N_2 g_2(h)]\zeta(k) = 0. \tag{11.26}
$$

The computation of $\Delta V_2(k)$ and $\Delta V_3(k)$ yields

$$
\begin{aligned}
\Delta V_2(k) &\le x^T(k)S_1x(k) - \lambda^{h_1}x^T(k-h_1)(S_1-S_2)x(k-h_1) - \lambda^{h_2}x^T(k-h_2)S_2x(k-h_2) \\
&= \zeta^T(k)\hat{S}\zeta(k)
\end{aligned}
\tag{11.27}
$$

and

$$
\begin{aligned}
\Delta V_3(k) &\le \eta^T(k+1)(h_1^2R_1 + h_{12}^2R_2)\eta(k+1) \\
&\quad - h_1\lambda^{h_1}\sum_{s=k-h_1+1}^{k}|\sqrt{R_1}\eta(s)|^2 - \lambda^{h_2}h_{12}\sum_{s=k-h_2+1}^{k-h_1}|\sqrt{R_2}\eta(s)|^2.
\end{aligned}
\tag{11.28}
$$

Then by Lemma 11.1, and the definition of the matrices G_2 and $\tilde{R}_1$ in (11.12) and (11.17), respectively, we arrive at the following upper bound of the first summation in (11.28):

$$-h_1\lambda^{h_1}\sum_{s=k-h_1+1}^{k}|\sqrt{R_1}\eta(s)|^2 \leq -\lambda^{h_1}\zeta^T(k)G_2^T\tilde{R}_1G_2\zeta(k). \tag{11.29}$$

The application of Lemma 11.1 to the last summation term of (11.28) yields

$$\begin{aligned}
&-\lambda^{h_2}h_{12}\sum_{s=k-h_2+1}^{k-h_1}|\sqrt{R_2}\eta(s)|^2 \\
&= -\lambda^{h_2}h_{12}\sum_{s=k-h+1}^{k-h_1}|\sqrt{R_2}\eta(s)|^2 - \lambda^{h_2}h_{12}\sum_{s=k-h_2+1}^{k-h}|\sqrt{R_2}\eta(s)|^2 \\
&\leq -\lambda^{h_2}\zeta^T(k)\Gamma^T\begin{bmatrix}\frac{h_{12}}{h-h_1}\tilde{R}_2 & 0\\ 0 & \frac{h_{12}}{h_2-h}\tilde{R}_2\end{bmatrix}\Gamma\zeta(k) \\
&\leq -\zeta^T(k)\Phi_1(h(k))\zeta(k),
\end{aligned} \tag{11.30}$$

where

$$\begin{aligned}
\Phi_1(h(k)) &= \lambda^{h_2}\Gamma^T\left(\begin{bmatrix}\tilde{R}_2 & X\\ X^T & \tilde{R}_2\end{bmatrix}+\begin{bmatrix}\frac{h_2-h}{h_{12}}T_1 & 0\\ 0 & \frac{h-h_1}{h_{12}}T_2\end{bmatrix}\right)\Gamma,\\
T_1 &= \tilde{R}_2 - X\tilde{R}_2^{-1}X^T,\ \ T_2 = \tilde{R}_2 - X^T\tilde{R}_2^{-1}X.
\end{aligned} \tag{11.31}$$

The latter inequality is guaranteed due to the refined reciprocally convex combination lemma [253].

From (11.25)–(11.30), it follows that

$$\Delta V(k) \leq \zeta^T(k)\Phi(h)\zeta(k),$$

where $\Phi(h) = \Phi_0(h) - \Phi_1(h) + (1-\lambda)G^T(h)PG(h) + (\Sigma-\rho_1)^T H(\Sigma-\rho_1)$ with $\Phi_0(h)$ and $\Phi_1(h)$ given in (11.17) and (11.31), respectively. By Schur complements, the two matrix inequalities of (11.15) are equivalent to $\Phi(h_i) \prec 0,\ i = 1, 2$, and, thus, guarantee $\Delta V(k) < 0$, implying the exponential stability with the decay rate λ of system (11.10) for all time-varying delays in the interval $[h_1,\ h_2]$.

Remark 11.2 In order to fully benefit from the summation inequality of (11.11), the augmented term V_1 in (11.19) not only includes the signals $x(k)$, $\sum_{s=k-h_1}^{k-1} x(s)$ and $\sum_{s=k-h_2}^{k-h_1-1} x(s)$ that were adopted in [202] but also includes two additional signals $\sum_{s=k-h_1}^{k-1} \delta_1(k,s)x(s)$ and $(h_{12}-1)\sum_{s=k-h_2}^{k-h_1-1} \delta_4(k,s)x(s)$. This state augmentation allows achieving less conservative stability criteria. Moreover, new summa-

Table 11.1 Example 11.1: admissible upper bound of h_2 for different h_1

h_1	2	4	6	10	15	20	25	NoVs
Reference [34]	21	21	21	22	24	27	31	$9.5n^2 + 5.5n$
Reference [114]	22	22	22	23	25	28	32	$27n^2 + 9n$
Reference [93]	26	27	28	31	34	35	36	$20n^2 + 5n$
Theorem 11.1	27	29	30	33	35	37	38	$79.5n^2 + 4.5n$

tion inequalities in double form were proposed in [93]. Therefore, the condition of Theorem 11.1 could be further improved by employing the generalized summation inequalities as well as the Lyapunov functional with triple summation terms [93].

Example 11.1 A widely used numerical example is taken from the literature to make a comparison with some results reported in the existing works. Consider the following much-studied system (11.10) with

$$A = \begin{bmatrix} 0.8 & 0.0 \\ 0.05 & 0.9 \end{bmatrix}, \quad A_1 = \begin{bmatrix} -0.1 & 0.0 \\ -0.2 & -0.1 \end{bmatrix}.$$

Table 11.1 shows that the maximum allowable delays h_2 for several values of h_1 obtained by Theorem 11.1 with $\lambda = 1$ are less conservative than those obtained by some methods from the literature. As usual, the reduction of the conservatism of Theorem 11.1 over existing results is at the price of additional decision variables, showing again a trade-off between the improvement and the numerical complexity [146].

11.4 Discrete-Time NCSs Under TOD Protocol

Based on the exponential delay-dependent analysis of Theorem 11.1, in this section we derive exponential stability criteria of (11.8)–(11.9) under (11.4), the definition of which is given as follows:

Definition 11.1 For any initial condition $x_{t_0} \in \underbrace{\mathbb{R}^n \times \cdots \times \mathbb{R}^n}_{\tau_M+1 \text{ times}}$, if there exist constants $b > 0$ and $0 < \kappa < 1$ such that the solutions of the hybrid system (11.8)–(11.9) under (11.4) satisfy

$$|x(k)|^2 \leq b\kappa^{2(k-t_0)}\{\|x_{t_0}\|_c^2 + |e(t_0)|^2\},$$

and

$$\{|e(k)|^2\} \leq b\kappa^{2(k-t_0)}\{\|x_{t_0}\|_c^2 + |e(t_0)|^2\},$$

where $\|x_{t_0}\|_c = \sup_{t_0-\tau_M \le s \le t_0} |x(s)|$, then the hybrid system (11.8)–(11.9) under (11.4) is said to be exponentially stable.

We apply the following discrete-time Lyapunov functional to system (11.8)–(11.9) under (11.4) for the exponential stability of systems with time-varying delays from the maximum delay interval $[h_m, \ \tau_M]$:

$$\begin{aligned} V_e(k) &= \tilde{V}(k) + \sum_{i=1}^{N} e_i^T(t_p) Q_i e_i(t_p), \\ \tilde{V}(k) &= V(k) + V_G(k), \ k \in [t_p, t_{p+1}-1], \ k, p \in \mathbb{Z}^+, \end{aligned} \tag{11.32}$$

where $V(k)$ is given by (11.19) with h_1, h_2 and h_{12} replaced by h_m, τ_M and $\tau_M - h_m$, respectively, and

$$V_G(k) = (\tau_M - h_m) \sum_{i=1}^{N} \sum_{s=s_p}^{k} \lambda^{k-s} |\sqrt{G_i} C_i \eta(s)|^2$$

with $\eta(k) = x(k) - x(k-1)$, $0 < \lambda < 1$, $G_i > 0$, $Q_i > 0$, $i = 1, \ldots, N$.

Following (9.26), we introduce the term V_G to deal with the delays in the reset conditions:

$$\begin{aligned} & V_G(t_{p+1}) - \lambda V_G(t_{p+1}-1) \\ & \le \sum_{i=1}^{N} \left[(\tau_M - h_m) |\sqrt{G_i} C_i \eta(t_{p+1})|^2 \right] - \sum_{i=1}^{N} \lambda^{\tau_M} \left| \sqrt{G_i} C_i [x(s_{p+1}) - x(s_p)] \right|^2 . \end{aligned}$$

To ease the presentation, for $i \in \mathscr{I}_N$, we will use in this section the following notations:

$$\begin{aligned} \tilde{\Sigma}^i &= [A - I \ 0_{n \times n} \ A_1 \ 0_{n \times 19n} \ \tilde{F}_0^i], \\ \tilde{F}_0^i &= [B_1 \ \cdots \ B_{j|j \ne i} \cdots B_N], \\ \tilde{H} &= h_m^2 R_0 + (\tau_M - h_m)^2 R_1 + (\tau_M - h_m) \sum_{l=1}^{N} C_l^T G_l C_l. \end{aligned} \tag{11.33}$$

Thanks to Theorem 11.1 for the exponential delay-dependent analysis, we prove the following theorem on the exponential stability of (11.8)–(11.9) under (11.4).

Theorem 11.2 *For any given scalar $0 < \lambda < 1$, integers $0 \le h_m < \tau_M$ and K_i, $i = 1, \ldots, N$, assume that there exist matrices $P \in \mathbb{S}_+^{5n}$, S_1, S_2, R_1, $R_2 \in \mathbb{S}_+^n$, Q_i, U_i, $G_i \in \mathbb{S}_+^{n_i}$, $i = 1, \ldots, N$, N_1, $N_2 \in \mathbb{R}^{14n \times 2n}$, and a matrix $X \in \mathbb{R}^{3n \times 3n}$ such that the following matrix inequalities*

$$\Omega_i \triangleq \begin{bmatrix} \Gamma_i & Q_i \\ * & Q_i - \lambda^{\tau_M} G_i \end{bmatrix} < 0, \ \ i = 1, \ldots, N, \tag{11.34}$$

$$\begin{bmatrix} \begin{bmatrix} \tilde{\Phi}_j & 0 \\ * & \phi_i \end{bmatrix} & (\tilde{\Sigma}^i)^T \tilde{H} \\ * & -\tilde{H} \end{bmatrix} < 0, \ j = 1, 2, \ i = 1, \ldots, N, \tag{11.35}$$

are feasible with $\tilde{\Phi}_j$, $j = 1, 2$, *defined in (11.16) and*

$$\begin{aligned} \Gamma_i &= -\frac{\lambda - (1-\lambda)(\tau_M - h_m)}{N-1} Q_i + \left(1 + \frac{1}{\tau_M - h_m}\right) U_i, \\ \phi_i &= \text{diag}\{W_1, \ldots, W_{j|_{j \neq i}}, \ldots, W_N\}, \\ W_i &= -\frac{1}{\tau_M - h_m} U_i + (1-\lambda) Q_i, \end{aligned} \tag{11.36}$$

and the other notations given by (11.33). Then, we have

(i) $V_e(k)$ *satisfies the following inequalities along (11.8)–(11.9) for* $k \in [t_p, t_{p+1} - 2]$*:*

$$\begin{aligned} \Theta_1(k) \triangleq V_e(k+1) - \lambda V_e(k) - \frac{1}{\tau_M - h_m} \sum_{i=1, i \neq i_p^*}^{N} |\sqrt{U_i} e_i(t_p)|^2 \\ -(1-\lambda) \left| \sqrt{Q_{i_p^*}} e_{i_p^*}(t_p) \right|^2 \leq 0; \end{aligned} \tag{11.37}$$

(ii) At $k = t_{p+1} - 1$*,*

$$\begin{aligned} \Theta_2 \triangleq V_e(t_{p+1}) - \lambda V_e(t_{p+1} - 1) + \sum_{i=1, i \neq i_p^*}^{N} |\sqrt{U_i} e_i(t_p)|^2 \\ +(1-\lambda)(\tau_M - h_m) \left| \sqrt{Q_{i_p^*}} e_{i_p^*}(t_p) \right|^2 \leq 0; \end{aligned} \tag{11.38}$$

(iii) The following bounds

$$\lambda_{\min}(P) |x(k)|^2 \leq \tilde{V}(k) \leq V_e(k) \leq \lambda^{k - t_0} V_e(t_0), \tag{11.39}$$

and

$$\sum_{i=1}^{N} |\sqrt{Q_i} e_i(k)|^2 \leq \tilde{c} \lambda^{k - t_0} V_e(t_0), \ k \geq t_0, \ k \in \mathbb{N}, \tag{11.40}$$

with $V_e(t_0) = \tilde{V}(t_0) + \sum_{i=1}^{N} |\sqrt{Q_i} e_i(t_0)|^2$ *and* $\tilde{c} = \lambda^{-(\tau_M - h_m)}$*, are valid for the solutions of (11.4), (11.8), (11.9) initialized by* $x_{t_0} \in \underbrace{\mathbb{R}^n \times \cdots \times \mathbb{R}^n}_{\tau_M + 1 \text{ times}}$, $e(t_0) \in \mathbb{R}^{n_y}$.

Moreover, the hybrid system (11.8)–(11.9) under (11.4) is exponentially stable.

Proof The proof consists in proving each item separately.

Proof of (i): First, from functional (11.32), it holds that for $k \in [t_p, t_{p+1} - 2]$

$$\begin{aligned}\Theta_1(k) \leq \tilde{V}(k+1) - \lambda\tilde{V}(k) + (\tau_M - h_m)\sum_{i=1}^{N} |\sqrt{G_i}C_i\eta(k+1)|^2 \\ + \sum_{i=1, i\neq i_p^*}^{N} |\sqrt{W_i}e_i(t_p)|^2 \triangleq \Psi(k).\end{aligned} \tag{11.41}$$

Therefore, $\Theta_1(k) \leq 0$ of (11.37) is satisfied if $\Psi(k) \leq 0$ for $k \in [t_p, t_{p+1} - 2]$.

Let $i_p^* = i \in \mathscr{I}_N$ and define $\xi_i(k) = [\zeta^T(k)\ \bar{\xi}_i^T(k)]^T$, where $\zeta(k)$ is defined in (11.20) and $\bar{\xi}_i(k) = \text{col}\{e_1(k), \ldots, e_j(k)_{|j\neq i}, \ldots, e_N(k)\}, i = 1, \ldots, N$. Following the arguments of Theorem 11.1 for exponential delay-dependent analysis, we arrive at $\Psi(k) \leq 0$ for $k \in [t_p, t_{p+1} - 2]$ if (11.35) is feasible. This completes the proof of (i).

Proof of (ii): From (11.32) and (11.41), we have

$$\begin{aligned}\Theta_2 \leq\ & \tilde{V}(t_{p+1}) - \lambda\tilde{V}(t_{p+1} - 1) + (\tau_M - h_m)\sum_{i=1}^{N} |\sqrt{G_i}C_i\eta(t_{p+1})|^2 \\ & - \sum_{i=1}^{N} \lambda^{\tau_M}\left|\sqrt{G_i}C_i[x(s_{p+1}) - x(s_p)]\right|^2 + \sum_{i=1}^{N}\left[|\sqrt{Q_i}e_i(t_{p+1})|^2 - \lambda|\sqrt{Q_i}e_i(t_p)|^2\right] \\ & + \sum_{i=1, i\neq i_p^*}^{N} |\sqrt{U_i}e_i(t_p)|^2 + (1-\lambda)(\tau_M - h_m)\left|\sqrt{Q_{i_p^*}}e_{i_p^*}(t_p)\right|^2 \\ \leq\ & \Psi(t_{p+1} - 1) + \frac{1}{\tau_M - h_m}\sum_{i=1, i\neq i_p^*}^{N} |\sqrt{U_i}e_i(t_p)|^2 - (1-\lambda)\sum_{i=1, i\neq i_p^*}^{N} |\sqrt{Q_i}e_i(t_p)|^2 \\ & - \sum_{i=1}^{N} \lambda^{\tau_M}\left|\sqrt{G_i}C_i[x(s_{p+1}) - x(s_p)]\right|^2 + \sum_{i=1}^{N}\left[|\sqrt{Q_i}e_i(t_{p+1})|^2 - \lambda|\sqrt{Q_i}e_i(t_p)|^2\right] \\ & + \sum_{i=1, i\neq i_p^*}^{N} |\sqrt{U_i}e_i(t_p)|^2 + (1-\lambda)(\tau_M - h_m)\left|\sqrt{Q_{i_p^*}}e_{i_p^*}(t_p)\right|^2.\end{aligned}$$

Note that under TOD protocol,

$$-\left|\sqrt{Q_{i_p^*}}e_{i_p^*}(t_p)\right|^2 \leq -\frac{1}{N-1}\sum_{i=1, i\neq i_p^*}^{N}\left|\sqrt{Q_i}e_i(t_p)\right|^2.$$

From (11.41) and (11.34), we have $\Psi(t_{p+1} - 1) \leq 0$ and $\lambda^{\tau_M}G_{i_p^*} - Q_{i_p^*}C_{i_p^*} > 0$, respectively. Denote $\varsigma_i = \text{col}\{e_i(t_p),\ C_i[x(s_{p+1}) - x(s_p)]\}$. Then employing (11.9), we arrive at

$$\Theta_2 \le \Psi(t_{p+1}-1) - \left|\sqrt{\lambda^{\tau_M} G_{i_p^*} - Q_{i_p^*}} C_{i_p^*}[x(s_{p+1}) - x(s_p)]\right|^2 + \sum_{i=1, i\neq i_p^*}^{N} \varsigma_i^T \Omega_i \varsigma_i \le 0,$$

that yields (11.38). This completes the proof of (ii).

Proof of (iii): The next step is to prove (11.39) and (11.40). By the comparison principle, for $k \in [t_p, t_{p+1} - 1]$ the inequality (11.37) implies

$$V_e(k) \le \lambda^{k-t_p} V_e(t_p) + \sum_{i=1, i\neq i_p^*}^{N} \left\{ |\sqrt{U_i} e_i(t_p)|^2 \right\} + (1-\lambda)(\tau_M - h_m) \left|\sqrt{Q_{i_p^*}} e_{i_p^*}(t_p)\right|^2. \tag{11.42}$$

Note that (11.34) guarantees $0 < (1-\lambda)(\tau_M - h_m) < \lambda < 1$ and $U_i < (1 + \frac{1}{\tau_M - h_m}) U_i < \frac{\lambda - (1-\lambda)(\tau_M - h_m)}{N-1} Q_i < Q_i$, $i = 1, \ldots, N$. Hence,

$$\tilde{V}(k) \le \lambda^{k-t_p} V_e(t_p), \ k \in [t_p, t_{p+1} - 1]. \tag{11.43}$$

On the other hand, the inequalities (11.38) and (11.42) with $k = t_{p+1} - 1$ imply

$$\begin{aligned} V_e(t_{p+1}) &\le \lambda V_e(t_{p+1} - 1) - \sum_{i=1, i\neq i_p^*}^{N} \left|\sqrt{U_i} e_i(t_p)\right|^2 - (1-\lambda)(\tau_M - h_m) \left|\sqrt{Q_{i_p^*}} e_{i_p^*}(t_p)\right|^2 \\ &\le \lambda^{t_{p+1}-t_p} V_e(t_p) - (1-\lambda) \sum_{i=1, i\neq i_p^*}^{N} \left|\sqrt{U_i} e_i(t_p)\right|^2 \\ &\quad -(1-\lambda)^2 (\tau_M - h_m) \left|\sqrt{Q_{i_p^*}} e_{i_p^*}(t_p)\right|^2 \\ &\le \lambda^{t_{p+1}-t_p} V_e(t_p). \end{aligned}$$

Then, a recursive argument allows us to conclude that, for all $p \in \mathbb{Z}$, we have

$$V_e(t_{p+1}) \le \lambda^{t_{p+1}-t_{p-1}} V_e(t_{p-1}) \le \lambda^{t_{p+1}-t_0} V_e(t_0). \tag{11.44}$$

Substituting $p + 1$ for p in (11.44) and taking into account (11.43), we arrive at (11.39), which yields the exponential stability of (11.8)–(11.9) under (11.4) since

$$\lambda_{\min}(P)|x(k)|^2 \le \tilde{V}(k), \ V(t_0) \le \delta \|x_{t_0}\|_c^2$$

for some scalar $\delta > 0$. Moreover, (11.44) with $p + 1$ replaced by p implies (11.40) since $\lambda^{t_p - t_0} = \lambda^{k-t_0} \lambda^{t_p - k} \le \bar{c} \lambda^{k-t_0}$ for $k \in [t_p, t_{p+1} - 1]$. This completes the proof of (iii).

Remark 11.3 The exponential stability of system (11.8)–(11.9) under (11.4) can be alternatively analyzed via the Lyapunov functional $\tilde{V}_e(k) = V_e(k) + V_W(k)$, $k \in [t_p, t_{p+1} - 1]$, where $V_e(k)$ is given by (11.32) and

$$V_W(k) = (1-\lambda)(t_p - k)\left|\sqrt{Q_{i_p^*}}e_{i_p^*}(t_p)\right|^2 + \frac{t_p - k}{t_{p+1} - t_p}\sum_{i=1, i\neq i_p^*}^{N}\left|\sqrt{U_i}e_i(t_p)\right|^2.$$

The negative term $V_W(k)$ is a discrete-time counterpart of piecewise-continuous in time term that was employed in (7.18) to simplify the exponential stability analysis of the hybrid system. Under the conditions (11.34) and (11.35), it holds that $\tilde{V}_e(k)$ is positive for $k \geq t_0$, $k \in \mathbb{Z}^+$, i.e., $\tilde{V}_e(k) \geq \rho(|x(k)|^2 + |e(k)|^2)$ with some $\rho > 0$, and that $\tilde{V}_e(k+1) - \lambda\tilde{V}_e(k) \leq 0$, $k \in [t_p, t_{p+1} - 1]$. These two inequalities imply the exponential stability of system (11.8)–(11.9) under (11.4).

Remark 11.4 For discrete-time NCSs under TOD scheduling protocol, Lemma 9.2 in Chap. 9 only guarantees the partial stability of the closed-loop system with $N = 2$ sensor nodes, while Theorem 11.2 in this chapter guarantees that (11.39) gives a bound not only on $x(k)$ but also on $e_i(k)$, $i = 1, \ldots, N$. That is why Theorem 11.2 assesses stability of system (11.8)–(11.9) under (11.4) with respect to the full state $\mathrm{col}\{x(k), e(k)\}$.

11.5 Discrete-Time NCSs Under Stochastic Protocol

In this section, we first reformulate system (11.8) and (11.9) under iid scheduling protocol (11.5) as a stochastic impulsive system with the system matrices having stochastic parameters with Bernoulli distributions and then derive exponential mean-square stability criteria by virtue of Theorem 11.1.

11.5.1 Stochastic Hybrid Time-Delay Model with Bernoulli-Distributed Parameters

Following [249], we introduce the indicator functions

$$\pi_{\{\sigma_p^*=i\}} = \begin{cases} 1, & \sigma_p^* = i \\ 0, & \sigma_p^* \neq i, \end{cases} \quad i \in \mathcal{I}_N, \; p \in \mathbb{Z}^+.$$

Thus, from (11.5) it follows that

$$\begin{aligned} &\mathbb{E}\{\pi_{\{\sigma_p^*=i\}}\} = \mathbb{E}\{[\pi_{\{\sigma_p^*=i\}}]^2\} = \mathrm{Prob}\{\sigma_p^* = i\} = \beta_i, \\ &\mathbb{E}\{[\pi_{\{\sigma_p^*=i\}} - \beta_i][\pi_{\{\sigma_p^*=j\}} - \beta_j]\} = \begin{cases} -\beta_i\beta_j, & i \neq j, \\ \beta_i(1-\beta_i), & i = j. \end{cases} \end{aligned}$$

Therefore, the stochastic impulsive system model (11.8)–(11.9) under (11.5) can be rewritten as

$$\begin{cases} x(k+1) = Ax(k) + A_1 x(s_p) + \sum_{i=1}^{N}(1 - \pi_{\{\sigma_p^*=i\}}) B_i e_i(t_p), \\ e(k+1) = e(k), \ k \in [t_p, t_{p+1} - 2], \ k \in \mathbb{Z}^+, \end{cases} \tag{11.45}$$

with the delayed reset system for $k = t_{p+1} - 1$

$$\begin{cases} x(t_{p+1}) = Ax(t_{p+1} - 1) + A_1 x(s_p) + \sum_{i=1}^{N}(1 - \pi_{\{\sigma_p^*=i\}}) B_i e_i(t_p), \\ e_i(t_{p+1}) = (1 - \pi_{\{\sigma_p^*=i\}}) e_i(t_p) + C_i[x(s_p) - x(s_{p+1})], \ i = 1, \ldots, N. \end{cases} \tag{11.46}$$

Definition 11.2 The hybrid system (11.45)–(11.46) is said to be exponentially mean-square stable if there exist constants $b > 0$ and $0 < \kappa < 1$ such that, for initial condition $x_{t_0} \in \underbrace{\mathbb{R}^n \times \cdots \times \mathbb{R}^n}_{\tau_M+1 \text{ times}}$, the solutions of the hybrid system (11.45)–(11.46) satisfy

$$\mathbb{E}\{|x(k)|^2\} \le b\kappa^{2(k-t_0)} \mathbb{E}\{\|x_{t_0}\|_c^2 + |e(t_0)|^2\}, \ k \ge t_0,$$

and

$$\mathbb{E}\{|e(k)|^2\} \le b\kappa^{2(k-t_0)} \mathbb{E}\{\|x_{t_0}\|_c^2 + |e(t_0)|^2\}, \ k \ge t_0.$$

The objective of this section is to derive condition for the exponential mean-square stability of the hybrid system (11.45)–(11.46).

11.5.2 Exponential Mean-Square Stability of NCSs Under Stochastic Protocol

The stability analysis of (11.45)–(11.46) will be based on discrete-time Lyapunov functional (11.32). The term V_G satisfies for $k = t_{p+1} - 1$

$$\begin{aligned} &\mathbb{E}\Big\{V_G(t_{p+1}) - \lambda V_G(t_{p+1} - 1)\Big\} \\ &\le (\tau_M - h_m) \sum_{i=1}^{N} \mathbb{E}\Big\{|\sqrt{G_i} C_i \eta(t_{p+1})|^2\Big\} - \sum_{i=1}^{N} \lambda^{\tau_M} \mathbb{E}\Big\{|\sqrt{G_i} C_i [x(s_{p+1}) - x(s_p)]|^2\Big\}. \end{aligned}$$

Following the arguments for network-based stabilization under TOD scheduling protocol, we arrive at:

Theorem 11.3 *For any given scalar* $0 < \lambda < 1$, *integers* $0 \le h_m < \tau_M$ *and* K_i, $i = 1, \ldots, N$, *assume that there exist matrices* $P \in \mathbb{S}_+^{5n}$, S_1, S_2, R_1, $R_2 \in \mathbb{S}_+^n$, Q_i, U_i, $G_i \in \mathbb{S}_+^{n_i}$, $i = 1, \ldots, N$, N_1, $N_2 \in \mathbb{R}^{14n \times 2n}$, *and a matrix* $X \in \mathbb{R}^{3n \times 3n}$ *such that the following matrix inequalities*

$$\hat{\Omega}_i \triangleq \begin{bmatrix} \hat{\Gamma}_i & (1-\beta_i)Q_i \\ * & Q_i - \lambda^{\tau_M} G_i \end{bmatrix} < 0, \quad i = 1, \ldots, N, \tag{11.47}$$

$$\begin{bmatrix} \begin{bmatrix} \tilde{\Phi}_j & 0 \\ * & -\psi \end{bmatrix} & \Xi^T \tilde{H} & \hat{\Xi}^T \tilde{H} \\ * & -\tilde{H} & 0 \\ * & * & -\beta \tilde{H} \end{bmatrix} < 0, \quad j = 1, 2, \tag{11.48}$$

are feasible for $k \in [t_p, t_{p+1} - 1]$, *where*

$$\begin{aligned} \hat{\Gamma}_i &= -\lambda Q_i + (1-\beta_i)Q_i + \Big(1 + \frac{1}{\tau_M - h_m}\Big)U_i. \\ \Xi &= [A - I \; 0_{n\times n} \; A_1 \; 0_{n\times 19n} \; \Xi_0], \; \Xi_0 = [(1-\beta_1)B_1 \; \cdots \; (1-\beta_N)B_N], \\ \Xi_1 &= [0_{n\times 22n} \; -B_1 \; 0_{n\times(n_y - n_1)}], \; \Xi_2 = [0_{n\times(22n+n_1)} \; -B_2 \; 0_{n\times(n_y-n_1-n_2)}], \ldots, \\ \Xi_N &= [0_{n\times(22n+n_y-n_N)} \; -B_N], \; \Xi_j \in \mathbb{R}^{n\times(22n+n_y)}, \; \hat{\Xi} = [\Xi_1^T \; \cdots \; \Xi_N^T]^T, \\ \beta &= \mathrm{diag}\{\beta_1^{-1}, \ldots, \beta_N^{-1}\}, \; \psi = \frac{1}{\tau_M - h_m} \mathrm{diag}\{U_1, \ldots, U_N\}, \end{aligned} \tag{11.49}$$

with the notations $\tilde{\Phi}_j$, $j = 1, 2$, *and* $\tilde{H}$ *given by (11.16) and (11.33), respectively. Then, we have*

(i) *For* $k \in [t_p, t_{p+1} - 2]$, $V_e(k)$ *satisfies the following inequalities along (11.45)–(11.46):*

$$\hat{\Theta}_1(k) \triangleq \mathbb{E}\Big\{V_e(k+1) - \lambda V_e(k) - \frac{1}{\tau_M - h_m}\sum_{i=1}^{N} |\sqrt{U_i} e_i(t_p)|^2\Big\} \le 0; \tag{11.50}$$

(ii) *At* $k = t_{p+1} - 1$,

$$\hat{\Theta}_2 \triangleq \mathbb{E}\Big\{V_e(t_{p+1}) - \lambda V_e(t_{p+1} - 1) + \sum_{i=1}^{N} |\sqrt{U_i} e_i(t_p)|^2\Big\} \le 0; \tag{11.51}$$

(iii) *For the solutions of (11.45)–(11.46) initialized by* $x_{t_0} \in \underbrace{\mathbb{R}^n \times \cdots \times \mathbb{R}^n}_{\tau_M + 1 \text{ times}}$, $e(t_0) \in \mathbb{R}^{n_y}$, *the following bounds*

$$\mathbb{E}\{\tilde{V}(k)\} \le \mathbb{E}\{V_e(k)\} \le \lambda^{k-t_0}\mathbb{E}\{V_e(t_0)\}, \; k \ge t_0, \; k \in \mathbb{N} \tag{11.52}$$

and

$$\sum_{i=1}^{N} \mathbb{E}\Big\{|\sqrt{Q_i} e_i(k)|^2\Big\} \le \tilde{c}\lambda^{k-t_0}\mathbb{E}\{V_e(t_0)\} \tag{11.53}$$

hold with $V_e(t_0) = \tilde{V}(t_0) + \sum_{i=1}^{N} |\sqrt{Q_i} e_i(t_0)|^2$ *and* $\tilde{c} = \lambda^{-(\tau_M - h_m)}$.

Consequently, the exponential mean-square stability of (11.45)–(11.46) is guaranteed.

Proof *Proof of (i)*: First, for $k \in [t_p, t_{p+1} - 2]$ it holds that from $V_e(k)$ of (11.32)

$$\begin{aligned}\hat{\Theta}_1(k) \leq \mathbb{E}\Big\{&\tilde{V}(k+1) - \lambda\tilde{V}(k) + (\tau_M - h_m)\sum_{i=1}^{N} |\sqrt{G_i} C_i \eta(k+1)|^2 \\ &- \frac{1}{\tau_M - h_m}\sum_{i=1}^{N} |\sqrt{U_i} e_i(t_p)|^2\Big\} \triangleq \hat{\Psi}(k).\end{aligned} \tag{11.54}$$

Therefore, $\hat{\Theta}_1(k) \leq 0$ (i.e., (11.50)) holds if $\hat{\Psi}(k) \leq 0$ is satisfied.

Consider $k \in [t_p, t_{p+1} - 1]$, $p \in \mathbb{Z}^+$, and define

$$\hat{\xi}(k) = [\zeta^T(k) \;\; 0_{1\times 8n} \;\; e_1^T(k) \;\; \cdots \;\; e_N^T(k)]^T.$$

It can be shown from (11.45) that

$$\begin{aligned}\eta(k+1) &= \varXi\hat{\xi}(k) + \sum_{i=1}^{N} [\pi_{\{\sigma_p^*=i\}} - \beta_i]\varXi_i\hat{\xi}(k), \\ |\sqrt{\tilde{H}}\eta(k+1)|^2 &= |\sqrt{\tilde{H}}\varXi\hat{\xi}(k)|^2 + 2\sum_{i=1}^{N} [\pi_{\{\sigma_p^*=i\}} - \beta_i]\hat{\xi}^T(k)\varXi^T\tilde{H}\varXi_i\hat{\xi}(k) \\ &\quad + \sum_{i,j=1,i\neq j}^{N} [\pi_{\{\sigma_p^*=i\}} - \beta_i][\pi_{\{\sigma_p^*=j\}} - \beta_j]\hat{\xi}^T(k)\varXi_i^T\tilde{H}\varXi_j\hat{\xi}(k) \\ &\quad + \sum_{i=1}^{N} [\pi_{\{\sigma_p^*=i\}} - \beta_i]^2 |\sqrt{\tilde{H}}\varXi_i\hat{\xi}(k)|^2\end{aligned} \tag{11.55}$$

with $\varXi$ and $\varXi_i$, $i = 1, \ldots, N$ given by (11.49).

Following the proof of Theorem 11.2, calculating the difference $\tilde{V}(k+1) - \lambda\tilde{V}(k)$ of (11.45) and taking the mathematical expectation, we have $\mathbb{E}\{\hat{\Psi}(k)\} \leq 0$ for $k \in [t_p, t_{p+1} - 2]$ if the matrix inequalities (11.48) with $j = 1, 2$ hold. This completes the proof of (i).

Proof of (ii): From $V_e(k)$ of (11.32) and (11.54), it follows that

$$\begin{aligned}
\hat{\Theta}_2 \leq \mathbb{E}\Big\{ & V(t_{p+1}) - \lambda V(t_{p+1}-1) + (\tau_M - h_m)\sum_{i=1}^{N} |\sqrt{G_i} C_i \eta(t_{p+1})|^2 \\
& - \sum_{i=1}^{N} \lambda^{\tau_M} \left|\sqrt{G_i} C_i [x(s_{p+1}) - x(s_p)]\right|^2 \\
& + \sum_{i=1}^{N} \Big[|\sqrt{Q_i} e_i(t_{p+1})|^2 - \lambda |\sqrt{Q_i} e_i(t_p)|^2 \Big] + \sum_{i=1}^{N} |\sqrt{U_i} e_i(t_p)|^2 \Big\} \\
= \hat{\Psi}(t_{p+1}-1) + \mathbb{E}\Big\{ & \frac{1}{\tau_M - h_m} \sum_{i=1}^{N} |\sqrt{U_i} e_i(t_p)|^2 - \sum_{i=1}^{N} \lambda^{\tau_M} \left|\sqrt{G_i} C_i [x(s_{p+1}) - x(s_p)]\right|^2 \\
& + \sum_{i=1}^{N} \Big[|\sqrt{Q_i} e_i(t_{p+1})|^2 - \lambda |\sqrt{Q_i} e_i(t_p)|^2 \Big] + \sum_{i=1}^{N} |\sqrt{U_i} e_i(t_p)|^2 \Big\}.
\end{aligned}$$

Note that $\hat{\Psi}(t_{p+1} - 1) \leq 0$ and

$$\begin{aligned}
& \mathbb{E}\{e_i^T(t_{p+1}) Q_i e_i(t_{p+1})\} \\
& = \mathbb{E}\Big\{ \left| \sqrt{Q_i} [(1 - \pi_{\{\sigma_p^* = i\}}) e_i(t_p) + C_i x(s_p) - C_i x(s_{p+1})] \right|^2 \Big\} \\
& = \mathbb{E}\Big\{ (1 - \beta_i) e_i^T(t_p) Q_i e_i(t_p) + 2(1 - \beta_i) e_i^T(t_p) Q_i C_i [x(s_p) - x(s_{p+1})] \\
& \qquad + \left| \sqrt{Q_i} C_i [x(s_p) - x(s_{p+1})] \right|^2 \Big\}, \quad i = 1, \ldots, N.
\end{aligned}$$

Denote $\hat{\varsigma}_i = \text{col}\{e_i(t_p), C_i[x(s_p) - x(s_{p+1})]\}$. Then employing (11.46), we arrive at

$$\hat{\Theta}_2 \leq \hat{\Psi}(t_{p+1} - 1) + \sum_{i=1}^{N} \mathbb{E}\{\hat{\varsigma}_i^T \hat{\Omega}_i \hat{\varsigma}_i\} \leq 0,$$

that yields (11.51). This completes the proof of (ii).

Proof of (iii): The objective of the next step is to prove (11.52) and (11.53). By the comparison principle, for $k \in [t_p, t_{p+1} - 1]$ the inequality (11.50) implies

$$\mathbb{E}\{V_e(k) \leq \lambda^{k - t_p} \mathbb{E}\{V_e(t_p) + \sum_{i=1}^{N} \mathbb{E}\{|\sqrt{U_i} e_i(t_p)|^2\}. \tag{11.56}$$

Note that (11.47) yields

$$\begin{aligned}
U_i & < \Big(1 + \frac{1}{\tau_M - h_m}\Big) U_i \\
& < [\lambda - (1 - \beta_i)] Q_i < Q_i, \; i = 1, \ldots, N.
\end{aligned}$$

Hence, for $k \in [t_p, t_{p+1} - 1]$ it holds that

$$\mathbb{E}\{\tilde{V}(k)\} \leq \lambda^{k - t_p} \mathbb{E}\{V_e(t_p)\}. \tag{11.57}$$

On the other hand, the inequalities (11.51) and (11.56) with $k = t_{p+1} - 1$ imply

$$\begin{aligned}\mathbb{E}\{V_e(t_{p+1})\} &\leq \lambda\mathbb{E}\{V_e(t_{p+1}-1)\} - \sum_{i=1}^{N}\mathbb{E}\{|\sqrt{U_i}e_i(t_p)|^2\}\\ &\leq \lambda^{t_{p+1}-t_p}\mathbb{E}\{V_e(t_p)\} - (1-\lambda)\sum_{i=1}^{N}\mathbb{E}\{|\sqrt{U_i}e_i(t_p)|^2\}\\ &\leq \lambda^{t_{p+1}-t_p}\mathbb{E}\{V_e(t_p)\}.\end{aligned}$$

Then, we have

$$\begin{aligned}\mathbb{E}\{V_e(t_{p+1})\} &\leq \lambda^{t_{p+1}-t_{p-1}}\mathbb{E}\{V_e(t_{p-1})\}\\ &\leq \lambda^{t_{p+1}-t_0}\mathbb{E}\{V_e(t_0)\}.\end{aligned} \tag{11.58}$$

Replacing in (11.58) $p+1$ by p and using (11.57), we arrive at (11.52), which yields the exponential mean-square stability of (11.45)–(11.46) since the inequalities

$$\lambda_{\min}(P)\mathbb{E}\{|x(k)|^2\} \leq \mathbb{E}\{\tilde{V}(k)\},\ \mathbb{E}\{V(t_0)\} \leq \delta\mathbb{E}\{\|x_{t_0}\|_c^2\}$$

hold for some scalar $\delta > 0$. Moreover, (11.58) with $p+1$ replaced by p implies (11.53) since $\lambda^{t_p-t_0} = \lambda^{k-t_0}\lambda^{t_p-k} \leq \tilde{c}\lambda^{k-t_0}$ for $k \in [t_p, t_{p+1}-1]$. This completes the proof of (iii).

Remark 11.5 To simplify the exponential mean-square stability analysis of the hybrid system (11.45)–(11.46), we can follow Remark 11.3 and adopt Lyapunov functional of the form $\hat{V}_e(k) = V_e(k) + V_U(k)$, $k \in [t_p, t_{p+1}-1]$, where $V_e(k)$ is shown in (11.32) and the term $V_U(k)$ is negative and is given by

$$V_U(k) = \frac{t_p - k}{t_{p+1} - t_p}\sum_{i=1}^{N}|\sqrt{U_i}e_i(t_p)|^2.$$

Then, the same conditions (11.47) and (11.48) guarantee that $\hat{V}_e(k)$ is positive for $k \geq t_0$, $k \in \mathbb{Z}$, i.e., $\mathbb{E}\{\hat{V}_e(k)\} \geq \hat{\rho}\ \mathbb{E}\{|x(k)|^2 + |e(k)|^2\}$ with some $\hat{\rho} > 0$, and that $\mathbb{E}\{\hat{V}_e(k+1) - \lambda\hat{V}_e(k)\} \leq 0$, $k \in [t_p, t_{p+1}-1]$, which can substitute for the conditions (i) and (ii) of Theorem 11.3.

Remark 11.6 To enlarge MATI and MAD (i.e., h_M), one can resort to Markovian scheduling (see, e.g., Chap. 6 and [28]) instead of iid scheduling. In this protocol, the value of σ_p^* is determined through a Markov chain and the closed-loop system can be modeled as a stochastic Markovian jump discrete-time impulsive system.

Remark 11.7 The modeling and stability analysis of discrete-time networked systems with multiple sensor nodes under TOD and iid protocols follow exactly the discrete-time counterpart of Lemmas 5.1 and 6.1 for TOD and iid protocols, respectively. Note that Lemmas 5.1 and 6.1 were derived by Jensen's inequality and the reciprocally convex combination approach [173], while Theorems 11.2 and 11.3 were obtained by virtue of the developed reciprocally convex combination inequality proposed in [253] and the discrete-time counterpart of augmented Lyapunov functional provided in [146] for the stability analysis of continuous-time systems with time-varying delays. Moreover, as stated in Remark 9.2, the stability analysis of discrete-time system (11.8)–(11.9) with time-varying interval delays under scheduling protocols can be alternatively analyzed by substituting the switched system transformation approach for the Lyapunov method.

Remark 11.8 The conditions of Theorems 11.2 and 11.3 are easily adapted to decentralized networked control of large-scale interconnected systems with local independent networks, where every plant is controlled under TOD or under iid stochastic scheduling protocol.

11.6 Numerical Example

In this section, we will verify the efficiency of the derived conditions in Theorems 11.2 and 11.3 through the widely used inverted pendulum system after discretization. The discrete-time system model is given by (9.49). The pendulum can be also stabilized by a state feedback $u(k) = Kx(k)$ with the gain $K = [K_1 \;\; K_2]$

$$\begin{aligned} K &= [K_1 \;\; K_2], \;\; K_1 = \begin{bmatrix} 7.7606 \; 14.6847 \end{bmatrix}, \\ K_2 &= \begin{bmatrix} -86.7306 \; -26.3029 \end{bmatrix}, \end{aligned} \tag{11.59}$$

which leads to the closed-loop system having eigenvalues

$$\{0.5374, 0.9860 + 0.0177i, \;\; 0.9860 - 0.0177i, 0.9924\}.$$

Suppose that the spatially distributed components of the state variables of system are not accessible simultaneously. We start with the case of $N = 2$ and consider two measurements $y_i(k) = C_i x(k)$, $k \in \mathbb{Z}^+$, where

$$C_1 = \begin{bmatrix} 1\,0\,0\,0 \\ 0\,1\,0\,0 \end{bmatrix}, \;\; C_2 = \begin{bmatrix} 0\,0\,1\,0 \\ 0\,0\,0\,1 \end{bmatrix}.$$

For the values of h_m given in Table 11.2, we apply Theorems 11.2 and 11.3 with $\lambda = 1$ and find the maximum allowable values of $\tau_M = \text{MATI} + h_M$ that preserve the stability of the hybrid time-delay system (11.8)–(11.9). From Table 11.2, it is observed that under TOD protocol the conditions of Theorem 11.2 stabilize the system for larger τ_M than the results in [131] and [143]. Note that the proposed time-dependent Lyapunov approach in [131] only guaranteed the partial stability of the resulting hybrid-delayed system. Moreover, it was restricted to $N = 2$ sensors and cannot be extended to the general case of $N \geq 2$. The condition of [143] was obtained by Jensen's inequality and the reciprocally convex combination approach

Table 11.2 Example ($N = 2$): maximum value of $\tau_M = \text{MATI} + h_M$ for different h_m

$\tau_M \setminus h_m$	0	2	5	10	15
Reference [131] (TOD)	19	20	23	27	30
Reference [143] (TOD)	17	19	22	25	28
Theorem 11.3 (iid, $\beta_1 = 0.3$)	18	20	23	26	29
Theorem 11.2 (TOD)	21	23	25	29	32

Table 11.3 Example ($N = 4$): maximum value of $\tau_M = \text{MATI} + h_M$ for different h_m

$\tau_M \setminus h_m$	0	1	2	4	8	10
Reference [143] (TOD)	1	2	3	5	9	–
Theorem 11.3 (iid)	2	3	4	6	10	11
Theorem 11.2 (TOD)	3	4	6	8	12	13

[173]. The improvement of Theorem 11.2 compared to [131] and [143] is achieved due to the application of both Lemma 11.1 and the developed reciprocally convex combination inequality proposed in [253]. Compared to TOD protocol in [131] and Theorem 11.2, the iid protocol of Theorem 11.3 with $\beta_1 = 0.3$ leads to conservative results, but can easily include the data packet dropouts or collisions in the presence of large communication delays [138].

We proceed next with the case of $N = 4$, where $C_1, \ldots, C_4$ are the rows of I_4 and $K_1, \ldots, K_4$ are the entries of K given by (11.59). In this case, the conditions of [131] are not applicable any more. By applying Theorems 11.2 and 11.3 with $\beta_i = 0.25, i = 1, \ldots, 4$, Table 11.3 shows the maximum value of $\tau_M = \text{MATI} + h_M$ that preserves the exponential stability of the hybrid system (11.8)–(11.9). Also, here Theorem 11.2 achieves the least conservative results. Moreover, when $h_m > \frac{\tau_M}{2}$ ($h_m = 8, 10$), the proposed method is still feasible representing the case where communication delays can be larger than the sampling intervals.

11.7 Notes

In this chapter, we addressed the stability problem of discrete-time NCSs under dynamic scheduling protocols in which the components communicate through a shared communication medium that introduces large but bounded time-varying transmission delays. The closed-loop system is modeled as a discrete-time hybrid system with time-varying delays in the dynamics and in the reset conditions. By a newly constructed augmented Lyapunov functional and the discrete-time counterpart of the second-order Bessel–Legendre integral inequality, an improved stability criterion to find MATI and MAD was derived such that the resulting closed-loop system is exponentially stable with respect to the full state. Numerical example illustrates the efficiency of our method.

Note that by reformulating an equivalent form of Moon et al.'s inequality [160], more insights on the relationship between some existing matrix inequalities were provided in [203]. It is demonstrated that some existing inequalities including the developed reciprocally convex combination inequality proposed in [253] can be captured as particular cases of Moon et al.'s inequality. Therefore, a less conservative analysis can be performed by substituting the developed reciprocally convex combination inequality for Moon et al.'s inequality.

Chapter 12
Decentralized Networked Control of Discrete-Time Systems with Local Networks

To manage with large communication delays in the presence of scheduling protocols from sensors to actuators, the time-delay approach to continuous-time decentralized NCSs has been developed in Chap. 7. The results of Chap. 7 are confined to stability analysis. This chapter is devoted to extend the time-delay approach to decentralized stabilization and H_∞ control of large-scale discrete-time systems with multiple local communication networks connecting sensors, controllers and actuators. The local networks operate asynchronously and independently of each other in the presence of variable sampling intervals, transmission delays and scheduling protocols from sensors to controllers.

The main challenge is in the stability and performance analysis of the closed-loop discrete-time NCS under TOD protocol. Note that as in Chaps. 5 and 7 for the continuous-time case, the closed-loop system in the discrete-time case is modeled as a hybrid large-scale system with delays. However, its structure is different, where the same state equation appears also in the reset system. This leads to a different from the continuous-time case Lyapunov technique (see Chap. 9 for the case of one plant). This chapter proposes a novel Lyapunov functional candidate explaining that its continuous-time counterpart leads to simplified LMIs comparatively to Chap. 7. Under Round-Robin protocol, the closed-loop system is modeled as a system with multiple delays, and the stability conditions naturally extend the conditions of Chap. 7 to the discrete-time case. The extension to l_2-gain analysis considers both, independent and coupled disturbances. A numerical example of decentralized control of two coupled inverted pendulums with local networks (see Fig. 12.1) illustrates the efficiency of the results. Throughout this chapter, by δ_{ij} we denote the Kronecker delta meaning $\delta_{nm} = 0,\ n \neq m$ and $\delta_{nn} = 1$ $(n, m \in \mathbb{N})$. The subscript or superscript j will stand for a subsystem index, while subscript i will stand for sensor index.

K. Liu et al., *Networked Control Under Communication Constraints*, Advances in Delays and Dynamics 11, https://doi.org/10.1007/978-981-15-4230-5_12

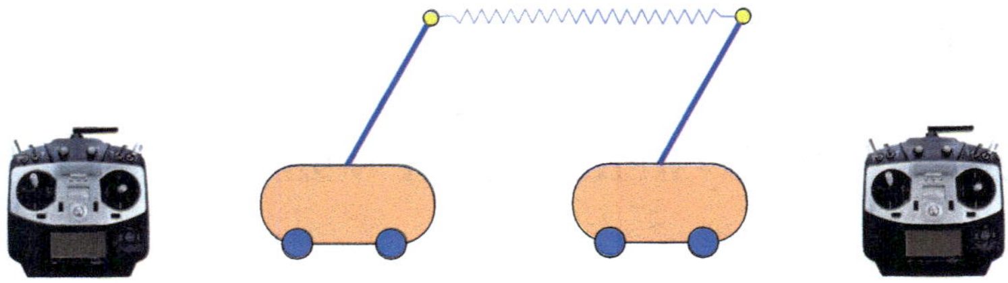

Fig. 12.1 Two coupled inverted pendulums

12.1 Problem Formulation and Preliminaries

Consider the large-scale system with M interconnected subsystems (see Fig. 7.1)

$$\begin{aligned} x_j(t+1) &= A_j x_j(t) + B_j u_j(t) + D_j w(t) + \sum_{l \neq j} F_j^l x_l(t), \\ y_{ij}(t) &= C_{ij} x_j(t) \in \mathbb{R}^{n_i^j}, \ i = 1, \ldots, N_j, \ t \in \mathbb{Z}^+, \end{aligned} \tag{12.1}$$

where $x_j(t) \in \mathbb{R}^{n^j}$ is the state, $w(t) \in l_2(\mathbb{R}^{n_w})$ is a perturbation, and F_j^l are the coupling matrices. The jth subsystem has N_j local sensors and a local controller, and $y_j(t) = [y_{1j}^T(t) \ldots y_{N_j j}^T(t)]^T \in \mathbb{R}^{n_y^j}$ is the local measurement vector.

The jth subsystem is assumed to have an independent sequence of sampling instants

$$0 = s_0^j < s_1^j < \cdots < s_k^j < \cdots, \ \lim_{k \to \infty} s_k^j = \infty$$

with bounded sampling intervals $1 \leq s_{k+1}^j - s_k^j \leq \text{MATI}_j$. At each sampling instant s_k^j, one of the sensors i_k^* is being chosen by a scheduling protocol and its output $y_{i_k^* j}(s_k^j)$ is being transmitted via a local sensor network to the local controller node. We assume that the local network is independent of other subsystem's networks.

Suppose that the data loss is not possible and that the transmission of the information over the networks from sensors to actuators (through controller) is subject to a variable roundtrip delay η_k^j. Then, $t_k^j = s_k^j + \eta_k^j$ is the updating time instant of the subsystem input $u_j(t)$. The communication delay is assumed to be bounded $\eta_k^j \in [\eta_m^j, \eta_M^j]$, where $\eta_M^j \equiv \text{MAD}_j$. We assume that an old sample cannot get to the same destination (same controller or same actuator) after the most recent one. Suppose that the controllers and the actuators are event-driven.

Following Chap. 7, we assume that there exist M gain matrices

$$K_j = \left[K_{1j} \cdots K_{N_j j} \right], \ K_{ij} \in \mathbb{R}^{m^j \times n_i^j}$$

such that the matrices $A_j + B_j K_j C_j$ are Schur, where $C_j = [C_{1j}^T \ \cdots \ C_{N_j j}^T]^T$.

Denote by

$$\hat{y}_j(s_k^j) = \left[\hat{y}_{1j}^T(s_k^j) \dots \hat{y}_{N_j j}^T(s_k^j)\right]^T \in \mathbb{R}^{n_y^j} \tag{12.2}$$

the most recent output information submitted to the scheduling protocol of the jth subsystem (i.e., the most recent information at the jth controller side) at the sampling instant s_k^j. Then, the static output feedbacks are given by

$$u_j(t) = \sum_{i=1}^{N_j} K_{ij}\hat{y}_{ij}(s_k^j),\ t_k^j \le t < t_{k+1}^j. \tag{12.3}$$

We assume $u_j(t) = 0$ for $t < t_0^j$.

Define the initial time

$$T_0 = \max_j t_0^j. \tag{12.4}$$

Definition 12.1 (*Exponential stability*) The closed-loop system (12.1), (12.3) with $w(t) \equiv 0$ is exponentially stable (with a decay rate $\lambda \in [0, 1)$) if there exists $\beta > 0$ such that for any initial condition x_0, the following inequality

$$|x(t)|^2 \le \beta \|x\|_{T_0}^2 \lambda^{t-T_0}$$

holds for $t > T_0$, where $x(t) = \text{col}\{x_1(t), \dots, x_M(t)\} \in \mathbb{R}^n$ and $\|x\|_{T_0}^2 \equiv \sum_{t=0}^{T_0} |x(t)|^2$.

In this chapter, the sufficient conditions are given for the exponential stability of the unperturbed closed-loop system, as well as for the l_2-gain analysis of the perturbed one.

12.2 Stability of Decentralized Networked Control Under TOD Protocol

Consider the system (12.1), (12.6) with M locally controlled subsystems, where the jth subsystem is under TOD protocol. In this section, we study the case where $w(t) \equiv 0$.

Following (5.5), we denote the error states $\{e_{ij}, i = 1, \dots, N_j\}$ by

$$e_{ij}(t) = \hat{y}_{ij}(s_{k-1}^j) - y_{ij}(s_k^j),\ t_k^j \le t < t_{k+1}^j, \tag{12.5}$$

and the error-state vectors by

$$e_j(t) = \text{col}\{e_{1j}(t) \dots e_{N_j j}(t)\},\ j = 1, \dots, M.$$

Then, the feedback (12.3) can be rewritten as

$$u_j(t) = \sum_{i=1}^{N_j} K_{ij} C_{ij} x_j(s_k) + \sum_{i=1, i \neq i_k^*}^{N_j} K_{ij} e_{ij}(t), \; t_k^j \leq t < t_{k+1}^j. \tag{12.6}$$

Denote

$$e_{ij}[k] \equiv e_{ij}(t_k^j), \; k = 0, 1, \ldots \tag{12.7}$$

for the error states in order to better express their piecewise-constant nature.

In TOD scheduling, the transmitted measurement sent at each instant is chosen according to a weighted error discriminant function

$$i_k^{j*} = \min \left\{ \arg \max_{i \in \{1, \ldots, N_j\}} e_{ij}^T[k] Q_{ij} e_{ij}[k] \right\} \tag{12.8}$$

with some weight matrices Q_{ij}. The conditions for choosing weight matrices will be given below. For simplicity, we will omit the subsystem index j in i_k^{j*} throughout the rest of this chapter. The error-state vector is given by

$$e_{ij}(t_{k+1}^j) = \begin{cases} -C_{ij}[x_j(s_{k+1}^j) - x_j(s_k^j)], & i = i_k^* \\ e_{ij}(t_k^j) - C_{ij}[x_j(s_{k+1}^j) - x_j(s_k^j)], & else \end{cases} \tag{12.9}$$

with $e_{ij}(t_0^j) = -C_{ij} x(s_0^j)$.

The jth subsystem closed loop is then given by

$$\begin{cases} x_j(t+1) = A_j x(t) + A_{1j} x_j(t - \tau_j(t)) + \sum_{i \neq i_{k,j}^*} B_{ij} e_{ij}(t) + \sum_{l \neq j} F_j^l x_l(t), \\ e_j(t+1) = e_j(t), \; t_k^j \leq t \leq t_{k+1}^j - 2 \end{cases} \tag{12.10}$$

with the reset equations (i.e., equations at reset times $t = t_{k+1}^j - 1$)

$$\begin{cases} x_j(t_{k+1}^j) = A_j x_j(t_{k+1}^j - 1) + A_{1j} x_j(s_k^j) + \sum_{i \neq i_k^*} B_{ij} e_{ij}(t_k^j) + \sum_{l \neq j} F_j^l x_l(t_{k+1}^j - 1), \\ e_{ij}(t_{k+1}^j) = (1 - \delta_{i_k^*}) e_{ij}(t_k^j) - C_{ij}[x_j(s_{k+1}^j) - x_j(s_k^j)], \; k \in \mathbb{Z}^+, \\ e_{ij}(t_0^j) = -C_{ij} x_j(s_0^j), \end{cases} \tag{12.11}$$

where $A_{1j} = B_j K_j C_j$ and $\tau_j(t) = t - s_k^j$ is the subsystem delay. We have

$$\begin{aligned} \tau_j(t) &\leq t_{k+1}^j - 1 - s_k^j \leq \text{MAD}_j + \text{MATI}_j - 1 \equiv \tau_M^j, \\ \tau_j(t) &\geq t_k^j - s_k^j \geq \eta_m^j \geq 0. \end{aligned} \tag{12.12}$$

Remark 12.1 Note that in the discrete-time case, the equation for $x_j(t)$ at reset times $t = t_k^j - 1$ is given by the same difference equation as for $t \in [t_{k-1}^j, t_k^j - 2]$. This is different from the continuous-time case, where $x_j(t_k^{j-}) = x_j(t_k^j)$.

We define a functional $V_j(t)$ for the jth subsystem. Consider the following LKF

$$V_j(t) = x_j^T(t)P_jx_j(t) + \tilde{V}_j(t) + V_Q(t,k) + V_G(t,k), \; t_k^j \le t \le t_{k+1}^j - 1, \tag{12.13}$$

where $\lambda \in (0,1)$ and

$$\begin{aligned}
V_Q(t,k) &= \lambda^{t-t_k^j}\sum_{i=1}^{N_j} e_{ij}^T(t)Q_{ij}e_{ij}(t) - \sum_{i\neq i_k^*}(t-t_k^j)e_{ij}^T(t)U_{ij}e_{ij}(t),\\
V_G(t,k) &= \sum_{s=s_k^j}^{t-1}\sum_{i=1}^{N_j}\lambda^{t-s-1}z_j^T(s)C_{ij}^TG_{ij}C_{ij}z_j(s),\\
\tilde{V}_j(t) &= V_{0j}(t) + V_{1j}(t),\\
V_{0j}(t) &= \sum_{s=t-\eta_m^j}^{t-1}\lambda^{t-s-1}x_j^T(s)S_{0j}x_j(s) + \eta_m^j\sum_{\theta=1}^{\eta_m}\sum_{s=t-\theta}^{t-1}\lambda^{t-s-1}z_j^T(s)R_{0j}z_j(s),\\
V_{1j}(t) &= \sum_{s=t-\tau_M^j}^{t-\eta_m-1}\lambda^{t-s-1}x_j^T(s)S_{1j}x_j(s) + h_j\sum_{\theta=\eta_m^j+1}^{\tau_M^j}\sum_{s=t-\theta}^{t-1}\lambda^{t-s-1}z_j^T(s)R_{1j}z_j(s),\\
z_j(t) &= x_j(t+1) - x_j(t), \; h_j = \tau_M^j - \eta_m^j.
\end{aligned} \tag{12.14}$$

Here $x_j^T(t)P_jx_j(t) + \tilde{V}_j(t)$ is a standard LKF for the stability of systems with interval delays [47], and the term $V_Q(t,k)$ is constructed in order to deal with the error-states that appear in (12.10). The term $V_G(t,k)$ allows to compensate the delayed terms $C_{ij}[x_j(s_{k+1}^j) - x_j(s_k^j)]$ in the reset Eq. (12.11).

Remark 12.2 In the continuous-time case, a LKF candidate V_j for the stability analysis of the resulting jth hybrid subsystem under TOD protocol has been proposed in Chap. 5 (for the uncoupled case, where $F_j^l = 0, l \neq j$) and has been modified in Chap. 7 for the coupled (large-scale) systems leading to simplified stability conditions. Thus, the stability of uncoupled jth subsystem has been guaranteed in Chap. 7 if the condition $\dot{V}_j + 2\alpha V_j \le 0$ holds for some $\alpha > 0$ along the continuous-time dynamics, and if V_j does not grow in the reset times. Different from this, in Chap. 9 for discrete-time case the stability of the uncoupled jth subsystem can be guaranteed if $V_j(t+1) - \lambda V_j(t) \le 0$ with some $\lambda \in (0,1)$ for all t (including the reset times $t = t_k^j - 1$). Note that the Lyapunov–Krasovskii method of Chap. 9 is confined to the case of two sensor nodes $N_j = 2$ and only partial x-stability is established. As in Chap. 7 for continuous-time case, an extension of results to large-scale systems is not a trivial result that will be presented in Proposition 12.1 and Theorem 12.1.

Remark 12.3 By introducing the term $V_Q(t,k)$ of (12.14), we avoid the positive terms containing $|\sqrt{Q_{i_k^*}}e_{i_k^*}|^2$ in the upper bound on $V_j(t+1) - \lambda V_j(t)$ for $t \neq t_k - 1$. The Q_{ij}-terms of V_Q in (12.14) is a discrete-time and modified version of the corresponding term in [38]

$$V_{conQ}(t) = \sum_{i=1}^{N_j} e_{ij}^T(t)Q_{i,j}e_{ij}(t) - 2\alpha(t - t_k^j)e_{i_k^*j}^T(t)Q_{i_k^*,j}e_{i_k^*j}(t),$$

where the negative term $-2\alpha(t_k^j - t)e_{i_k^*j}^T(t)Q_{i_k^*,j}e_{i_k^*j}(t)$ allows to compensate a positive term of the form $2\alpha e_{i_k^*j}^T(t)Q_{i_k^*,j}e_{i_k^*j}(t)$ that arises in $\dot{V}_j(t) + 2\alpha V_j(t)$.

The continuous-time counterpart of Q-terms of (12.14) has the form

$$V_{cQ}(t) = e^{-2\alpha(t-t_k)} \sum_{i=1}^{N_j} e_{ij}^T(t)Q_{i,j}e_{ij}(t). \tag{12.15}$$

The term (12.15) improves and simplifies the condition for $\dot{V} + 2\alpha V$ of [38] based on V_{conQ}:

$$\begin{aligned} &\dot{V}_{cQ}(t) + 2\alpha V_{cQ}(t) \le 0, \\ &\dot{V}_{conQ}(t) + 2\alpha V_{conQ}(t) = \sum_{i \ne i_k^*} e_{ij}^T(t)Q_{i,j}e_{ij}(t). \end{aligned}$$

However, in the jump condition $t = t_{k+1}^j$ these functionals lead to different results:

$$V_{cQ}(t_{k+1}^{j+}) - V_{cQ}(t_{k+1}^{j-}) = \sum_{i=1}^{N_j} e_{ij}^T(t_{k+1}^j)Q_{i,j}e_{ij}(t_{k+1}^j) - e^{-2\alpha(t_{k+1}^j - t_k^j)} \sum_{i=1}^{N_j} e_{ij}^T(t_k^j)Q_{i,j}e_{ij}(t_k^j),$$

whereas

$$\begin{aligned} V_{conQ}(t_{k+1}^{+j}) - V_{conQ}(t_{k+1}^{-j}) = &\sum_{i=1}^{N_j} e_{ij}^T(t_{k+1}^j)Q_{i,j}e_{ij}(t_{k+1}^j) - \sum_{i=1}^{N_j} e_{ij}^T(t_k^j)Q_{i,j}e_{ij}(t_k^j) \\ &+2\alpha(t_{k+1}^j - t_k^j)e_{ij}^T(t_k^j)Q_{i,j}e_{ij}(t_k^j)_{|i=i_k^*}. \end{aligned}$$

Hence, V_{cQ} may lead to different from [38] (complementary) results. Note that in the example from [38], V_{cQ} does not change the numerical result, but $Q_{i,j}$-terms are eliminated from part of LMIs (simplifying them).

Proposition 12.1 *Consider the jth closed-loop subsystem* (12.10)–(12.11). *Given parameters* $0 < \lambda < 1$ *and* $\varepsilon \le 1 - \lambda$ *(* $\sum_{j \ne l} \varepsilon_{jl} \le \varepsilon$, $l = 1, \ldots, M$*), and matrices* $P_l > 0$, $l = 1 \ldots M$, *assume that there exist matrices* $S_{0j} > 0$, $R_{0j} > 0$, $S_{1j} > 0$, $R_{1j} > 0$, $Q_{ij} > 0$, $U_{ij} > 0$, $G_{ij} > 0$ *and* W_j *that satisfy LMIs*

$$\Omega_j = \begin{bmatrix} R_{1j} & W_j \\ * & R_{1j} \end{bmatrix} \ge 0 \tag{12.16}$$

and

$$\begin{aligned} &\Sigma_i^j - \tilde{\Upsilon}_i^j + \tilde{E}_{ij2}^T G_j \tilde{E}_{ij2} < 0, \\ &\Psi_i^j + [I\ 0]^T U_{ij} [I\ 0] < 0,\ i = 1, \ldots, N_j, \end{aligned} \tag{12.17}$$

where

$$
\begin{aligned}
\Sigma_i^j &= \tilde{E}_{ij1}^T P_j \tilde{E}_{ij1} + \tilde{E}_{ij2}^T H_j \tilde{E}_{ij2} - \tilde{E}_{ij3}^T \lambda^{\tau_M^j} \Omega_j \tilde{E}_{ij3} - \tilde{E}_{ij4}^T \lambda^{\eta_m^j} R_{0j} \tilde{E}_{ij4} + \tilde{E}_{ij5}, \\
h_j &= \tau_M^j - \eta_m^j, \quad \mathscr{F}^j = [F_j^1 \cdots F_j^{l \neq j} \cdots F_j^M], \\
H_j &= \eta_m^{j2} R_{0j} + h_j^2 R_{1j}, \quad G_j = \sum_{i=1}^{N_j} C_{ij}^T G_{ij} C_{ij}, \\
E_{ij1} &= \left[A_j \ \ [0\ 1\ 0] \otimes A_{1j} \ \ \Theta_{ij} \right], \\
E_{ij2} &= \left[(A_j - I_{n^j}) \ \ [0\ 1\ 0] \otimes A_{1j} \ \ \Theta_{ij} \right], \\
\Theta_{ij} &= \left[B_{1j} \cdots B_{r \neq ij} \cdots B_{N_j j} \right], \\
E_{ij3} &= \left[0_{2n^j \times n^j} \ \ \left[\begin{smallmatrix} 1 & -1 & 0 \\ 0 & 1 & -1 \end{smallmatrix}\right] \otimes I_{n^j} \ \ 0_{2n^j \times (n_y^j - n_i^j)} \right], \\
E_{ij4} &= [I_{n^j} \ \ -I_{n^j} \ \ 0], \\
E_{i5} &= \mathrm{diag}\{S_{0j} - \lambda P_j, -(S_{0j} - S_{1j})\lambda^{\eta_m^j}, 0_{n^j}, -S_{1j}\lambda^{\tau_M^j}, 0\}, \\
\Psi_i^j &= \begin{bmatrix} \lambda h_j U_{ij} + \left(1 - \dfrac{N_j \lambda^{h_j+1}}{N_j - 1}\right) Q_{ij} & Q_{ij} \\ * & Q_{ij} - \dfrac{\lambda^{\tau_M^j}}{h_j + 1} G_{ij} \end{bmatrix}, \\
\Upsilon_i^j &= \mathrm{diag}\{0_{4n^j}, U_{1j}, \cdots, U_{r \neq i,j}, \cdots, U_{N_j j}\}, \\
\tilde{E}_{ij1} &= [E_{ij1} \ \mathscr{F}^j], \quad \tilde{E}_{ij2} = [E_{ij2} \ \mathscr{F}^j], \\
\tilde{E}_{ij3} &= \left[E_{ij3} \ \ 0 \cdot \left[\begin{smallmatrix} \mathscr{F}^j \\ \mathscr{F}^j \end{smallmatrix}\right] \right], \quad \tilde{E}_{ij4} = [E_{ij4} \ \ 0 \cdot \mathscr{F}^j], \\
\tilde{E}_{ij5} &= \mathrm{diag}\{E_{ij5}, \ -\mathscr{P}^j\}, \\
\mathscr{P}^j &= \mathrm{diag}\{\varepsilon_{jl} P_l, l \neq j\}, \\
\tilde{\Upsilon}_i^j &= \mathrm{diag}\{\Upsilon_i^j, 0_{n-n_j}\}.
\end{aligned}
\tag{12.18}
$$

Then, V_j *given by* (12.13) *satisfies the following inequalities along* (12.10)–(12.11) *(with some* $\beta_j > 0$*):*

$$
V_j(t) \geq \beta_j (|x_j(t)|^2 + |e_j(t)|^2), \ t \geq t_0^j, \tag{12.19}
$$

$$
V_j(t+1) - \lambda V_j(t) \leq \sum_{l \neq j} \varepsilon_{jl} x_l^T(t) P_l x_l(t), \ t \geq t_0^j. \tag{12.20}
$$

Proof We prove first the positivity condition (12.19). Note that

$$
t_{k+1}^j - t_k^j - 1 = (t_{k+1}^j - 1 - s_k^j) - \eta_k^j \leq \tau_M^j - \eta_m^j = h_j. \tag{12.21}
$$

We have

$$
\begin{aligned}
V_Q(t,k) &\geq \lambda^{t_{k+1}^j-1-t_k^j} \sum_{i=1}^{N_j} e_{ij}^T(t) Q_{ij} e_{ij}(t) - (t_{k+1}^j - 1 - t_k^j) \sum_{i \neq i_k^*} e_{ij}^T(t) U_{ij} e_{ij}(t) \\
&\geq \lambda^{h_j} \sum_{i=1}^{N_j} e_{ij}^T(t) Q_{ij} e_{ij}(t) - h_j \sum_{i \neq i_k^*} e_{ij}^T(t) U_{ij} e_{ij}(t), \ t_k^j \leq t \leq t_{k+1}^j - 1.
\end{aligned}
\tag{12.22}
$$

From the second inequality in (12.17) it follows that

$$
\lambda h_j U_{ij} < \Big(\frac{N_j \lambda^{h_j+1}}{N_j - 1} - 1\Big) Q_{ij} < \lambda^{h_j+1}\Big(\frac{N_j}{N_j-1} - 1\Big) Q_{ij} = \frac{\lambda^{h_j+1}}{N_j - 1} Q_{ij} \leq \lambda^{h_j+1} Q_{ij}.
$$

The latter inequality together with (12.22) implies

$$
V_Q(t,k) \geq \beta_j |e_j(t)|^2, \ t_k^j \leq t \leq t_{k+1}^j - 1
$$

for some $\beta_j > 0$, yielding (12.19).

We prove next (12.20). We find

$$
\begin{aligned}
V_{0j}(t+1) - \lambda V_{0j}(t) &\leq \eta_m^{j2} z_j^T(t) R_{0j} z_j(t) + x_j^T(t) S_{0j} x_j(t) \\
&\quad - \lambda^{\eta_m^j} x_j^T(t - \eta_m^j) S_{0j} x_j(t - \eta_m^j) \\
&\quad - \eta_m^j \sum_{\theta=1}^{\eta_m^j} \lambda^\theta z_j^T(t-\theta) R_{0j} z_j(t-\theta), \\
V_{1j}(t+1) - \lambda V_{1j}(t) &\leq h_j^2 z_j^T(t) R_{1j} z_j(t) + \lambda^{\eta_m^j} x_j^T(t - \eta_m^j) S_{1j} x_j(t - \eta_m^j) \\
&\quad - \lambda^{\tau_M^j} x_j^T(t - \tau_M^j) S_{1j} x_j(t - \tau_M^j) \\
&\quad - h_j \sum_{\theta=\eta_m^j+1}^{\tau_M^j} \lambda^\theta z_j^T(t-\theta) R_{1j} z_j(t-\theta).
\end{aligned}
\tag{12.23}
$$

By using Jensen's inequality, we obtain

$$
\begin{aligned}
\eta_m^j \sum_{\theta=1}^{\eta_m^j} \lambda^\theta z_j^T(t-\theta) R_{0j} z_j(t-\theta) \geq \quad & \lambda^{\eta_m^j} \Big| \sqrt{R_{0j}} (x_j(t) - x_j(t - \eta_m^j)) \Big|^2 \\
= & \begin{bmatrix} \eta_j(t) \\ \xi_{ij}(t) \\ X_j^c(t) \end{bmatrix} \tilde{E}_{ij4}^T \lambda^{\eta_m^j} R_{0j} \tilde{E}_{ij4} \begin{bmatrix} \eta_j(t) \\ \xi_{ij}(t) \\ X_j^c(t) \end{bmatrix}.
\end{aligned}
\tag{12.24}
$$

Under (12.16), the following inequality

$$
\begin{aligned}
h_j \sum_{\theta=\eta_m^j+1}^{\tau_M^j} \lambda^{\theta} z_j^T(t-\theta) R_{1j} z_j(t-\theta) &\geq \lambda^{\tau_M^j} \begin{bmatrix} x_j(t-\eta_m^j) - x_j(t-\tau_j(t)) \\ x_j(t-\tau_j(t)) - x_j(t-\tau_M^j) \end{bmatrix}^T \Omega_j \\
&\quad \times \begin{bmatrix} x_j(t-\eta_m^j) - x_j(t-\tau_j(t)) \\ x_j(t-\tau_j(t)) - x_j(t-\tau_M^j) \end{bmatrix} \\
&= \begin{bmatrix} \eta_j(t) \\ \xi_{ij}(t) \\ X_j^c(t) \end{bmatrix}^T \tilde{E}_{ij3}^T \lambda^{\tau_M^j} \Omega_j \tilde{E}_{ij3} \begin{bmatrix} \eta_j(t) \\ \xi_{ij}(t) \\ X_j^c(t) \end{bmatrix}
\end{aligned} \tag{12.25}
$$

holds [173]. Denote

$$
\begin{aligned}
\eta_j(t) &= \mathrm{col}\{x_j(t), x_j(t-\eta_m^j), x_j(t-\tau_j(t)), x_j(t-\tau_M^j)\}, \\
X_j^c(t) &= \mathrm{col}\{x_l(t), l \neq j\}, \\
\xi_{ij}(t) &= \mathrm{col}\{e_{rj}(t), r \neq i\}, \\
\sigma_{ij}[k] &= C_{ij}(x_j(s_{k+1}^j) - x_j(s_k^j)).
\end{aligned}
$$

Employing the relations

$$
z_j(t) = \tilde{E}_{i_k^* j2} \begin{bmatrix} \eta_j(t) \\ \xi_{i_k^* j}(t) \\ X_j^c(t) \end{bmatrix}, \quad x_j(t+1) = \tilde{E}_{i_k^* j1} \begin{bmatrix} \eta_j(t) \\ \xi_{i_k^* j}(t) \\ X_j^c(t) \end{bmatrix} \tag{12.26}
$$

and

$$
\sum_{l \neq j} \varepsilon_{jl} x_l^T(t) P_l x_l(t) = X_j^{cT}(t) \mathscr{P}^j X_j^c(t), \tag{12.27}
$$

from (12.23)–(12.25), we obtain for $t_k \leq t \leq t_{k+1} - 1$

$$
\begin{aligned}
&x_j^T(t+1) P_j x_j(t+1) + \tilde{V}_j(t+1) - \lambda(x_j(t)^T P_j x_j(t) + \tilde{V}_j(t)) \\
&\quad - \sum_{l \neq j} \varepsilon_{jl} x_l^T(t) P_l x_l(t) \leq \begin{bmatrix} \eta_j^T(t) \ \xi_{i_k^* j}^T(t) \ \ X_j^{cT}(t) \end{bmatrix} \Sigma_{i_k^*}^j \begin{bmatrix} \eta_j(t) \\ \xi_{i_k^* j}(t) \\ X_j^c(t) \end{bmatrix}.
\end{aligned} \tag{12.28}
$$

Consider $t_k^j \leq t \leq t_{k+1}^j - 2$. From (12.21) and definition of $V_Q(t)$, we have

$$
\begin{aligned}
V_Q(t+1,k) - \lambda V_Q(t,k) &= \Big[-1 + (\lambda - 1)(t - t_k^j)\Big] \sum_{i \neq i_k^*} |\sqrt{U_{ij}} e_{ij}(t)|^2 \\
&\leq -\sum_{i \neq i_k^*} |\sqrt{U_{ij}} e_{ij}(t)|^2 \\
&\leq -\begin{bmatrix} \eta_j^T(t) \ \xi_{i_k^* j}^T(t) \end{bmatrix} \Upsilon_{i_k^*}^j \begin{bmatrix} \eta_j(t) \\ \xi_{i_k^* j}(t) \end{bmatrix} \\
&= -\begin{bmatrix} \eta_j(t) \\ \xi_{i_k^* j}(t) \\ X_j^c(t) \end{bmatrix}^T \tilde{\Upsilon}_{i_k^*}^j \begin{bmatrix} \eta_j(t) \\ \xi_{i_k^* j}(t) \\ X_j^c(t) \end{bmatrix}.
\end{aligned} \tag{12.29}
$$

Note that the terms $|\sqrt{Q_{ij}}e_{ij}(t)|^2$ vanish in (12.29) due to the multiplication of them by $\lambda^{t-t_k^j}$. Therefore, we obtain

$$V_j(t+1)-\lambda V_j(t)-\sum_{l\neq j}\varepsilon_{jl}x_l^T(t)P_lx_l(t) \leq \begin{bmatrix}\eta_j^T(t) & \xi_{i_k^*j}^T(t) & X_j^{cT}(t)\end{bmatrix}\left(\Sigma_{i_k^*}^j-\tilde{\Upsilon}_{i_k^*}^j\right)\begin{bmatrix}\eta_j(t)\\ \xi_{i_k^*j}(t)\\ X_j^c(t)\end{bmatrix}$$
$$+z_j^T(t)G_jz_j(t),\ t_k^j \leq t \leq t_{k+1}^j-2.$$

Substituting (12.26) into the latter inequality, we see that the first inequality (12.17) implies the inequality (12.20) for $t_k^j \leq t \leq t_{k+1}^j - 2$.

Consider now the case of reset times, where $t = t_{k+1}^j - 1$. Using the notation (12.7), we have

$$\begin{aligned}&V_Q(t_{k+1}^j, k+1) - \lambda V_Q(t_{k+1}^j - 1, k)\\ &\leq \sum_{i=1}^{N_j}\left[e_{ij}^T[k+1]Q_{ij}e_{ij}[k+1] - e_{ij}^T[k]\lambda^{t_{k+1}^j-t_k^j}Q_{ij}e_{ij}[k]\right] + \lambda h_j\sum_{i\neq i_k^*}e_{ij}^T[k]U_{ij}e_{ij}[k].\end{aligned} \tag{12.30}$$

Exploiting the reset equations (12.9), we find

$$\begin{aligned}\sum_{i=1}^{N_j}e_{ij}^T[k+1]Q_{ij}e_{ij}[k+1] &= \sigma_{i_k^*j}^T[k]Q_{i_k^*j}\sigma_{i_k^*j}^T[k]\\ &+\sum_{i\neq i_k^*}(e_{ij}[k]+\sigma_{ij}[k])^TQ_{ij}(e_{ij}[k]+\sigma_{ij}[k]).\end{aligned} \tag{12.31}$$

Since under TOD

$$\left|\sqrt{Q_{i_k^*j}}e_{i_k^*j}(t)\right|^2 \geq \frac{1}{N_j-1}\sum_{i\neq i_k^*}\left|\sqrt{Q_{i,j}}e_{ij}(t)\right|^2, \tag{12.32}$$

we arrive at

$$\begin{aligned}\sum_{i=1}^{N_j}e_{ij}^T[k]Q_{ij}e_{ij}[k] &= e_{i_k^*j}^T[k]Q_{i_k^*j}e_{i_k^*j}^T[k] + \sum_{i\neq i_k^*}e_{ij}[k]^TQ_{ij}e_{ij}[k]\\ &\geq \sum_{i\neq i_k^*}e_{ij}[k]^T\left(1+\frac{1}{N_j-1}\right)Q_{ij}e_{ij}[k].\end{aligned} \tag{12.33}$$

Substituting (12.31), (12.33) into (12.30) and using (12.21), we obtain

$$
\begin{aligned}
V_Q(t^j_{k+1}, k+1) - \lambda V_Q(t^j_{k+1}-1, k) \le & \sum_{i \ne i^*_k} e^T_{ij}[k]\left[1 - \lambda^{h_j+1}\left(1 + \frac{1}{N_j - 1}\right)\right] Q_{ij} e_{ij}[k] \\
& + \lambda h_j \sum_{i \ne i^*_k} e^T_{ij}[k] U_{ij} e_{ij}[k] + 2 \sum_{i \ne i^*_k} e^T_{ij}[k] Q_{ij} \sigma_{ij}[k] \\
& + \sum_{i \ne i^*_k} \sigma_{ij}[k]^T Q_{ij} \sigma_{ij}[k] + \sigma^T_{i^*_k j}[k] Q_{i^*_k j} \sigma^T_{i^*_k j}[k].
\end{aligned}
\tag{12.34}
$$

For V_G, consider two cases. In the case where $t = t^j_{k+1} - 1 \le s^j_k$, we have $t^j_k = s^j_k$, $t^j_{k+1} = s^j_{k+1}$. Thus,

$$
\begin{aligned}
& V_G(t^j_{k+1}, k+1) - \lambda V_G(t^j_{k+1} - 1, k) = 0 \\
& = \sum_{i=1}^{N_j} z^T_j(t^j_{k+1} - 1) C^T_{ij} G_{ij} C_{ij} z_j(t^j_{k+1} - 1) - \sum_{i=1}^{N_j} z^T_j(s^j_k) C^T_{ij} G_{ij} C_{ij} z_j(s^j_k) \\
& \le \sum_{i=1}^{N_j} z^T_j(t) C^T_{ij} G_{ij} C_{ij} z_j(t) - \frac{\lambda^{\tau^j_M}}{h_j + 1} \sum_{i=1}^{N_j} \sigma^T_{ij}[k] G_{ij} \sigma_{ij}[k].
\end{aligned}
\tag{12.35}
$$

Here, the latter inequality is due to Jensen's inequality.

Otherwise, $t^j_{k+1} - 1 > s^j_k$ and similar to (12.35), we have

$$
\begin{aligned}
& V_G(t^j_{k+1}, k+1) - \lambda V_G(t^j_{k+1} - 1, k) \\
& \le \sum_{i=1}^{N_j} z^T_j(t^j_{k+1} - 1) C^T_{ij} G_{ij} C_{ij} z_j(t^j_{k+1} - 1) - \sum_{s=s^j_k}^{s^j_{k+1}-1} \sum_{i=1}^{N_j} \lambda^{t^j_{k+1}-s-1} z^T_j(s) C^T_{ij} G_{ij} C_{ij} z_j(s) \\
& \le \sum_{i=1}^{N_j} z^T_j(t) C^T_{ij} G_{ij} C_{ij} z_j(t) - \frac{\lambda^{\tau^j_M}}{h_j + 1} \sum_{i=1}^{N_j} \sigma^T_{ij}[k] G_{ij} \sigma_{ij}[k].
\end{aligned}
\tag{12.36}
$$

To sum up, we obtain

$$
\begin{aligned}
& V_Q(t^j_{k+1}, k+1) + V_G(t^j_{k+1}, k+1) - \lambda[V_Q(t^j_{k+1} - 1, k) + V_G(t^j_{k+1} - 1, k)] \\
& \qquad \le z_j(t)^T G_j z_j(t) + \sum_{i \ne i^*_k} \begin{bmatrix} e_{ij}[k] \\ \sigma_{ij}[k] \end{bmatrix}^T \Psi^j_i \begin{bmatrix} e_{ij}[k] \\ \sigma_{ij}[k] \end{bmatrix} \\
& \qquad + \begin{bmatrix} e_{i^*_k j}[k] \\ \sigma_{i^*_k j}[k] \end{bmatrix}^T \left(Q_{i^*_k j} - \frac{\lambda^{\tau^j_M}}{h_j + 1} G_{i^*_k j} \right) \begin{bmatrix} e_{i^*_k j}[k] \\ \sigma_{i^*_k j}[k] \end{bmatrix},
\end{aligned}
\tag{12.37}
$$

and together with (12.28), we then have

$$
\begin{aligned}
&V_j(t^j_{k+1}) - \lambda V_j(t^j_{k+1}-1) - \sum_{l\neq j} \varepsilon_{jl}|\sqrt{P_l}x_l(t_{k+1}-1)|^2 \\
&\quad \le \left[\eta_j^T(t)\ \xi^T_{i^*_k j}(t)\ X_j^{cT}(t)\right] \Sigma^j_{i^*_k} \begin{bmatrix} \eta_j(t) \\ \xi_{i^*_k j}(t) \\ X^c_j(t) \end{bmatrix} + z_j^T(t) G_j z_j(t) \\
&\qquad + \sum_{i\neq i^*_k} \begin{bmatrix} e_{ij}[k] \\ \sigma_{ij}[k] \end{bmatrix}^T \Psi^j_i \begin{bmatrix} e_{ij}[k] \\ \sigma_{ij}[k] \end{bmatrix} + \begin{bmatrix} e_{i^*_k j}[k] \\ \sigma_{i^*_k j}[k] \end{bmatrix}^T \left(Q_{i^*_k j} - \frac{\lambda^{\tau^j_M}}{h_j+1} G_{i^*_k j} \right) \begin{bmatrix} e_{i^*_k j}[k] \\ \sigma_{i^*_k j}[k] \end{bmatrix}.
\end{aligned}
\tag{12.38}
$$

From the second inequality in (12.17), it follows that

$$
Q_{i^*_k j} - \frac{\lambda^{\tau^j_M}}{h_j+1} G_{i^*_k j} < 0.
$$

Therefore, we obtain

$$
\begin{aligned}
&V_j(t^j_{k+1}) - \lambda V_j(t^j_{k+1}-1) - \sum_{l\neq j} \varepsilon_{jl}|\sqrt{P_l}x_l(t_{k+1}-1)|^2 \\
&\le \left[\eta_j^T(t)\ \xi^T_{i^*_k j}(t)\ X_j^{cT}(t)\right] \Sigma^j_{i^*_k} \begin{bmatrix} \eta_j(t) \\ \xi_{i^*_k j}(t) \\ X^c_j(t) \end{bmatrix} + z_j^T(t) G_j z_j(t) + \sum_{i\neq i^*_k} \begin{bmatrix} e_{ij}[k] \\ \sigma_{ij}[k] \end{bmatrix}^T \Psi^j_i \begin{bmatrix} e_{ij}[k] \\ \sigma_{ij}[k] \end{bmatrix}.
\end{aligned}
\tag{12.39}
$$

Note that

$$
\left[\eta_j^T(t)\ \xi^T_{i^*_k j}(t)\ X_j^{cT}(t)\right] \tilde{\Upsilon}^j_{i^*_k} \begin{bmatrix} \eta_j(t) \\ \xi_{i^*_k j}(t) \\ X^c_j(t) \end{bmatrix} - \sum_{i\neq i^*_k} e^T_{ij}[k] U_{ij} e_{ij}[k] = 0. \tag{12.40}
$$

Hence, by adding (12.40) to (12.39), we arrive at

$$
\begin{aligned}
&V_j(t^j_{k+1}) - \lambda V_j(t^j_{k+1}-1) - \sum_{l\neq j} \varepsilon_{jl}|\sqrt{P_l}x_l(t_{k+1}-1)|^2 \\
&\quad \le \left[\eta_j^T(t)\ \xi^T_{i^*_k j}(t)\ X_j^{cT}(t)\right] \left(\Sigma^j_{i^*_k} - \tilde{\Upsilon}^j_{i^*_k}\right) \begin{bmatrix} \eta_j(t) \\ \xi_{i^*_k j}(t) \\ X^c_j(t) \end{bmatrix} + z_j^T(t) G_j z_j(t) \\
&\qquad + \sum_{i\neq i^*_k} \begin{bmatrix} e_{ij}[k] \\ \sigma_{ij}[k] \end{bmatrix}^T \Psi^j_i \begin{bmatrix} e_{ij}[k] \\ \sigma_{ij}[k] \end{bmatrix} + \sum_{i\neq i^*_k} e^T_{ij}[k] U_{ij} e_{ij}[k].
\end{aligned}
\tag{12.41}
$$

Therefore, taking into account (12.26), the inequality (12.17) implies (12.20) for $t = t^j_{k+1} - 1$.

We are in a position to formulate the stability result.

Theorem 12.1 *Consider the large-scale system* (12.10)–(12.11), $j = 1, \ldots, M$. *Given tuning parameters* $0 < \lambda < 1$ *and* $\varepsilon \le 1 - \lambda$ *(* $\sum_{j\neq l} \varepsilon_{jl} \le \varepsilon$, $l = 1, \ldots, M$*),* *assume that there exist matrices* $P_j > 0$, $S_{0j} > 0$, $R_{0j} > 0$, $S_{1j} > 0$, $R_{1j}>0$, $Q_{ij}>0$, $U_{ij}>0$, $G_{ij} > 0$ *and* W_j *that satisfy* (12.16) *and* (12.17) *for all* $j = 1, \dots, M$. *Then, the system* (12.10)–(12.11) *is exponentially stable with a decay rate* $\lambda + \varepsilon$.

Proof From Proposition 12.1, (12.16) and (12.17) imply the inequality (12.19) for $j = 1, \dots, M$. Then, there exist positive $\{\beta_j\}_{j=1}^M$ such that

$$V_j(t) \geq x_j^T(t)P_jx_j(t) + \beta_j|e_j(t)|^2. \tag{12.42}$$

Consider now the LKF

$$V(t) = \sum_{j=1}^{M} V_j(t). \tag{12.43}$$

Since $x_j^T(t)P_jx_j(t) \leq V_j(t)$ and $\sum_{j\neq l} \varepsilon_{jl} \leq \varepsilon, l = 1, \dots, M$, summing (12.20) over $j = 1, \dots, M$ for $t \geq T_0$, we obtain

$$V(t+1) - (\lambda + \varepsilon)V(t) \leq 0$$

that implies

$$V(t) \leq (\lambda + \varepsilon)^{t-T_0} V(T_0), \ t > T_0. \tag{12.44}$$

The inequalities (12.20) for $j = 1, \dots, M$, yield

$$V_j(T_0) \leq \lambda^{T_0 - t_0^j} V_j(t_0^j) + \sum_{s=t_0^j}^{T_0} \sum_{l\neq j} \lambda^{T_0 - s} \varepsilon_{jl} x_l^T(t) P_l x_l(t).$$

Then, there exist some constants c_j such that

$$V_j(T_0) \leq V_j(t_0^j) + \frac{1}{2} c_j \|x\|_{T_0}^2 \leq c_j \|x\|_{T_0}^2, \ \forall j.$$

Hence, for some constant c, we have

$$V(T_0) \leq c\|x\|_{T_0}^2,$$

implying (for some $\beta^- > 0$)

$$\beta^- \Big(|x(t)|^2 + \sum_{j=1}^{M} |e_j(t)|^2 \Big) \leq V(t) \leq c\|x\|_{T_0}^2 (\lambda + \varepsilon)^{t-T_0}$$

with $\lambda + \varepsilon < 1$.

Remark 12.4 Note that different from [38], the LMIs (12.17) are not affine in the system matrices. This is due to substitution of (12.26) into the positive terms $z_j^T(t)H_jz_j^T(t)$, $z_j^T(t)G_jz_j^T(t)$ and $x_j^T(t+1)P_jx_j(t+1)$ of $V_j(t+1) - \lambda V_j(t)$. However, by applying Schur complements to these terms, one can arrive at equivalent to (12.17) LMIs that are affine in the systems matrices. Hence, the result of Theorem 12.1 is applicable to the case of system matrices from an uncertain time-varying polytope, where one can solve the LMIs (12.17) simultaneously for all the vertices of the polytope applying the same decision matrices.

12.3 l_2-Gain Analysis of the Large-Scale System

Consider now the large-scale system (12.1) under the controller (12.6), where all subsystems are orchestrated by TOD protocol. The closed-loop is then given by

$$x_j(t+1) = A_jx(t) + A_{1j}x_j(t-\tau_j(t)) + \sum_{i\neq i^*_{k,j}} B_{ij}e_{ij}[k] + \sum_{l\neq j} F^l_jx_l(t) + D_jw_j(t), \tag{12.45}$$

where $t^j_k \le t \le t^j_{k+1} - 1$, $j = 1, \ldots, M$, $w(t) = \text{col}\{w_j(t), j = 1, \ldots, M\} \in l_2([T_0, \infty), \mathbb{R}^{n_w})$ is a disturbance. Let $z(t) = \text{col}\{z_j(t), j = 1, \ldots, M\}$ be the controlled output, where

$$Z_j(t) = \Lambda_{1j}x_j(t) + \Lambda_{2j}u_j(t) \in \mathbb{R}^{n_z}.$$

Given $\gamma > 0$, define the following performance index

$$J = \sum_{t=t_0}^{\infty}\left[\sum_j z_j^T(t)z_j(t) - \gamma^2w^T(t)w(t)\right].$$

Definition 12.2 The closed-loop large-scale system (12.45) with initial time $T_0 = \max_j\{t^j_0\}$ is said to have an induced l_2-gain less than γ, if

$$J < V(T_0), \ \forall w \in l_2(\mathbb{R}^{n_w}), \ w \neq 0$$

holds, where V is given by (12.43).

It is well known (see, e.g., [47]) that $J < V(T_0)$ if for some $\alpha > 0$ the following condition holds along (12.45)

$$V(t+1) - V(t) + z^T(t)z(t) - \gamma^2w^T(t)w(t) \le -\alpha[|x(t)|^2 + |w(t)|^2], \ t \ge T_0. \tag{12.46}$$

Lemma 1 *Consider* $\{V_j(t)\}_{j=1}^M$ *given by* (12.13). *Let there exist positive tuning parameters* $\varepsilon < 1$ *and* $\{\varepsilon_{jl}\}_{j,l=1}^M$ *such that* $\sum_{j\neq l}\varepsilon_{jl} \le \varepsilon$, $l = 1, \ldots, M$ *and*

$$V_j(t+1) - (1-\varepsilon)V_j(t) + z_j^T(t)z_j(t) - \gamma^2 w_j^T(t)w_j(t) \le \sum_{l\neq j} \varepsilon_{jl} x_l^T(t)P_l x_l(t) \tag{12.47}$$

for $t \ge T_0$, *then* (12.46) *holds along* (12.45) *with* $V(t) = \sum_{j=1}^M V_j(t)$.

Taking into account (12.42), we obtain that the result of Lemma 1 follows from summation in (12.47). By extending derivations of Proposition 12.1, we arrive at the following LMI conditions that guarantee (12.47).

Theorem 12.2 *Given* $\gamma > 0$, *consider the closed-loop system* (12.45) *with a tuning parameter* $0 < \varepsilon < 1$ $(\sum_{j\neq l} \varepsilon_{jl} \le \varepsilon,\ l = 1, \ldots, M)$. *Let there exist matrices* $P_j > 0$, $S_{0j} > 0$, $R_{0j} > 0$, $S_{1j} > 0$, $R_{1j} > 0$, $Q_{ij} > 0$, $U_{ij} > 0$, $G_{ij} > 0$ *and* W_j *such that* (12.16) *holds for all* $j = 1, \ldots, M$, *and*

$$\begin{aligned} &\Sigma_i^j - \tilde{\Upsilon}_i^j + \tilde{E}_{ij2}^T G_j \tilde{E}_{ij2} + \zeta_{ij}^T \zeta_{ij} < 0, \\ &\Psi_i^j + [I\ \ 0]^T U_{ij} [I\ \ 0] < 0,\ i = 1, \ldots, N_j,\ j = 1, \ldots, M, \end{aligned} \tag{12.48}$$

where $\lambda = 1 - \varepsilon$ *and*

$$\begin{aligned} \Sigma_i^j &= \tilde{E}_{ij1}^T P_j \tilde{E}_{ij1} + \tilde{E}_{ij2}^T H_j \tilde{E}_{ij2} - \tilde{E}_{ij3}^T \lambda^{\tau_M} \Omega_j \tilde{E}_{ij3} \\ &\quad - \tilde{E}_{ij4}^T \lambda^{\eta_m^j} R_{0j} \tilde{E}_{ij4} + \tilde{E}_{ij5}, \\ h_j &= \tau_M^j - \eta_m^j, \quad \mathscr{F}^j = [F_j^1 \cdots F_j^{l\neq j} \cdots F_j^M], \\ \tilde{E}_{ij1} &= [E_{ij1}\ \ \mathscr{F}^j\ \ D_j], \quad \tilde{E}_{ij2} = [E_{ij2}\ \ \mathscr{F}^j\ \ D_j], \\ \tilde{E}_{ij3} &= \begin{bmatrix} E_{ij3} & 0 \cdot \begin{bmatrix} \mathscr{F}^j & D_j \\ \mathscr{F}^j & D_j \end{bmatrix} \end{bmatrix}, \\ \tilde{E}_{ij4} &= [E_{ij4}\ \ 0\cdot\mathscr{F}^j\ \ 0\cdot D_j], \\ \tilde{E}_{ij5} &= \mathrm{diag}\{E_{ij5}, -\mathscr{P}^j, -\gamma^2 I_{n_w^j}\}, \\ \mathscr{P}^j &= \mathrm{diag}\{\varepsilon_{jl} P_{l\neq j}\}, \quad \tilde{\Upsilon}_i^j = \mathrm{diag}\{\Upsilon_i^j, 0_{n-n_j}, 0_{n_w^j}\}, \\ \mathscr{K}_i^j &= row\{K_{rj}, r \neq i\}, \\ \zeta_{ij} &= [\Lambda_{1j}\ [0\ \ 1\ \ 0] \otimes \Lambda_{2j} K_j C_j\ \ \ \Lambda_{2j}\mathscr{K}_i^j\ \ 0] \end{aligned} \tag{12.49}$$

with the notations given by (12.18). *Then, the large-scale system* (12.45) *has an induced* l_2*-gain less than* γ, *where* $V(t) = \sum_j V_j(t)$ *with* $V_j(t)$ *given by* (12.13).

Remark 12.5 In the general case where the disturbance is common for all subsystems (i.e., $w_j(t) \equiv w(t)$, $j = 1, \ldots, M$), we modify condition (12.47) as

$$V_j(t+1) - (1-\varepsilon)V_j(t) + z_j^T(t)z_j(t) - w^T(t)\Gamma_j w(t) \le \sum_{l\neq j} \varepsilon_{jl} x_l^T(t) P_l x_l(t),$$

where $n_w \times n_w$ positive definite matrices Γ_j are subject to $\gamma^2 I_{n_w} \ge \sum_j \Gamma_j$. Then, Theorem 12.2 still holds if for every j, $\gamma^2 I_{n_w^j}$ in (12.49) is replaced by Γ_j. Here $\{\Gamma_j\}_{j=1}^M$ are additional decision variables of LMIs. The resulting bound γ_{Rem} on the l_2-gain is given by $\gamma_{Rem}^2 = \gamma_{Rem}^2(\Gamma_j) = \gamma^2$. Simpler LMIs, where $\gamma^2 I_{n_w^j}$ in (12.49) is replaced by (common for all j) $\gamma^2 I_{n_w}/M$, may lead to a larger bound $\gamma_{Rem}^2 = \gamma_{Rem}^2(\gamma) = \gamma^2$ (as shown in the example below).

12.4 Decentralized Control Under Round-Robin Protocol

Consider next the stabilization of (12.1) ($w_j = 0$), where the Round-Robin protocol orchestrates the transmitted measurements sent at each instant from sensors to a controller for every subsystem. Under Round-Robin protocol, the transmitted sensor measurement is chosen in a periodic manner. In this case, the control law (12.6) can be expressed as

$$u_j(t) = \sum_{i=1}^{N_j} K_{ij} C_{ij} x_j(t - \tau_j(t)),$$

where $\tau_j(t)$ are piecewise-linear delays with known bounds

$$\eta_m^j \le \tau_j(t) \le N_j \cdot \mathrm{MATI}_j + \eta_M^j - 1 = \tau_M^j.$$

The closed-loop system is then given by

$$x_j(t+1) = A_j x_j(t) + \sum_{i=1}^{N_j} A_{ij} x_j(t - \tau_j(t)) + \sum_{l \ne j} F_j^l x_l(t),\ t \ge t_{N_j}^j,\ j = 1, \ldots, M, \tag{12.50}$$

where $A_{ij} = B_j K_{ij} C_{ij}$. We use Lyapunov functional of the form

$$V_j(t) = x_j^T(t) P_j x_j(t) + V_0^j(t) + \sum_{i=1}^{N_j} V_{1i}^j(t),$$

where

$$\begin{aligned} V_0^j(t) = &\sum_{s=t-\eta_m^j}^{t-1} \lambda^{t-s-1} x_j^T(s) S_{0j} x_j(s) + \sum_{s=t-\tau_M^j}^{t-\eta_m^j-1} \lambda^{t-s-1} x_j^T(s) S_{1j} x_j(s) \\ &+ \eta_m^j \sum_{\theta=1}^{\eta_m^j} \sum_{s=t-\theta}^{t-1} \lambda^{t-s-1} z_j^T(s) R_{0j} z_j(s), \\ V_{1i}^j(t) = &(\tau_M^j - \eta_m^j) \sum_{\theta=\eta_m^j+1}^{\tau_M^j} \sum_{s=t-\theta}^{t-1} \lambda^{t-s-1} z_j^T(s) R_{1ij} z_j(s). \end{aligned}$$

By using the arguments of Theorem 12.1, we arrive at the following result.

Theorem 12.3 *Consider the closed-loop large-scale system* (12.50). *Given tuning parameters* $0 < \lambda < 1$ *and* $\varepsilon \le 1 - \lambda$ ($\sum_{j \neq l} \varepsilon_{jl} \le \varepsilon$, $l = 1, \dots, M$), *assume that there exist matrices* $P_j > 0$, $S_{0j} > 0$, $R_{0j} > 0$, $S_{1j} > 0$, $R_{1ij} > 0$ *and* W_{ij}, $i = 1, \dots, N_j$, $j = 1, \dots, M$, *that satisfy*

$$\begin{aligned} &\begin{bmatrix} R_{1ij} & W_{ij} \\ * & R_{1ij} \end{bmatrix} \ge 0, \ i = 1, \dots, N_j, \\ &\Sigma^j < 0, \ j = 1, \dots, M, \end{aligned} \tag{12.51}$$

where

$$\begin{aligned}
\Sigma^j &= E_{1j}^T P_j E_{1j} + E_{2j}^T H_j E_{2j} - \sum_{i=1}^{N_j} E_{3ij}^T \lambda^{\tau_M^j} \Omega_{ij} E_{3ij} - E_{4j}^T \lambda^{\eta_m^j} R_{0j} E_{4j} + E_{5j}, \\
\mathcal{F}^j &= [F_j^1 \cdots F_j^{l \neq j} \cdots F_j^M], \\
H_j &= {\eta_m^j}^2 R_{0j} + (\tau_M^j - \eta_m^j)^2 \sum_{i=1}^{N_j} R_{1ij}, \\
E_{1j} &= \Big[\begin{bmatrix} 1\,0\,0 \end{bmatrix} \otimes A_j \ \ A_{ij} \cdots A_{N_j j} \ \ \mathcal{F}^j \Big], \\
E_{2j} &= \Big[\begin{bmatrix} 1\,0\,0 \end{bmatrix} \otimes (A_j - I_{n^j}) \ A_{ij} \cdots A_{N_j j} \ \ \mathcal{F}^j \Big], \\
E_{3ij} &= \left[\begin{bmatrix} 0 \\ 0 \end{bmatrix} \otimes A_j \ \begin{bmatrix} 1 & 0 \\ 0 & -1 \end{bmatrix} \otimes I_{n^j} \ \ \chi_1^i \cdots \chi_{N_j}^i \ \begin{bmatrix} 0 \\ 0 \end{bmatrix} \otimes \mathcal{F}^j \right], \\
\chi_r^i &= \delta_{i,r} \begin{bmatrix} -1 \\ 1 \end{bmatrix} \otimes I_{n^j}, \\
E_{4j} &= [I_{n^j} \ - I_{n^j} \ 0], \\
E_{5j} &= \mathrm{diag}\{S_{0j} - \lambda P_j, -(S_{0j} - S_{1j})\lambda^{\eta_m^j}, -S_{1j}\lambda^{\tau_M^j}, 0_{n_y^j}, -\mathcal{P}^j\}, \\
\mathcal{P}^j &= \mathrm{diag}\{\varepsilon_{jl} P_l, l \neq j\}.
\end{aligned}$$

Then, the closed-loop system (12.50) *is exponentially stable with a decay rate* $\lambda + \varepsilon$, *where* $T_0^j = t_{N_j}^j$.

The derivation of LMIs consists of the arguments similar to Proposition 12.1, where

$$\eta_j(t) = \mathrm{col}\left\{ x_j(t), x_j(t - \eta_m^j), x_j(t - \tau_M^j), x_j(t - \tau_1^j(t)), \dots, x_j(t - \tau_{N_j}^j(t)) \right\}.$$

Remark 12.6 In the numerical Example 12.5 below, the results that follow from Theorem 12.3 (under Round-Robin protocol) are less conservative than the results based on Theorem 12.1. However, the improvement is achieved on the account of the numerical complexity:

- The conditions of Theorem 12.1 possess LMIs of $N_j n + (4N_j + 7)n^j + 4n_y^j$ total rows and $5T_{n^j} + {n^j}^2 + 3\sum_{i=1}^{N_j} T_{n_i^j}$ scalar decision variables for each subsystem j, where $T_n = \frac{n^2+n}{2}$ is the nth triangular number;

- The conditions of Theorem 12.3 possess LMIs of $(6+4N_j)n^j+n$ total rows and $(4+N_j)T_{n^j}+N_j {n^j}^2$ scalar decision variables for each subsystem j.

Comparatively to LMIs of Theorem 12.1, the number of lines in LMIs of Theorem 12.2 is enlarged by $N_j n_w^j$, whereas the number of scalar decision variables remains the same.

Remark 12.7 There is a trade-off between enlarging the decay rate and the values of MATI / MAD. The choice of a decay rate λ close to 1 enlarges the values of MATI / MAD. Moreover, fast convergence of subsystems without coupling, where $\lambda << 1$ if $F_j^l = 0$, allows stronger coupling.

12.5 Example: Two Coupled Inverted Pendulums

Consider the example of two coupled inverted pendulums on carts (Fig. 12.1), under the scenario of decentralized networked control. We discretize the continuous-time model of [14] with the sampling time $T_s = 10^{-4}$. Here, $M = 2$, $N_j = 2$ or $N_j = 4$, $j = 1, 2$. The resulting system matrices of (12.1), (12.3) are given by

$$\begin{aligned}
A_j &= A = I_4 + 10^{-4} \cdot \begin{bmatrix} 0 & 1 & 0 & 0 \\ 2.9156 & 0 & -0.0005 & 0 \\ 0 & 0 & 0 & 1 \\ -1.6663 & 0 & 0.0002 & 0 \end{bmatrix}, \\
B_j^T &= B^T = 10^{-4} \cdot \begin{bmatrix} 0 & -0.0042 & 0 & 0.0167 \end{bmatrix}, \\
F_1^2 &= F_2^1 = A_{12} = 10^{-4} \cdot \begin{bmatrix} 0 & 0 & 0 & 0 \\ 0.0011 & 0 & 0.0005 & 0 \\ 0 & 0 & 0 & 0 \\ -0.0003 & 0 & -0.0002 & 0 \end{bmatrix}, \\
K_1 &= [k_{11}\ k_{21}\ k_{31}\ k_{41}] = [11396 \;\; 7196.2 \;\; 573.96 \;\; 1199.0], \\
K_2 &= [k_{12}\ k_{22}\ k_{32}\ k_{42}] = [29241 \;\; 18135 \;\; 2875.3 \;\; 3693.9].
\end{aligned}$$

In the case of $N_j = 2$, we consider

$$\begin{aligned}
C_{1j} &= \begin{bmatrix} 1 & 0 & 0 & 0 \\ 0 & 1 & 0 & 0 \end{bmatrix}, \quad C_{2j} = \begin{bmatrix} 0 & 0 & 1 & 0 \\ 0 & 0 & 0 & 1 \end{bmatrix}, \\
K_{1j} &= [k_{1j}\ k_{2j}], \quad K_{2j} = [k_{3j}\ k_{4j}], \; j = 1, 2.
\end{aligned}$$

In the case of $N_j = 4$, $C_{1j}, \ldots, C_{4j}$ are the rows of I_4 and $K_{1j}, \ldots, K_{4j}$ are the entries of K_j.

We start with the stability analysis and apply the conditions of Theorems 12.1 and 12.3. Choosing the tuning parameters $\lambda = e^{-3\times 10^{-6}}$, $\varepsilon_{12} = \varepsilon_{21} = 1 - \lambda$, we achieve maximal delays preserving the stability that are given in Table 12.1. For constant communication delays $\text{MAD}_1 = \text{MAD}_2 = 1$, MATI_1 and MATI_2 are given in Table 12.2.

Table 12.1 Maximum value of τ_M^1, τ_M^2 preserving stability

$\eta_m^1 = \eta_m^2$	$N_1 = N_2$	Theorem 12.1 (TOD)		Theorem 12.3 (Round-Robin)	
		τ_M^1	τ_M^2	τ_M^1	τ_M^2
1	2	102	39	339	133
1	4	29	10	327	130

Table 12.2 MATI_1, MATI_2 preserving stability for MAD = 1

$\eta_m^1 = \eta_m^2$	$N_1 = N_2$	Theorem 12.1 (TOD)		Theorem 12.3 (Round-Robin)	
		MATI_1	MATI_2	MATI_1	MATI_2
1	2	101	38	169	66
1	4	28	9	81	32

Table 12.3 Results of [38]: maximum value of τ_M^j preserving stability ($j = 1, 2$) for $\eta_m^1 = \eta_m^2 = 10^{-4}$

N_j	2		4	
	τ_M^1	τ_M^2	τ_M^1	τ_M^2
Reference [38] (TOD)	0.0103	0.004	0.003	0.001

It is seen that in this example, the results under TOD are more conservative (see Remark 12.6).

Table 12.3 illustrates that in this example under TOD protocol, the discrete-time results with $T_s = 10^{-4}$ are similar to the continuous-time ones obtained in [38]. The same is true for the results under Round-Robin protocol.

Consider now the perturbed model of two coupled inverted pendulums given as above, where each system input is perturbed such that

$$u_j(t) = \tilde{u}_j(t) + v_j(t) + \frac{9}{10} v_{3-j}(t), \quad j = 1, 2.$$

Here, $\tilde{u}_j(t)$ are the controllers as considered above, and $v_1, v_2 \in l_2$ are the perturbations. The system is now given by (12.45) where $w_1^T(t) = w_2^T(t) = \left[v_1^T(t) \;\; v_2^T(t)\right]$ and

$$D_1 = \left[B_1 \;\; \frac{9}{10} B_1\right], \quad D_2 = \left[\frac{9}{10} B_2 \;\; B_2\right].$$

The controlled output is given by

$$z_1(t) = [500\ 0\ 0\ 0] x_1(t), \quad z_2(t) = [100\ 0\ 0\ 0] x_2(t).$$

Table 12.4 Minimum values of γ^2_{Rem}

$N_1 = N_2$	$\eta^1_m = \eta^2_m$	τ^1_M	τ^2_M	$\gamma^2_{Rem}(\Gamma_j)$	$\gamma^2_{Rem}(\gamma)$
2	1	2	2	0.0262	0.0482
4	1	2	2	0.0461	0.0791

Choosing λ, ε_{12} and ε_{21} as above, we find bounds on l_2-gain by using Remark 12.5 with different Γ_j (denoted by $\gamma_{Rem}(\Gamma_j)$), and common $\gamma^2 I_{n_w}/M$ (denoted by $\gamma_{Rem}(\gamma)$) (see Table 12.4). Here, for $N_j = 2$, the following gain matrices are obtained:

$$\Gamma_1 = \text{diag}\{0.0224, 0.0216\}, \quad \Gamma_2 = \text{diag}\{0.0038, 0.0045\}$$

and for $N_j = 4$

$$\Gamma_1 = \text{diag}\{0.04, 0.0388\}, \quad \Gamma_2 = \text{diag}\{0.0061, 0.0073\}$$

It is seen that the bounds on γ_{Rem} achieved by using different Γ_j are less restrictive. However, the corresponding LMIs possess more decision variables.

12.6 Notes

In this chapter, a time-delay approach was developed for the decentralized networked control of large-scale discrete-time systems with local networks, where asynchronous variable sampling intervals, large variable communication delays and scheduling protocols from sensors to controllers were taken into account. The proposed Lyapunov–Krasovskii method leads to efficient LMI conditions for the exponential stability and l_2-gain analysis of the closed-loop large-scale system.

Appendix

A.1 Proof of Theorem 5.1

Proof Consider $t \in [t_k, t_{k+1})$, $k \in \mathbb{Z}^+$ and define $\xi_i(t) = \text{col}\{x(t), x(t-\eta_m), x(t-\tau(t)), x(t-\tau_M), \bar{\xi}_i(t), \omega(t)\}$, $i = 1, \ldots, N$, where

$$\begin{aligned}
\bar{\xi}_i(t) &= \text{col}\{e_2(t), \cdots, e_N(t)\},\ i = 1,\\
\bar{\xi}_i(t) &= \text{col}\{e_1(t), \cdots, e_{N-1}(t)\},\ i = N,\\
\bar{\xi}_i(t) &= \text{col}\{e_1(t), \cdots, e_j(t)_{|j\neq i}, \cdots, e_N(t)\},\ i = 2, \ldots, N-1.
\end{aligned}$$

Let $i_k^* = i \in \mathbb{N}$. Differentiating $V_e(t)$ along (5.8) and applying Jensen's inequality, we have

$$\begin{aligned}
\eta_m \int_{t-\eta_m}^{t} \dot{x}^T(s) R_0 \dot{x}(s) ds &\geq \int_{t-\eta_m}^{t} \dot{x}^T(s) ds R_0 \int_{t-\eta_m}^{t} \dot{x}(s) ds\\
&= \xi_i^T(t)(F_2^i)^T R_0 F_2^i \xi_i(t),
\end{aligned}$$

and

$$\begin{aligned}
&-(\tau_M - \eta_m) \int_{t-\tau_M}^{t-\eta_m} \dot{x}^T(s) R_1 \dot{x}(s) ds\\
&= -(\tau_M - \eta_m) \int_{t-\tau(t)}^{t-\eta_m} \dot{x}^T(s) R_1 \dot{x}(s) ds - (\tau_M - \eta_m) \int_{t-\tau_M}^{t-\tau(t)} \dot{x}^T(s) R_1 \dot{x}(s) ds\\
&\leq -\frac{\tau_M - \eta_m}{\tau(t) - \eta_m} \xi_i^T(t) \Big[[I_n\ 0_{n\times n}] F^i\Big]^T R_1 [I_n\ 0_{n\times n}] F^i \xi_i(t)\\
&\quad -\frac{\tau_M - \eta_m}{\tau_M - \tau(t)} \xi_i^T(t) \Big[[0_{n\times n}\ I_n] F^i\Big]^T R_1 [0_{n\times n}\ I_n] F^i \xi_i(t)\\
&\leq -\xi_i^T(t) (F^i)^T \Phi F^i \xi_i(t).
\end{aligned}$$

The latter inequality holds if (5.22) is feasible [173]. Then, we have

$$\begin{aligned}
&\dot{V}_e(t) + 2\alpha V_e(t) - \frac{1}{\tau_M - \eta_m} \sum_{l=1, l\neq i}^{N} |\sqrt{U_l} e_l(t)|^2 - 2\alpha |\sqrt{Q_i} e_i(t)|^2 - b|\omega(t)|^2\\
&\leq \xi_i^T(t)[\Sigma_i + \Xi_i^T H \Xi_i - (F^i)^T \Phi F^i e^{-2\alpha\tau_M}]\xi_i(t) \leq 0,
\end{aligned}$$

K. Liu et al., *Networked Control Under Communication Constraints*, Advances in Delays and Dynamics 11, https://doi.org/10.1007/978-981-15-4230-5

if $\Sigma_i + \Xi_i^T H \Xi_i - (F^i)^T \Phi F^i e^{-2\alpha\tau_M} < 0$, i.e., by Schur complements, if (5.23) is feasible. Thus due to Lemma 5.1, inequalities (5.15), (5.22) and (5.23) imply (5.17).

A.2 Proof of Lemma 5.3

Proof Since $|\omega(t)| \le \Delta$, (5.28) implies

$$
\begin{aligned}
V(t, x_t, \dot{x}_t) &\le V_e(t) \\
&\le e^{-2\alpha(t-t_k)} V_e(t_k) + b\Delta^2 \int_{t_k}^{t} e^{-2\alpha(t-s)} ds, \ t \in [t_k, t_{k+1}).
\end{aligned} \tag{A.1}
$$

Note that $V_e(t_{k+1}) = \tilde{V}_{|t=t_{k+1}} + V_{Q_{|t=t_{k+1}}} + V_{G_{|t=t_{k+1}}}$. Taking into account (5.32) and the relations $\tilde{V}_{t=t_{k+1}} = \tilde{V}_{t=t_{k+1}^-}$, $e(t_{k+1}^-) = e(t_k)$, we obtain due to (5.35), (5.36) and (5.39) for $k > N - 1$

$$
\begin{aligned}
\Theta_{k+1} &\overset{\Delta}{=} V_e(t_{k+1}) - V_e(t_{k+1}^-) \\
&= [V_Q + V_G]_{t=t_{k+1}} - [V_Q + V_G]_{t=t_{k+1}^-} \\
&\le (\tau_M - \eta_m) e^{2\alpha[\tau_M + (N-2)(\tau_M - \eta_m)]} \Big[\sum_{j=0}^{N-2} \int_{s_{k-j}}^{s_{k+1}} e^{2\alpha(s-t_{k+1})} \Big| \sqrt{Q^*_{i_{k-j}}} C^*_{i_{k-j}} \dot{x}(s) \Big|^2 ds \\
&\quad - \sum_{i=1}^{N} (N-1) \int_{s_k}^{s_{k+1}} e^{2\alpha(s-t_{k+1})} |\sqrt{Q_i} C_i \dot{x}(s)|^2 ds \Big],
\end{aligned} \tag{A.2}
$$

whereas for $k = N - 1$ due to (5.37) and (5.39)

$$
\begin{aligned}
\Theta_N &\le \sum_{j=0}^{N-2} (\tau_M - \eta_m) e^{2\alpha[\tau_M + (N-2)(\tau_M - \eta_m)]} \int_{s_{N-1-j}}^{s_N} e^{2\alpha(s-t_N)} \Big| \sqrt{Q^*_{i_{N-1-j}}} C^*_{i_{N-1-j}} \dot{x}(s) \Big|^2 ds \\
&\quad - \sum_{i=1}^{N} (\tau_M - \eta_m) \int_{s_0}^{s_N} e^{2\alpha(s-t_N)} |\sqrt{G_i} C_i \dot{x}(s)|^2 ds \\
&\le -(\tau_M - \eta_m) e^{2\alpha[\tau_M + (N-2)(\tau_M - \eta_m)]} \\
&\quad \times \int_{s_0}^{s_N} e^{2\alpha(s-t_N)} \Big[(N-2) \sum_{i=1}^{N} |\sqrt{Q_i} C_i \dot{x}(s)|^2 + |\sqrt{Q_l} C_l \dot{x}(s)|^2_{|l=i_N^*} \Big] ds.
\end{aligned}
$$

We will prove (5.40) by induction. For $k = N - 1$, we have

$$
\begin{aligned}
V_e(t_N) &\le \Theta_N + V_e(t_N^-) \\
&\le -(\tau_M - \eta_m) e^{2\alpha[\tau_M + (N-2)(\tau_M - \eta_m)]} \\
&\quad \times \int_{s_0}^{s_N} e^{2\alpha(s-t_N)} \Big[(N-2) \sum_{i=1}^{N} |\sqrt{Q_i} C_i \dot{x}(s)|^2 + |\sqrt{Q_l} C_l \dot{x}(s)|^2_{|l=i_N^*} \Big] ds \\
&\quad + e^{-2\alpha(t_N - t_{N-1})} V_e(t_{N-1}) + b\Delta^2 \int_{t_{N-1}}^{t_N} e^{-2\alpha(t_N - s)} ds,
\end{aligned}
$$

which implies (5.40).

Assume that (5.40) holds for $k-1$ $(k \geq N-1)$:

$$V_e(t_k) \leq e^{-2\alpha(t_k - t_{N-1})} V_e(t_{N-1}) + \Psi_k + b\Delta^2 \int_{t_{N-1}}^{t_k} e^{-2\alpha(t_k - s)} ds.$$

Then due to (A.1) for $t = t_{k+1}^-$, we obtain

$$V_e(t_{k+1}) \leq \Theta_{k+1} + e^{-2\alpha(t_{k+1} - t_k)} \Psi_k + e^{-2\alpha(t_{k+1} - t_{N-1})} V_e(t_{N-1}) \\ + b\Delta^2 \int_{t_{N-1}}^{t_{k+1}} e^{-2\alpha(t_{k+1} - s)} ds.$$

We have

$$\begin{aligned} e^{-2\alpha(t_{k+1}-t_k)}\Psi_k &= -(\tau_M - \eta_m) e^{2\alpha[\tau_M + (N-2)(\tau_M - \eta_m)]} \\ &\quad \times \Big[\sum_{l=0}^{N-3} (N-2-l) \int_{s_{k-l-2}}^{s_k} e^{2\alpha(s - t_{k+1})} \Big| \sqrt{Q_{i^*_{k-1-l}}} C_{i^*_{k-1-l}} \dot{x}(s) \Big|^2 ds \\ &\quad + (N-1) \int_{s_{k-1}}^{s_k} e^{2\alpha(s - t_{k+1})} \Big| \sqrt{Q_{i^*_k}} C_{i^*_k} \dot{x}(s) \Big|^2 ds \Big] \\ &= -(\tau_M - \eta_m) e^{2\alpha[\tau_M + (N-2)(\tau_M - \eta_m)]} \\ &\quad \times \Big[\sum_{j=0}^{N-2} (N-1-j) \int_{s_{k-j-1}}^{s_k} e^{2\alpha(s - t_{k+1})} \Big| \sqrt{Q_{i^*_{k-j}}} C_{i^*_{k-j}} \dot{x}(s) \Big|^2 ds \Big]. \end{aligned}$$

Then taking into account (A.2), we find

$$\begin{aligned} &\Theta_{k+1} + e^{-2\alpha(t_{k+1}-t_k)}\Psi_k \\ &\leq (\tau_M - \eta_m) e^{2\alpha[\tau_M + (N-2)(\tau_M - \eta_m)]} \Big[\sum_{j=0}^{N-2} \int_{s_{k-j-1}}^{s_{k+1}} e^{2\alpha(s - t_{k+1})} \Big| \sqrt{Q^*_{i_{k-j}}} C^*_{i_{k-j}} \dot{x}(s) \Big|^2 ds \\ &\quad - \sum_{i=1}^{N} (N-1) \int_{s_k}^{s_{k+1}} e^{2\alpha(s - t_{k+1})} |\sqrt{Q_i} C_i \dot{x}(s)|^2 ds \\ &\quad - \sum_{j=0}^{N-2} (N-1-j) \int_{s_{k-j-1}}^{s_k} e^{2\alpha(s - t_{k+1})} \Big| \sqrt{Q_{i^*_{k-j}}} C_{i^*_{k-j}} \dot{x}(s) \Big|^2 ds \Big] \\ &\leq -(\tau_M - \eta_m) e^{2\alpha[\tau_M + (N-2)(\tau_M - \eta_m)]} \\ &\quad \times \Big[\sum_{j=0}^{N-2} (N-2-j) \int_{s_{k-j-1}}^{s_{k+1}} e^{2\alpha(s - t_{k+1})} \Big| \sqrt{Q_{i^*_{k-j}}} C_{i^*_{k-j}} \dot{x}(s) \Big|^2 ds \\ &\quad + (N-1) \int_{s_k}^{s_{k+1}} e^{2\alpha(s - t_{k+1})} \Big| \sqrt{Q_{i^*_{k+1}}} C_{i^*_{k+1}} \dot{x}(s) \Big|^2 ds \Big] \\ &= \Psi_{k+1}, \end{aligned}$$

which implies (5.40). Hence, (5.40) and (A.1) yield (5.42).

A.3 Proof of Lemma 8.1

Proof Consider $t \in [t_k, t_{k+1})$, $k \in \mathbb{Z}^+$ and define $\xi(t) = \text{col}\{x(t), x(t-\eta_m), x(t-\tau(t)), x(t-\tau_M), \omega_1(t), \cdots, \omega_N(t)\}$. Differentiating V along (8.5) and applying Jensen's inequality, we have

$$\begin{aligned}\eta_m \int_{t-\eta_m}^{t} \dot{x}^T(s) R_0 \dot{x}(s) ds &\geq \int_{t-\eta_m}^{t} \dot{x}^T(s) ds R_0 \int_{t-\eta_m}^{t} \dot{x}(s) ds \\ &= \xi^T(t) F_2^T R_0 F_2 \xi(t),\end{aligned}$$

and

$$\begin{aligned}&-(\tau_M - \eta_m) \int_{t-\tau_M}^{t-\eta_m} \dot{x}^T(s) R_1 \dot{x}(s) ds \\ &= -(\tau_M - \eta_m) \int_{t-\tau(t)}^{t-\eta_m} \dot{x}^T(s) R_1 \dot{x}(s) ds - (\tau_M - \eta_m) \int_{t-\tau_M}^{t-\tau(t)} \dot{x}^T(s) R_1 \dot{x}(s) ds \\ &\leq -\frac{\tau_M - \eta_m}{\tau(t) - \eta_m} \xi^T(t) \Big[[I_n \; 0_{n\times n}] F \Big]^T R_1 [I_n \; 0_{n\times n}] F \xi(t) \\ &\quad - \frac{\tau_M - \eta_m}{\tau_M - \tau(t)} \xi^T(t) \Big[[0_{n\times n} \; I_n] F \Big]^T R_1 [0_{n\times n} \; I_n] F \xi(t) \\ &\leq -\xi^T(t) F^T \Phi F \xi(t).\end{aligned}$$

The latter inequality holds if (8.7) is feasible [173]. Then, we have

$$\frac{d}{dt} V + 2\alpha V - \sum_{i=1}^{N} b_i |\omega_i(t)|^2 \leq \xi^T(t) [\Sigma + \Xi^T H \Xi - F^T \Phi F e^{-2\alpha \tau_M}] \xi(t) \leq 0, \tag{A.3}$$

if $\Sigma + \Xi^T H \Xi - F^T \Phi F e^{-2\alpha\tau_M} < 0$, i.e., by Schur complements, if (8.8) is feasible.

Since $|\omega_i(t)| \leq \mu \Delta_i$, $i = 1, \ldots, N$, by the comparison principle [108], (A.3) implies for $t \in [t_k, t_{k+1})$

$$\begin{aligned}V(t, x_t, \dot{x}_t) &\leq e^{-2\alpha(t-t_k)} V(t_k, x_{t_k}, \dot{x}_{t_k}) + \mu^2 \sum_{i=1}^{N} b_i \Delta_i^2 \int_{t_k}^{t} e^{-2\alpha(t-s)} ds \\ &\leq e^{-2\alpha(t-t_{k-1})} V(t_{k-1}, x_{t_{k-1}}, \dot{x}_{t_{k-1}}) + \mu^2 \sum_{i=1}^{N} b_i \Delta_i^2 \int_{t_{k-1}}^{t} e^{-2\alpha(t-s)} ds \\ &\;\;\vdots \\ &\leq e^{-2\alpha(t-t_0)} V(t_0, x_{t_0}, \dot{x}_{t_0}) + \mu^2 \sum_{i=1}^{N} b_i \Delta_i^2 \int_{t_0}^{t} e^{-2\alpha(t-s)} ds \\ &\leq e^{-2\alpha(t-t_0)} V(t_0, x_{t_0}, \dot{x}_{t_0}) + \frac{\mu^2}{2\alpha} \sum_{i=1}^{N} b_i \Delta_i^2,\end{aligned}$$

that completes the proof.

A.4 Proof of Lemma 8.2

Proof For all $x_t \in \mathscr{X}_{t_0,(1+\beta\nu^2)\mu^2 M_0^2}$ starting from $\mathscr{W}_{\mu^2 M_0^2}$, we have

$$x^T(t)Px(t) \leq V(t, x_t, \dot{x}_t) < (1+\beta\nu^2)\mu^2 M_0^2,$$

and thus, (8.12) guarantees that

$$x^T(s_k)C_i^T C_i x(s_k) < \frac{x^T(s_k)Px(s_k)\cdot \mu^2 M_i^2}{(1+\beta\nu^2)\mu^2 M_0^2} < \mu^2 M_i^2,\ k \in \mathbb{Z}^+.$$

Hence, there is no saturation for the sensor node $y_i(s_k)=C_i x(s_k)$, $k \in \mathbb{Z}^+$ ($|C_i x(s_k)| < \mu M_i$), which implies $|\omega_i(t)| \leq \mu \Delta_i$, $t \geq t_0$, $i = 1, \ldots, N$, whenever $x_t \in \mathscr{X}_{t_0,(1+\beta\nu^2)\mu^2 M_0^2}$ and $x_{t_0} \in \mathscr{X}_{\mu^2 M_0^2}$.

Let $x_{t_0} \in \mathscr{X}_{\mu^2 M_0^2}$, then the solutions of (8.5) satisfy $V(t, x_t, \dot{x}_t) < (1+\beta\nu^2)\mu^2 M_0^2$ for $t \in [t_0, t')$ for some $t' > t_0$. We will show next that the solutions of (8.5) with $x_{t_0} \in \mathscr{X}_{\mu^2 M_0^2}$ stay in $\mathscr{X}_{t_0,(1+\beta\nu^2)\mu^2 M_0^2}$ for all $t \geq t_0$ if LMIs (8.7)–(8.8), (8.12)–(8.13) are feasible.

Assume, on the contrary, that there exists a finite time $t' > t_0$ such that $V(t, x_t, \dot{x}_t) < (1+\beta\nu^2)\mu^2 M_0^2$ for $t \in [t_0, t')$ and $V(t', x_{t'}, \dot{x}_{t'}) = (1+\beta\nu^2)\mu^2 M_0^2$. Then under (8.12), we have $|\omega_i(t)| \leq \mu\Delta_i$, $i = 1, \ldots, N$ for $t \in [t_0, t']$. From (8.10)–(8.13), it follows that

$$\begin{aligned} V(t, x_t, \dot{x}_t) &\leq e^{-2\alpha(t-t_0)} V(t_0, x_{t_0}, \dot{x}_{t_0}) + \frac{\mu^2}{2\alpha}\sum_{i=1}^{N} b_i \Delta_i^2 \\ &< (1+\beta\nu^2)\mu^2 M_0^2,\ t \in [t_0, t'], \end{aligned}$$

which contradicts to $V(t', x_{t'}, \dot{x}_{t'}) = (1+\beta\nu^2)\mu^2 M_0^2$.

Then, (8.10)–(8.14) yield

$$\begin{aligned} V(t, x_t, \dot{x}_t) &\leq e^{-2\alpha \mathrm{T}} V(t_0, x_{t_0}, \dot{x}_{t_0}) + \frac{\mu^2}{2\alpha}\sum_{i=1}^{N} b_i \Delta_i^2 \\ &< (1-\beta)\nu^2 \cdot \mu^2 M_0^2 + \beta\nu^2\mu^2 M_0^2 \\ &= \nu^2\mu^2 M_0^2,\ t \geq t_0 + \mathrm{T}, \end{aligned}$$

that completes the proof.

A.5 Proof of Lemma 10.1

Proof Given $i \in \mathscr{I}$, consider $k \in [t_p, t_{p+1} - 1]$, $k \in \mathbb{Z}^+$ and define $\xi(k) = \mathrm{col}\{x(k), x(k-\eta_m), x(k-\tau_N(k)), \cdots, x(k-\tau_1(k)), x(k-\tau_M^1), \omega_1(k), \cdots, \omega_N(k)\}$.

Applying Cauchy–Schwarz inequality, taking advantage of the ordered delays and using the convex analysis [173], we have

$$\begin{aligned} &V(x_{k+1}) - \lambda V(x_k) - \sum_{i=1}^{N} b_i |\omega_i(k)|^2 \\ &\leq \xi^T(k)[\Psi + (F_0^i)^T P F_0^i + (F_0^i - F_1)^T H (F_0^i - F_1)]\xi(k) \leq 0, \end{aligned} \tag{A.4}$$

if $\Psi + (F_0^i)^T P F_0^i + (F_0^i - F_1)^T H (F_0^i - F_1) < 0$, i.e., by Schur complements, if (10.14) is feasible.

Since $|\omega_i(k)| \leq \mu \Delta_i$, $i = 1, \ldots, N$, the inequality (A.4) implies for $k \in [t_p, t_{p+1} - 1]$

$$\begin{aligned} V(x_k) &\leq \lambda V(x_{k-1}) + \mu^2 \sum_{i=1}^{N} b_i \Delta_i^2 \\ &\;\vdots \\ &\leq \lambda^{k-t_{N-1}} V(x_{t_{N-1}}) + \frac{\mu^2}{1-\lambda} \sum_{i=1}^{N} b_i \Delta_i^2, \end{aligned}$$

that completes the proof.

References

1. A. Adaldo, D. Liuzza, D.V. Dimarogonas, K.H. Johansson, Cloud-supported formation control of second-order multi-agent systems. IEEE Trans. Control Netw. Syst. **5**(4), 1563–1574 (2018)
2. J. Almeida, C. Silvestre, A. Pascoal, Synchronization of multiagent systems using event-triggered and self-triggered broadcasts. IEEE Trans. Autom. Control **62**(9), 4741–4746 (2017)
3. S. Amin, G.A. Schwartz, S.S. Sastry, Security of interdependent and identical networked control systems. Automatica **49**(1), 186–192 (2013)
4. A. Anta, P. Tabuada, To sample or not to sample: self-triggered control for nonlinear systems. IEEE Trans. Autom. Control **55**(9), 2030–2042 (2010)
5. P. Antsaklis, J. Baillieul, Special issue on technology of networked control systems. Proc. IEEE **95**(1), 5–8 (2007)
6. B. Bamieh, J. Pearson, B. Francis, A. Tannenbaum, A lifting technique for linear periodic systems with applications to sampled-data control. Syst. Control Lett. **17**(2), 79–88 (1991)
7. N. Bar Am, E. Fridman, Network-based H_∞ filtering of parabolic systems. Automatica **50**(12), 3139–3146 (2014)
8. J.E.F. Baruch, M.J. Cox, Remote control and robots: an Internet solution. Comput. Control Eng. J. **7**, 39–45 (1996)
9. T. Başar, P. Bernhard, *H_∞ Optimal Control and Related Minimax Design Problems: A Dynamic Game Approach* (Birkhäuser, Boston, 1995)
10. N. Bauer, P. Maas, W. Heemels, Stability analysis of networked control systems: a sum of squares approach. Automatica **48**(8), 1514–1524 (2012)
11. D.S. Bernstein, A.N. Michel, Special issue on saturating actuators. Int. J. Robust Nonlinear Control **5**(5), 375–540 (1995)
12. F. Blanchini, S. Miani, Stabilization of LPV systems: state feedback, state estimation, and duality. SIAM J. Control Optim. **42**(1), 76–97 (2003)
13. D.P. Borgers, W.P.M.H. Heemels, Event-separation properties of event-triggered control systems. IEEE Trans. Autom. Control **59**(10), 2644–2656 (2014)
14. D.P. Borgers, W.P.M.H. Heemels, Stability analysis of large-scale networked control systems with local networks: a hybrid small-gain approach, in *Proceedings of the 17th International Conference on Hybrid Systems: Computation and Control*, Berlin, Germany (2014)
15. C. Briat, A. Seuret, A looped-functional approach for robust stability analysis of linear impulsive systems. Syst. Control Lett. **61**(10), 980–988 (2012)
16. R. Brockett, D. Liberzon, Quantized feedback stabilization of linear systems. IEEE Trans. Autom. Control **45**(7), 1279–1289 (2000)

K. Liu et al., *Networked Control Under Communication Constraints*, Advances in Delays and Dynamics 11, https://doi.org/10.1007/978-981-15-4230-5

17. D. Carnevale, A.R. Teel, D. Nesic, A Lyapunov proof of an improved maximum allowable transfer interval for networked control systems. IEEE Trans. Autom. Control **52**(5), 892–897 (2007)
18. H. Chan, Ü. Özgüner, Closed-loop control of systems over a communications network with queues. Int. J. Control **62**(3), 493–510 (1995)
19. T.W. Chen, B.A. Francis, *Optimal Sampled-Data Control Systems* (Springer, London, 1995)
20. W.H. Chen, W.X. Zheng, Input-to-state stability for networked control systems via an improved impulsive system approach. Automatica **47**(4), 789–796 (2011)
21. D. Christmann, R. Gotzhein, S. Siegmund, F. Wirth, Realization of Try-Once-Discard in wireless multihop networks. IEEE Trans. Ind. Inf. **10**(1), 17–26 (2014)
22. M.B.G. Cloosterman, L. Hetel, N. van de Wouw, W.P.M.H. Heemels, J. Daafouz, H. Nijmeijer, Controller synthesis for networked control systems. Automatica **46**(10), 1584–1594 (2010)
23. M.B.G. Cloosterman, N. van de Wouw, W.P.M.H. Heemels, H. Nijmeijer, Stability of networked control systems with large delays, in *Proceedings of the 46th IEEE Conference on Decision and Control*, New Orleans, USA (2007)
24. M.B.G. Cloosterman, N. van de Wouw, W.P.M.H. Heemels, H. Nijmeijer, Stability of networked control systems with uncertain time-varying delays. IEEE Trans. Autom. Control **54**(7), 1575–1580 (2009)
25. B. Cui, Y.Q. Xia, K. Liu, J.H. Zhang, Cooperative tracking control for general linear multiagent systems with directed intermittent communications: an artificial delay approach. Int. J. Robust Nonlinear Control **29**(10), 3063–3077 (2019)
26. D.B. Dačić, D. Nešić, Quadratic stabilization of linear networked control systems via simultaneous protocol and controller design. Automatica **43**(7), 1145–1155 (2007)
27. V.S. Dolk, P. Tesi, C. De Persis, W.P.M.H. Heemels, Event-triggered control systems under denial-of-service attacks. IEEE Trans. Control Netw. Syst. **4**(1), 93–105 (2017)
28. M.C.F. Donkers, W.P.M.H. Heemels, D. Bernardini, A. Bemporad, V. Shneer, Stability analysis of stochastic networked control systems. Automatica **48**(5), 917–925 (2012)
29. M.C.F. Donkers, W.P.M.H. Heemels, N. van de Wouw, L. Hetel, Stability analysis of networked control systems using a switched linear systems approach. IEEE Trans. Autom. Control **56**(9), 2101–2115 (2011)
30. M.C.F. Donkers, L. Hetel, W.P.M.H. Heemels, N. van de Wouw, M. Steinbuch, Stability analysis of networked control systems using a switched linear systems approach. Hybrid Syst.: Comput. Control Lect. Notes Comput. Sci. **5469**, 150–164 (2009)
31. G. Dullerud, K. Glover, Robust stabilization of sampled-data systems to structured LTI perturbations. IEEE Trans. Autom. Control **38**(10), 1497–1508 (1993)
32. N. Elia, S.K. Mitter, Stabilization of linear systems with limited information. IEEE Trans. Autom. Control **46**(9), 1384–1400 (2001)
33. B. Etkin, L.D. Reid, *Dynamics of Flight: Stability and Control* (Wiley, New York, 1996)
34. Z.G. Feng, J. Lam, G.H. Yang, Optimal partitioning method for stability analysis of continuous/discrete delay systems. Int. J. Robust Nonlinear Control **25**(4), 559–574 (2015)
35. Z.G. Feng, P. Shi, Sliding mode control of singular stochastic Markov jump systems. IEEE Trans. Autom. Control **62**(8), 4266–4273 (2017)
36. Z.G. Feng, P. Shi, Two equivalent sets: application to singular systems. Automatica **77**, 198–205 (2017)
37. C. Fiter, L. Hetel, W. Perruquetti, J.P. Richard, A state dependent sampling for linear state feedback. Automatica **48**(8), 1860–1867 (2012)
38. D. Freirich, E. Fridman, Decentralized networked control of systems with local networks: A time-delay approach. Automatica **69**, 201–209 (2016)
39. D. Freirich, E. Fridman, Decentralized networked control of discrete-time systems with local networks. Int. J. Robust Nonlinear Control **28**(1), 365–380 (2018)
40. E. Fridman, Use of models with aftereffect in the problem of the design of optimal digital-control systems. Autom. Remote Control **53**(10), 1523–1528 (1992)
41. E. Fridman, New Lyapunov-Krasovskii functionals for stability of linear retarded and neutral type systems. Syst. Control Lett. **43**(4), 309–319 (2001)

42. E. Fridman, Stability of linear descriptor systems with delay: a Lyapunov-based approach. J. Math. Anal. Appl. **273**(1), 24–44 (2002)
43. E. Fridman, Discussion on: stabilization of networked control systems with data packet dropout and transmission delays: continuous-time case. Eur. J. Control **11**(1), 53–55 (2005)
44. E. Fridman, Descriptor discretized Lyapunov functional method: analysis and design. IEEE Trans. Autom. Control **51**(5), 890–897 (2006)
45. E. Fridman, A new Lyapunov technique for robust control of systems with uncertain non-small delays. IMA J. Math. Control Inf. **23**(2), 165–179 (2006)
46. E. Fridman, A refined input delay approach to sampled-data control. Automatica **46**(2), 421–427 (2010)
47. E. Fridman, *Introduction to Time-Delay Systems* (Birkhäuser, Basel, 2014)
48. E. Fridman, Tutorial on Lyapunov-based methods for time-delay systems. Eur. J. Control **20**(6), 271–283 (2014)
49. E. Fridman, A. Blighovsky, Robust sampled-data control of a class of semilinear parabolic systems. Automatica **48**(5), 826–836 (2012)
50. E. Fridman, M. Dambrine, Control under quantization, saturation and delay: an LMI approach. Automatica **45**(10), 2258–2264 (2009)
51. E. Fridman, Y. Orlov, Exponential stability of linear distributed parameter systems with time-varying delays. Automatica **45**(1), 194–201 (2009)
52. E. Fridman, A. Pila, U. Shaked, Regional stabilization and H_∞ control of time-delay systems with saturating actuators. Int. J. Robust Nonlinear Control **13**(9), 885–907 (2003)
53. E. Fridman, A. Seuret, J.P. Richard, Robust sampled-data stabilization of linear systems: an input delay approach. Automatica **40**(8), 1441–1446 (2004)
54. E. Fridman, L. Shaikhet, Delay-induced stability of vector second-order systems via simple Lyapunov functionals. Automatica **74**, 288–296 (2016)
55. E. Fridman, L. Shaikhet, Stabilization by using artificial delays: an LMI approach. Automatica **81**, 429–437 (2017)
56. E. Fridman, U. Shaked, Delay-dependent stability and H_∞ control: constant and time-varying delays. Int. J. Control **76**(1), 48–60 (2003)
57. E. Fridman, U. Shaked, Stability and guaranteed cost control of uncertain discrete delay systems. Int. J. Control **78**(4), 235–246 (2005)
58. E. Fridman, U. Shaked, Input-output approach to stability and L_2-gain analysis of systems with time-varying delays. Syst. Control Lett. **55**(12), 1041–1053 (2006)
59. E. Fridman, G. Tsodik, H_∞ control of distributed and discrete delay systems via discretized Lyapunov functional. Eur. J. Control **15**(1), 84–96 (2009)
60. M.Y. Fu, L.H. Xie, The sector bound approach to quantized feedback control. IEEE Trans. Autom. Control **50**(11), 1698–1711 (2005)
61. H. Fujioka, Stability analysis for a class of networked/embedded control systems: a discrete-time approach, in *Proceedings of the American Control Conference*, USA, Washington (2008)
62. H. Fujioka, Stability analysis of systems with aperiodic sample-and-hold devices. Automatica **45**(3), 771–775 (2009)
63. H.J. Gao, T.W. Chen, Network-based H_∞ output tracking control. IEEE Trans. Autom. Control **53**(3), 655–667 (2008)
64. H.J. Gao, T.W. Chen, T.Y. Chai, Passivity and passification for networked control systems. SIAM J. Control Optim. **46**(4), 1299–1322 (2007)
65. H.J. Gao, T.W. Chen, J. Lam, A new delay system approach to network-based control. Automatica **44**(1), 39–52 (2008)
66. H.J. Gao, X.Y. Meng, T.W. Chen, J. Lam, Stabilization of networked control systems via dynamic output-feedback controllers. SIAM J. Control Optim. **48**(5), 3643–3658 (2010)
67. H.J. Gao, J. Wu, P. Shi, Robust sampled-data H_∞ control with stochastic sampling. Automatica **45**(7), 1729–1736 (2009)
68. J.C. Geromel, R. Korogui, J. Bernussou, H_2 and H_∞ robust output feedback control for continuous time polytopic systems. IET Control Theory Appl. **1**(5), 1541–1549 (2007)

69. R.H. Gielen, S. Olaru, M. Lazar, W.P.M.H. Heemels, N. van de Wouw, S.I. Niculescu, On polytopic inclusions as a modeling framework for systems with time-varying delays. Automatica **46**(3), 615–619 (2010)
70. R. Goebel, R.G. Sanfelice, A. Teel, Hybrid dynamical systems. IEEE Control Syst. Mag. **29**(2), 28–93 (2009)
71. K. Gu, Discretized LMI set in the stability problem of linear uncertain time-delay systems. Int. J. Control **68**(4), 923–934 (1997)
72. K. Gu, Stability problem of systems with multiple delay channels. Automatica **46**, 743–751 (2010)
73. K. Gu, V. Kharitonov, J. Chen, *Stability of Time-Delay Systems* (Birkhäuser, Boston, 2003)
74. K. Gu, S.I. Niculescu, Further remarks on additional dynamics in various model transformations of linear delay systems. IEEE Trans. Autom. Control **46**(3), 497–500 (2001)
75. J.K. Hale, *Theory of Functional Differential Equations* (Springer, New York, 1977)
76. X.R. Han, E. Fridman, S.K. Spurgeon, Sampled-data sliding mode observer for robust fault reconstruction: a time-delay approach. J. Franklin Inst. **351**(4), 2125–2142 (2014)
77. G. Hardy, J. Littlewood, G. Pólya, *Inequalitites* (Cambridge University Press, Cambridge, 1934)
78. K. Hashimoto, S. Adachi, D.V. Dimarogonas, Event-triggered intermittent sampling for nonlinear model predictive control. Automatica **81**, 148–155 (2017)
79. K. Hashimoto, S. Adachi, D.V. Dimarogonas, Self-triggered model predictive control for nonlinear input-affine dynamical systems via adaptive control samples selection. IEEE Trans. Autom. Control **62**(1), 177–189 (2017)
80. Y. He, Q.G. Wang, C. Lin, M. Wu, Delay-range-dependent stability for systems with time-varying delay. Automatica **43**(2), 371–376 (2007)
81. W.P.M.H. Heemels, D.P. Borgers, N. van de Wouw, D. Nesic, A.R. Teel, Stability analysis of nonlinear networked control systems with asynchronous communication: a small-gain approac, in *Proceedings of the 52nd IEEE Conference on Decision and Control*, Florence, Italy (2013)
82. W.P.M.H. Heemels, K.H. Johansson, P. Tabuada, An introduction to event-triggered and self-triggered control, in *Proceedings of the 51st IEEE Conference on Decision and Control*, Hawaii, USA (2012)
83. W.P.M.H. Heemels, D. Nesic, A.R. Teel, N. van de Wouw, Networked and quantized control systems with communication delays, in *Proceedings of the 48th IEEE Conference on Decision and Control*, Shanghai, China (2009)
84. W.P.M.H. Heemels, A. Teel, N. van de Wouw, D. Nesic, Networked control systems with communication constraints: tradeoffs between transmission intervals, delays and performance. IEEE Trans. Autom. Control **55**(8), 1781–1796 (2010)
85. W.P.M.H. Heemels, N. van de Wouw, R.H. Gielen, M.C.F. Donkers, L. Hetel, S. Olaru, M. Lazar, J. Daafouz, S.I. Niculescu, Comparison of overapproximation methods for stability analysis of networked control systems, in *Proceedings of the 13th International Conference on Hybrid Systems: Computation and Control*, Stockholm, Sweden (2010)
86. J.M. Hendrickx, K.H. Johansson, R.M. Jungers, H. Sandberg, K.C. Sou, Efficient computations of a security index for false data attacks in power networks. IEEE Trans. Autom. Control **59**(12), 3194–3208 (2014)
87. J. Hespanha, A. Morse, Stability of switched systems with average dwell-time, in *Proceedings of the 38th IEEE Conference on Decision and Control*, Phoenix, Arizona, USA (1999)
88. J. Hespanha, P. Naghshtabrizi, Y. Xu, A survey of recent results in networked control systems. Proc. IEEE **95**(1), 138–162 (2007)
89. L. Hetel, J. Daafouz, C. Iung, Stabilization of arbitrary switched linear systems with unknown time-varying delays. IEEE Trans. Autom. Control **51**(10), 1668–1674 (2006)
90. L. Hetel, J. Daafouz, C. Iung, Equivalence between the Lyapunov-Krasovskii functionals approach for discrete delay systems and that of the stability conditions for switched systems. Nonlinear Anal.: Hybrid Syst. **2**(3), 697–705 (2008)

91. L. Hetel, C. Fiter, H. Omran, A. Seuret, E. Fridman, J.P. Richard, S.I. Niculescu, Recent developments on the stability of systems with aperiodic sampling: an overview. Automatica **76**, 309–335 (2017)
92. L. Hetel, A. Kruszewski, W. Perruquetti, J.P. Richard, Discrete and intersample analysis of systems with aperiodic sampling. IEEE Trans. Autom. Control **56**(7), 1696–1701 (2011)
93. L.V. Hien, H. Trinh, New finite-sum inequalities with applications to stability of discrete time-delay systems. Automatica **71**, 197–201 (2016)
94. D. Hristu-Varsakelis, W.S. Levine, *Handbook of Networked and Embedded Control Systems* (Birkhäuser, Boston, 2005)
95. J. Hu, J.L. Liang, D.Y. Chen, D.H. Ji, J.H. Du, A recursive approach to non-fragile filtering for networked systems with stochastic uncertainties and incomplete measurements. J. Franklin Inst. **352**(5), 1946–1962 (2015)
96. S. Hu, W.Y. Yan, Stability robustness of networked control systems with respect to packet loss. Automatica **43**(7), 1243–1248 (2007)
97. S.L. Hu, D. Yue, L_2-gain analysis of event-triggered networked control systems: a discontinuous Lyapunov functional approach. Int. J. Robust Nonlinear Control **23**(11), 1277–1300 (2013)
98. S.L. Hu, D. Yue, X.P. Xie, X.L. Chen, X.X. Yin, Resilient event-triggered controller synthesis of networked control systems under periodic DoS jamming attacks. IEEE Trans. Cybern. **49**(12), 4271–4281 (2018)
99. S.S. Hu, Q.X. Zhu, Stochastic optimal control and analysis of stability of networked control systems with long delay. Automatica **39**(11), 1877–1884 (2003)
100. T. Hu, Z. Lin, *Control Systems with Actuator Saturation: Analysis and Design* (Birkhäuser, Boston, 2001)
101. C.C. Hua, C. Ge, X.P. Guan, Synchronization of chaotic Lur'e systems with time delays using sampled-data control. IEEE Trans. Neural Netw. Learn. Syst. **26**(6), 1214–1221 (2015)
102. D. Huang, S.K. Nguang, State feedback control of uncertain networked control systems with random time delays. IEEE Trans. Autom. Control **53**(3), 829–834 (2008)
103. D. Huang, S.K. Nguang, *Robust Control for Uncertain Networked Control Systems with Random Delays* (Springer, Berlin, 2009)
104. O.C. Imer, S. Yüksel, T. Başar, Optimal control of LTI systems over unreliable communication links. Automatica **42**(9), 1429–1439 (2006)
105. H. Ishii, H_∞ control with limited communication and message losses. Syst. Control Lett. **57**(4), 322–331 (2008)
106. K.H. Johansson, The quadruple-tank process: A multivariable laboratory process with an adjustable zero. IEEE Trans. Control Syst. Technol. **8**(3), 456–465 (2000)
107. W. Kang, E. Fridman, Distributed sampled-data control of Kuramoto-Sivashinsky equation. Automatica **95**, 514–524 (2018)
108. H. Khalil, J. Grizzle, *Nonlinear Systems* (Prentice Hall, Upper Saddle River, 2002)
109. V.L. Kharitonov, D. Melchor-Aguilar, On delay-dependent stability conditions. Syst. Control Lett. **40**(1), 71–76 (2000)
110. V.L. Kharitonov, S.I. Niculescu, On the stability of linear systems with uncertain delay. IEEE Trans. Autom. Control **48**(1), 127–132 (2003)
111. V.L. Kharitonov, S.I. Niculescu, J. Moreno, W. Michiels, Static output feedback stabilization: necessary conditions for multiple delay controllers. IEEE Trans. Autom. Control **50**(1), 82–86 (2005)
112. V.B. Kolmanovskii, J. Richard, Stability of some linear systems with delays. IEEE Trans. Autom. Control **44**(5), 984–989 (1999)
113. G.V. Kondraske, R.A. Volz, D.H. Johnson, D. Tesar, J.C. Trinkle, C.R. Price, Network-based infrastructure for distributed remote operations and robotics research. IEEE Trans. Robot. Autom. **9**(5), 702–704 (1993)
114. O. Kwon, M. Park, J. Park, S. Lee, E. Cha, Stability and stabilization for discrete-time systems with time-varying delays via augmented Lyapunov-Krasovskii functional. J. Franklin Inst. **350**(3), 521–540 (2013)

115. J.W. Lee, On uniform stabilization of discrete-time linear parameter-varying control systems. IEEE Trans. Autom. Control **51**(10), 1714–1721 (2006)
116. K.B. Lee, R.D. Schneeman, Internet-based distributed measurement and control applications. IEEE Instrum. Meas. Mag. **2**(2), 23–27 (1999)
117. S.Y. Lee, W.I. Lee, P.G. Park, Polynomials-based integral inequality for stability analysis of linear systems with time-varying delays. J. Franklin Inst. **354**(4), 2053–2067 (2017)
118. T.H. Lee, J.H. Park, S.M. Lee, O.M. Kwon, Robust synchronisation of chaotic systems with randomly occurring uncertainties via stochastic sampled-data control. Int. J. Control **86**(1), 107–119 (2013)
119. H.Y. Li, C.W. Wu, P. Shi, Y.B. Gao, Control of nonlinear networked systems with packet dropouts: interval type-2 fuzzy model-based approach. IEEE Trans. Cybern. **45**(11), 2378–2389 (2015)
120. X. Li, C.E. de Souza, Criteria for robust stability and stabilization of uncertain linear systems with state delay. Automatica **33**(9), 1657–1662 (1997)
121. Y.L. Li, K. Liu, W. He, Y.X. Yin, R. Johansson, K. Zhang, Bilateral teleoperation of multiple robots under scheduling communication. IEEE Trans. Control Syst. Technol. (2019). https://doi.org/10.1109/TCST.2019.2923788
122. Y.L. Li, K. Zhang, K. Liu, R. Johansson, Y.X. Yin, Neural-network-based adaptive control for bilateral teleoperation with multiple slaves under Round-Robin scheduling protocol. Int. J. Control (2019). https://doi.org/10.1080/00207179.2019.1652853
123. D. Liberzon, Hybrid feedback stabilization of systems with quantized signals. Automatica **39**(9), 1543–1554 (2003)
124. D. Liberzon, *Switching in Systems and Control. Systems & Control: Foundations & Applications* (Birkhuser, Boston, 2003)
125. D. Liberzon, D. Nesic, Input-to-state stabilization of linear systems with quantized state measurements. IEEE Trans. Autom. Control **52**(5), 767–781 (2007)
126. F. Liu, H.J. Gao, J.B. Qiu, S. Yin, J. Fan, T.Y. Chai, Networked multirate output feedback control for setpoints compensation and its application to rougher flotation process. IEEE Trans. Ind. Electron. **61**(1), 460–468 (2014)
127. G.P. Liu, Y.Q. Xia, J. Chen, D. Rees, W.S. Hu, Networked predictive control of systems with random network delays in both forward and feedback channels. IEEE Trans. Ind. Electron. **54**(3), 1282–1297 (2007)
128. K. Liu, E. Fridman, Networked-based stabilization via discontinuous Lyapunov functionals. Int. J. Robust Nonlinear Control **22**(4), 420–436 (2012)
129. K. Liu, E. Fridman, Wirtinger's inequality and Lyapunov-based sampled-data stabilization. Automatica **48**(1), 102–108 (2012)
130. K. Liu, E. Fridman, Delay-dependent methods and the first delay interval. Syst. Control Lett. **64**(1), 57–63 (2014)
131. K. Liu, E. Fridman, Discrete-time network-based control under scheduling and actuator constraints. Int. J. Robust Nonlinear Control **25**(12), 1816–1830 (2015)
132. K. Liu, E. Fridman, L. Hetel, Network-based control via a novel analysis of hybrid systems with time-varying delays, in *Proceedings of the 51st IEEE Conference on Decision and Control*, Hawaii, USA (2012)
133. K. Liu, E. Fridman, L. Hetel, Stability and L_2-gain analysis of networked control systems under Round-Robin scheduling: a time-delay approach. Syst. Control Lett. **61**(5), 666–675 (2012)
134. K. Liu, E. Fridman, L. Hetel, Networked control systems in the presence of scheduling protocols and communication delays. SIAM J. Control Optim. **53**(4), 1768–1788 (2015)
135. K. Liu, E. Fridman, L. Hetel, J.P. Richard, Sampled-data stabilization via Round-Robin scheduling: a direct Lyapunov–Krasovskii approach, in *Proceedings of the 18th World Congress of the International Federation of Automatic Control*, Milano, Italy (2011)
136. K. Liu, E. Fridman, K.H. Johansson, Networked control with a stochastic scheduling: a time-delay approach, in *Proceedings of the 19th World Congress of the International Federation of Automatic Control*, Cape Town, South Africa (2014)

137. K. Liu, E. Fridman, K.H. Johansson, Dynamic quantization of uncertain linear networked control systems. Automatica **59**, 248–255 (2015)
138. K. Liu, E. Fridman, K.H. Johansson, Networked control with stochastic scheduling. IEEE Trans. Autom. Control **60**(11), 3071–3076 (2015)
139. K. Liu, E. Fridman, K.H. Johansson, Y.Q. Xia, Generalized jensen inequalities with application to stability analysis of systems with distributed delays over infinite time-horizons. Automatica **69**, 222–231 (2016)
140. K. Liu, E. Fridman, K.H. Johansson, Y.Q. Xia, Quantized control under Round-Robin communication protocol. IEEE Trans. Ind. Electron. **63**(7), 4461–4471 (2016)
141. K. Liu, H. Guo, Q.R. Zhang, Y.Q. Xia, Distributed secure filtering for discrete-time systems under Round-Robin protocol and deception attacks. IEEE Trans. Cybern. (2019). https://doi.org/10.1109/TCYB.2019.2897366
142. K. Liu, A.Y. Ma, Y.Q. Xia, Z.Q. Sun, K.H. Johansson, Network scheduling and control co-design for multi-loop MPC. IEEE Trans. Autom. Control **64**(12), 5238–5245 (2019)
143. K. Liu, X. Pan, Y. Q. Xia, E. Fridman, J. Lam, Networked control under communication constraints: the discrete-time case, in *Proceedings of the 55th IEEE Conference on Decision and Control*, Las Vegas, USA (2016)
144. K. Liu, A. Selivanov, E. Fridman, Survey on time-delay approach to networked control. Ann. Rev. Control **48**, 57–79 (2019)
145. K. Liu, A. Seuret, Comparison of bounding methods for stability analysis of systems with time-varying delays. J. Franklin Inst. **354**(7), 2979–2993 (2017)
146. K. Liu, A. Seuret, Y.Q. Xia, Stability analysis of systems with time-varying delays via the second-order Bessel-Legendre inequality. Automatica **76**, 138–142 (2017)
147. K. Liu, A. Seuret, Y.Q. Xia, F. Gouaisbaut, Y. Ariba, Bessel-Laguerre inequality and its application to systems with infinite distributed delays. Automatica **109**, 108562 (2019)
148. K. Liu, V. Suplin, E. Fridman, Stability of linear systems with general sawtooth delay. IMA J. Math. Control Inf. **27**(4), 419–436 (2010)
149. J. Louisell, New examples of quenching in delay differential equations having time-varying delay, in *Proceedings of the 4th European Control Conference*, Karlshruhe, Germany (1999)
150. J. Louisell, Delay differential systems with time-varying delay: new directions for stability theory. Kybernetika **37**(3), 239–251 (2001)
151. R.Q. Lu, P. Shi, H.Y. Su, Z.G. Wu, J.Q. Lu, Synchronization of general chaotic neural networks with nonuniform sampling and packet missing: a switched system approach. IEEE Trans. Neural Netw. Learn. Syst. **29**(3), 523–533 (2018)
152. J. Lunze, *Feedback Control of Large Scale Systems* (Prentice Hall PTR, Upper Saddle River, 1992)
153. J. Lunze, D. Lehmann, A state-feedback approach to event-based control. Automatica **46**(1), 211–215 (2010)
154. A.Y. Ma, K. Liu, Q.R. Zhang, Y.Q. Xia, Distributed MPC for linear discrete-time systems with disturbances and coupled states. Syst. Control Lett. **135**, 104578 (2020)
155. M. Mazo, A. Anta, P. Tabuada, An ISS self-triggered implementation of linear controllers. Automatica **46**(8), 1310–1314 (2010)
156. Y. Mikheev, V. Sobolev, E. Fridman, Asymptotic analysis of digital control systems. Autom. Remote Control **49**(9), 1175–1180 (1988)
157. L. Mirkin, Some remarks on the use of time-varying delay to model sample-and-hold circuits. IEEE Trans. Autom. Control **52**(6), 1109–1112 (2007)
158. R. Mitchell, R. Chen, Adaptive intrusion detection of malicious unmanned air vehicles using behavior rule specifications. IEEE Trans. Syst. Man. Cybern.: Syst. **44**(5), 593–604 (2014)
159. M. Moarref, L. Rodrigues, On exponential stability of linear networked control systems. Int. J. Robust Nonlinear Control **24**(7), 1221–1240 (2014)
160. Y.S. Moon, P. Park, W.H. Kwon, Y.S. Lee, Delay-dependent robust stabilization of uncertain state-delayed systems. Int. J. Control **74**(14), 1447–1455 (2001)
161. P. Naghshtabrizi, J. Hespanha, A.R. Teel, Stability of delay impulsive systems with application to networked control systems. Trans. Inst. Meas. Control **32**(5), 511–528 (2010)

162. P. Naghshtabrizi, J.P. Hespanha, A.R. Teel, Exponential stability of impulsive systems with application to uncertain sampled-data systems. Syst. Control Lett. **57**(5), 378–385 (2008)
163. D. Nesic, D. Liberzon, A unified framework for design and analysis of networked and quantized control systems. IEEE Trans. Autom. Control **54**(4), 732–747 (2009)
164. D. Nesic, A.R. Teel, Input-output stability properties of networked control systems. IEEE Trans. Autom. Control **49**(10), 1650–1667 (2004)
165. S.I. Niculescu, W. Michiels, Stabilizing a chain of integrators using multiple delays. IEEE Trans. Autom. Control **49**(5), 802–807 (2004)
166. J. Nilsson, *Real-time control systems with delays*. Ph.D. thesis, Lund Institute of Technology (1998)
167. J. Nilsson, B. Bernhardsson, B. Wittenmark, Stochastic analysis and control of real-time systems with random time delays. Automatica **34**(1), 57–64 (1998)
168. C. Nowzari, J. Cortés, Self-triggered coordination of robotic networks for optimal deployment. Automatica **48**(6), 1077–1087 (2012)
169. P.G. Otanez, J.R. Moyne, D.M. Tilbury, Using deadbands to reduce communication in networked control systems, in *Proceedings of the American Control Conference*, Alaska, USA (2002)
170. A. Papachristodoulou, M.M. Peet, S.I. Niculescu, Stability analysis of linear systems with time-varying delays: delay uncertainty and quenching, in *Proceedings of the 46th IEEE Conference on Decision and Control*, New Orleans, USA (2007)
171. P.G. Park, A delay-dependent stability criterion for systems with uncertain time-invariant delays. IEEE Trans. Autom. Control **44**(4), 876–877 (1999)
172. P.G. Park, J.W. Ko, Stability and robust stability for systems with a time-varying delay. Automatica **43**(10), 1855–1858 (2007)
173. P.G. Park, J.W. Ko, C. Jeong, Reciprocally convex approach to stability of systems with time-varying delays. Automatica **47**(1), 235–238 (2011)
174. P.G. Park, W.I. Lee, S.Y. Lee, Auxiliary function-based integral inequalities for quadratic functions and their applications to time-delay systems. J. Franklin Inst. **352**(4), 1378–1396 (2015)
175. C. Peng, Q.L. Han, A novel event-triggered transmission scheme and L_2 control co-design for sampled-data control systems. IEEE Trans. Autom. Control **58**(10), 2620–2626 (2013)
176. C. Peng, S.D. Ma, X.P. Xie, Observer-based non-PDC control for networked T-S fuzzy systems with an event-triggered communication. IEEE Trans. Cybern. **47**(8), 2279–2287 (2017)
177. C. Peng, T.C. Yang, Event-triggered communication and H_∞ control co-design for networked control systems. Automatica **49**(5), 1326–1332 (2013)
178. A. Polyakov, D. Efimov, W. Perruquetti, J.P. Richard, Output stabilization of time-varying input delay systems using interval observation technique. Automatica **49**(11), 3402–3410 (2013)
179. A. Ray, Distributed data communication networks for real-time process control. Chem. Eng. Commun. **65**(1), 139–154 (1988)
180. O. Roesch, H. Roth, S.I. Niculescu, Remote control of mechatronic systems over communication networks, in *Proceedings of the IEEE International Conference on Mechatronics & Automation*, Niagara Falls, Canada (2005)
181. A. Saberi, Z.L. Lin, A.R. Teel, Control of linear systems with saturating actuators. IEEE Trans. Autom. Control **41**(3), 368–378 (1996)
182. A. Saberi, A.A. Stoorvogel, Special issue on saturating actuators. Int. J. Robust Nonlinear Control **9**(10), 583–584 (1999)
183. M.F. Sågfors, H.T. Toivonen, H_∞ and LQG control of asynchronous sample-data systems. Automatica **33**(9), 1663–1668 (1997)
184. H. Sandberg, S. Amin, K.H. Johansson, Special issue on cyberphysical security in networked control systems. IEEE Control Syst. Mag. **35**(1), 20–23 (2015)
185. L. Schenato, To zero or to hold control inputs with lossy links? IEEE Trans. Autom. Control **54**(5), 1093–1099 (2009)

186. P. Seiler, R. Sengupta, An H_∞ approach to networked control. IEEE Trans. Autom. Control **50**(3), 356–364 (2005)
187. A. Selivanov, E. Fridman, Distributed event-triggered control of diffusion semilinear PDEs. Automatica **68**, 344–351 (2016)
188. A. Selivanov, E. Fridman, Event-triggered H_∞ control: a switching approach. IEEE Trans. Autom. Control **61**(10), 3221–3226 (2016)
189. A. Selivanov, E. Fridman, Observer-based input-to-state stabilization of networked control systems with large uncertain delays. Automatica **74**, 63–70 (2016)
190. A. Selivanov, E. Fridman, An improved time-delay implementation of derivative-dependent feedback. Automatica **98**, 269–276 (2018)
191. A. Selivanov, E. Fridman, Sampled-data implementation of derivative-dependent control using artificial delays. IEEE Trans. Autom. Control **63**(10), 3594–3600 (2018)
192. A. Seuret, A novel stability analysis of linear systems under asynchronous samplings. Automatica **48**(1), 177–182 (2012)
193. A. Seuret, C. Briat, Stability analysis of uncertain sampled-data systems with incremental delay using looped-functionals. Automatica **55**, 274–278 (2015)
194. A. Seuret, E. Fridman, Wirtinger-like Lyapunov-Krasovskii functionals for discrete-time delay systems. IMA J. Math. Control Inf. **35**(3), 861–876 (2018)
195. A. Seuret, E. Fridman, J.P. Richard, Sampled-data exponential stabilization of neutral systems with input and state delays, in *Proceedings of the 13th IEEE Mediterranean Conference on Control and Automation, Cyprus*, Limassol, Cyprus (2005)
196. A. Seuret, J.M. Gomes da Silva, Taking into account period variations and actuator saturation in sampled-data systems. Syst. Control Lett. **61**(12), 1286–1293 (2012)
197. A. Seuret, F. Gouaisbaut, Wirtinger-based integral inequality: application to time-delay systems. Automatica **49**(9), 2860–2866 (2013)
198. A. Seuret, F. Gouaisbaut, Hierarchy of LMI conditions for the stability analysis of time-delay systems. Syst. Control Lett. **81**, 1–7 (2015)
199. A. Seuret, F. Gouaisbaut, Stability of linear systems with time-varying delays using Bessel-Legendre inequalities. IEEE Trans. Autom. Control **63**(1), 225–232 (2018)
200. A. Seuret, F. Gouaisbaut, Y. Ariba, Complete quadratic Lyapunov functionals for distributed delay systems. Automatica **62**, 168–176 (2015)
201. A. Seuret, F. Gouaisbaut, E. Fridman, Stability of systems with fast-varying delay using improved Wirtinger's inequality, in *Proceedings of the 52nd IEEE Conference on Decision and Control*, Florence, Italy (2013)
202. A. Seuret, F. Gouaisbaut, E. Fridman, Stability of discrete-time systems with time-varying delays via a novel summation inequality. IEEE Trans. Autom. Control **60**(10), 2740–2745 (2015)
203. A. Seuret, K. Liu, F. Gouaisbaut, Generalized reciprocally convex combination lemmas and its application to time-delay systems. Automatica **95**, 488–493 (2018)
204. A. Seuret, J.P. Richard, Control of a remote system over network including delays and packet dropout, in *Proceedings of the 17th World Congress of the International Federation of Automatic Control*, Seoul, Korea (2008)
205. D.H. Shah, A. Mehta, *Discrete-Time Sliding Mode Control for Networked Control System* (Springer Nature, Singapore, 2018)
206. H.Y. Shao, New delay-dependent stability criteria for systems with interval delay. Automatica **45**(3), 744–749 (2009)
207. Y. Shi, B. Yu, Output feedback stabilization of networked control systems with random delays modeled by Markov chains. IEEE Trans. Autom. Control **54**(7), 1668–1674 (2009)
208. Y. Shi, B. Yu, Robust mixed H_2/H_∞ control of networked control systems with random time delays in both forward and backward communication links. Automatica **47**(4), 754–760 (2011)
209. N. Sivashankar, P.P. Khargonekar, Characterization of the L_2-induced norm for linear systems with jumps with applications to sampled-data systems. SIAM J. Control Optim. **32**(4), 1128–1150 (1994)

210. J. Skaf, S. Boyd, Analysis and synthesis of state-feedback controllers with timing jitter. IEEE Trans. Autom. Control **54**(3), 652–657 (2009)
211. O. Solomon, E. Fridman, New stability conditions for systems with distributed delays. Automatica **49**(11), 3467–3475 (2013)
212. M. Souza, G.S. Deaecto, J.C. Geromel, J. Daafouz, Self-triggered linear quadratic networked control. Optim. Control Appl. Methods **35**(5), 524–538 (2014)
213. Y.S. Suh, Stability and stabilization of nonuniform sampling systems. Automatica **44**(12), 3222–3226 (2008)
214. X.M. Sun, G.P. Liu, W. Wang, D. Rees, Stability analysis for networked control systems based on average dwell time method. Int. J. Robust Nonlinear Control **20**(15), 1774–1784 (2010)
215. X.M. Sun, J. Zhao, D. Hill, Stability and L_2-gain analysis for switched delay systems: a delay-dependent method. Automatica **42**(10), 1769–1774 (2006)
216. V. Suplin, E. Fridman, U. Shaked, H_∞ control of linear uncertain time-delay systems-a projection approach. IEEE Trans. Autom. Control **51**(4), 680–685 (2006)
217. V. Suplin, E. Fridman, U. Shaked, Sampled-data H_∞ control and filtering: nonuniform uncertain sampling. Automatica **43**(6), 1072–1083 (2007)
218. V. Suplin, E. Fridman, U. Shaked, H_∞ sampled-data control of systems with time-delays. Int. J. Control **82**(2), 298–309 (2009)
219. V.L. Syrmos, C.T. Abdallah, P. Dorato, K. Grigoriadis, Static output feedback-a survey. Automatica **33**(2), 125–137 (1997)
220. M. Tabbara, D. Nesic, Input-output stability of networked control systems with stochastic protocols and channels. IEEE Trans. Autom. Control **53**(5), 1160–1175 (2008)
221. P. Tabuada, Event-triggered real-time scheduling of stabilizing control tasks. IEEE Trans. Autom. Control **52**(9), 1680–1685 (2007)
222. S. Tarbouriech, J.J.M. Gomes da Silva, Synthesis of controllers for continuous-time delay systems with saturating controls via LMIs. IEEE Trans. Autom. Control **45**(1), 105–111 (2000)
223. A. Teixeira. *Toward Cyber-Secure and Resilient Networked Control Systems*. Ph.D. thesis, KTH Royal Institute of Technology (2014)
224. E.G. Tian, W.K. Wong, D. Yue, Robust H_∞ control for switched systems with input delays: a sojourn-probability-dependent method. Inf. Sci. **283**, 22–35 (2014)
225. E.G. Tian, W.K. Wong, D. Yue, T.C. Yang, H_∞ filtering for discrete-time switched systems with known sojourn probabilities. IEEE Trans. Autom. Control **60**(9), 2446–2451 (2015)
226. V. Ugrinovskii, E. Fridman, A Round-Robin type protocol for distributed estimation with H_∞ consensus. Syst. Control Lett. **69**, 103–110 (2014)
227. N. van de Wouw, P. Naghshtabrizi, M.B.G. Cloosterman, J.P. Hespanha, Tracking control for sampled-data systems with uncertain time-varying sampling intervals and delays. Int. J. Robust Nonlinear Control **20**(4), 387–411 (2009)
228. S. van Loon, M. Donkers, N. van de Wouw, W.P. M.H. Heemels, Stability analysis of networked control systems with periodic protocols and uniform quantizers, in *Proceedings of the 4th IFAC Conference on Analysis and Design of Hybrid Systems*, Eindhoven, Netherlands (2012)
229. G. Walsh, H. Ye, L. Bushnell, Stability analysis of networked control systems. IEEE Trans. Control Syst. Technol. **10**(3), 438–446 (2002)
230. G.C. Walsh, O. Beldiman, L.G. Bushnell, Asymptotic behavior of nonlinear networked control systems. IEEE Trans. Autom. Control **46**(7), 1093–1097 (2001)
231. R. Wang, G.P. Liu, W. Wang, D. Rees, Y.B. Zhao, H_∞ control for networked predictive control systems based on the switched Lyapunov function method. IEEE Trans. Ind. Electron. **57**(10), 3565–3571 (2010)
232. X.F. Wang, M.D. Lemmon, Self-triggered feedback control systems with finite-gain L_2 stability. IEEE Trans. Autom. Control **54**(3), 452–467 (2009)
233. X.F. Wang, M.D. Lemmon, Event-triggering in distributed networked control systems. IEEE Trans. Autom. Control **56**(3), 586–601 (2011)

234. Y.L. Wang, S.H. Yu, An improved dynamic quantization scheme for uncertain linear networked control systems. Automatica **92**, 244–248 (2018)
235. Z.D. Wang, F. Yang, D.W.C. Ho, X. Liu, Robust H_∞ filtering for stochastic time-delay systems with missing measurements. IEEE Trans. Signal Process. **54**(7), 2579–2587 (2006)
236. G.H. Wen, Y. Wan, J.D. Cao, T.W. Huang, W.W. Yu, Master-slave synchronization of heterogeneous systems under scheduling communication. IEEE Trans. Syst. Man Cybern.: Syst. **48**(3), 473–484 (2018)
237. Y.Q. Wu, R.Q. Lu, H.Y. Su, P. Shi, Z.G. Wu, *Synchronization Control for Large-Scale Network Systems* (Springer, Berlin, 2017)
238. Z.G. Wu, J.H. Park, H.Y. Su, J. Chu, Discontinuous Lyapunov functional approach to synchronization of time-delay neural networks using sampled-data. Nonlinear Dyn. **69**(4), 2021–2030 (2012)
239. Y.Q. Xia, Cloud control system. IEEE/CAA J. Autom. Sin. **2**(2), 134–142 (2015)
240. Y.Q. Xia, M.Y. Fu, G.P. Liu, *Analysis and Synthesis of Networked Control Systems* (Springer, Berlin, 2011)
241. Y.Q. Xia, G.P. Liu, M.Y. Fu, D. Rees, Predictive control of networked systems with random delay and data dropout. IET Control Theory Appl. **3**(11), 1476–1486 (2009)
242. J.L. Xiong, J. Lam, Stabilization of linear systems over networks with bounded packet loss. Automatica **43**(1), 80–87 (2007)
243. J.L. Xiong, J. Lam, Stabilization of networked control systems with a logic ZOH. IEEE Trans. Autom. Control **54**(2), 358–363 (2009)
244. M. Yu, L. Wang, T. Chu, F. Hao, Stabilization of networked control systems with data packet dropout and transmission delays: continuous-time case. Eur. J. Control **11**(1), 40–49 (2005)
245. W.W. Yu, G.R. Chen, M. Cao, W. Ren, Delay-induced consensus and quasi-consensus in multi-agent dynamical systems. IEEE Trans. Circuits Syst. I Regul. Pap. **60**(10), 2679–2687 (2013)
246. D. Yue, Q.L. Han, J. Lam, Network-based robust H_∞ control of systems with uncertainty. Automatica **41**(6), 999–1007 (2005)
247. D. Yue, C. Peng, G.Y. Tang, Guaranteed cost control of linear systems over networks with state and input quantisations. IEE Proc.: Control Theory Appl. **153**(6), 658–664 (2006)
248. D. Yue, E.G. Tian, Q.L. Han, A delay system method for designing event-triggered controllers of networked control systems. IEEE Trans. Autom. Control **58**(2), 475–481 (2013)
249. D. Yue, E.G. Tian, Y. Zhang, C. Peng, Delay-distribution-dependent robust stability of uncertain systems with time-varying delay. Int. J. Robust Nonlinear Control **19**(4), 377–393 (2009)
250. S. Yüksel, T. Başar, *Stochastic Networked Control Systems: Stabilization and Optimization under Information Constraints* (Birkhäuser, Basel, 2013)
251. H.B. Zeng, Y. He, M. Wu, J.H. She, Free-matrix-based integral inequality for stability analysis of systems with time-varying delay. IEEE Trans. Autom. Control **60**(10), 2768–2772 (2015)
252. H.B. Zeng, Y. He, M. Wu, J.H. She, New results on stability analysis for systems with discrete distributed delay. Automatica **60**, 189–192 (2015)
253. C.K. Zhang, Y. He, L. Jiang, M. Wu, H.B. Zeng, Stability analysis of systems with time-varying delay via relaxed integral inequalities. Syst. Control Lett. **92**, 52–61 (2016)
254. H.G. Zhang, J.Y. Wang, Z.S. Wang, H.J. Liang, Sampled-data synchronization analysis of markovian neural networks with generally incomplete transition rates. IEEE Trans. Neural Netw. Learn. Syst. **28**(3), 740–752 (2017)
255. J.H. Zhang, D.Y. Zhao, W.X. Zheng, Output feedback control of discrete-time systems with self-triggered controllers. Int. J. Robust Nonlinear Control **25**, 3698–3713 (2015)
256. L. Zhang, Y. Shi, T. Chen, B. Huang, A new method for stabilization of networked control systems with random delays. IEEE Trans. Autom. Control **50**(8), 1177–1181 (2005)
257. Q.R. Zhang, K. Liu, Y.Q. Xia, A.Y. Ma, Optimal stealthy deception attack against cyber-physical systems. IEEE Trans. Cybern. (2019). https://doi.org/10.1109/TCYB.2019.2912622
258. W. Zhang. *Stability analysis of networked control systems*. Ph.D. thesis, Case Western Reserve University (2001)

259. W. Zhang, M. Branicky, S.M. Phillips, Stability of networked control systems. IEEE Control Syst. Mag. **21**(1), 84–99 (2001)
260. W.A. Zhang, L. Yu, Modelling and control of networked control systems with both network-induced delay and packet-dropout. Automatica **44**(12), 3206–3210 (2008)
261. W.A. Zhang, L. Yu, A robust control approach to stabilization of networked control systems with time-varying delays. Automatica **45**(10), 2440–2445 (2009)
262. X.M. Zhang, Q.L. Han, Abel lemma-based finite-sum inequality and its application to stability analysis for linear discrete time-delay systems. Automatica **57**, 199–202 (2015)
263. X.M. Zhang, Q.L. Han, Event-based H_∞ filtering for sampled-data systems. Automatica **51**, 55–69 (2015)
264. X.M. Zhang, Q.L. Han, X.H. Yu, Survey on recent advances in networked control systems. IEEE Trans. Ind. Inf. **12**(5), 1740–1752 (2016)
265. L. Zou, Z.D. Wang, H.J. Gao, Observer-based H_∞ control of networked systems with stochastic communication protocol: the finite-horizon case. Automatica **63**, 366–373 (2016)
266. L. Zou, Z.D. Wang, H.J. Gao, Set-membership filtering for time-varying systems with mixed time-delays under Round-Robin and weighted Try-Once-Discard protocols. Automatica **74**, 341–348 (2016)

Index

K. Liu et al., *Networked Control Under Communication Constraints*, Advances in Delays and Dynamics 11, https://doi.org/10.1007/978-981-15-4230-5

MIX
Papier aus verantwortungsvollen Quellen
Paper from responsible sources
FSC® C105338

Printed by Books on Demand, Germany